The Radio Station

The Radio Station

FOURTH EDITION

Michael C. Keith

Foreword by

Ed Shane

Focal Press
Boston   Oxford
Johannesburg   Melbourne
New Delhi   Singapore

Focal Press is an imprint of Butterworth–Heinemann.

A member of the Reed Elsevier group

 Recognizing the importance of preserving what has been written, Butterworth–Heinemann prints its books on acid-free paper whenever possible.

Library of Congress Cataloging-in-Publication Data

Keith, Michael C., date
    The radio station / Michael C. Keith. — 4th ed.
        p.   cm.
    Includes bibliographical references and index.
    ISBN 0-240-80261-6
    1. Radio broadcasting—United States.   I. Title.
HE8698.K45 1997
384.54'53'069—dc20   96-22433
                                             CIP

British Library Cataloguing-in-Publication Data
A catalogue record for this book is available from the British Library.

The publisher offers special discounts on bulk orders of this book.
For information, please contact:
Manager of Special Sales
Butterworth–Heinemann
313 Washington Street
Newton, MA 02158–1626
Tel: 617-928-2500
Fax: 617-928-2620

For information on all Focal Press publications available, contact our World Wide Web home page at: http://www.bh.com/fp

10 9 8 7 6 5 4 3 2 1

Printed in the United States of America

CONTENTS

Foreword
by Ed Shane

Ed Shane is a broadcast adviser and founder of Houston-based Shane Media Services, which provides management, programming, and research consultation to radio. Since 1977 he has helped broadcast companies, both large and small, achieve success for their stations in a variety of formats. A trend-watcher, Shane combines insights on the world at large with research conducted specifically for radio. His writing has been published in both consumer and trade press. His first radio book, Programming Dynamics *(1984, Kansas City, KS: Globecom), was an industry bestseller. He is also the author of* Cutting Through *(1991, Houston, TX: Shane Media).*

Welcome to the Golden Age of Radio. Revenues are the highest ever, setting records in the mid-1990s. Profits are up, too. Advertiser demand is high. Rule changes have allowed ownerships to consolidate.

Most people refer to "the Golden Age" as some period in the past, usually the "theater of the mind" days of the 1930s and 40s when listeners turned words and sound into mental pictures that commanded the attention of the nation. That was a unique period. Radio was the only free-entertainment medium for a nation emerging from a disastrous depression. For that reason alone, the phrase "Golden Age" applies.

The fifties and sixties are also described as a "Golden Age." That was a time when the deejay was as much a king as Elvis because the deejay spun the soundtrack to the lives of Baby Boomers as teenagers.

If you're using this book to learn radio, chances are you have no direct experience of those long-gone Golden Ages. That's why I feel that today's radio—and for that matter, tomorrow's—are Golden Ages, too.

To hear some people talk, the end of the nineties is hardly the stuff of a "Golden Age." They say radio is endangered, threatened, about to die. New media will invade radio's turf, they say, replacing radio as an information provider and music provider.

I'll be the first to admit that there's no holding back the interactive and digital future. Yet radio has already faced challenges, even threats, from other media: television, the long play album, CDs and cassettes, multichannel cable television, cable audio. Each time, radio survived and adapted.

Radio even survived the announcement of its own demise. Henry Morgan, a radio performer and satirist during the fifties, once claimed, "Radio actually died when 'Stop the Music' got higher ratings than Fred Allen." While the event was noteworthy for lovers of the original Golden Age, it hardly spelled death. It did spell change and challenge. When Morgan made his acerbic comment, network radio faced more than a challenge. The medium was truly threatened by television.

Before television, the "pictures" were on the radio. A man named "Raymond" opened a squeaking door on "Inner Sanctum" and the stories told behind the door made spines tingle for a half hour. That age brought "The Shadow," a Gothic thriller whose main character was a mental projection against a foggy night full of smoke from coal-burning furnaces.

The lighter side of that era was Fibber McGee's closet, a packed-to-the-gills jumble that fell with a crash to the floor once per episode in an avalanche that usually included samples of the sponsor's product. After the cacophony there was a pause. Finally, a dinner bell crash-tinkled to the floor as punctuation.

These images, all formed in the mind of the listener, could not stand the transition to television. Even NBC-TV's attempt at a revival of "Fibber McGee and Molly" left the famous closet out of sight—in the hands of sound-effects experts and in the imagination of the viewer.

The transition was easier for other radio shows. Jack Benny simply added scenery and sets to the comedy scripts his crew was already performing. It was the same for Burns and Allen and for scores of other radio performers who moved from their radio desks to their TV desks and continued their careers in the visual medium.

While radio was disrupted by the introduction of television, it survived the challenge by finding new applications for its medium. Panic began to subside when audio broadcasters stretched individual music programs into 24-hour formats. Thus were born Top 40, Middle of the Road, Beautiful Music. Each station became something distinct. The days of being "all things to all people" on one radio station virtually disappeared.

The stories of radio's rejuvenation have been told plenty of times. The names Todd Storz, Gordon McLendon, and Rick Sklar are the stuff of legend because those men did not view the future of radio in terms of its past. The change to format radio stimulated another "Golden Age" as teenagers of the Baby Boom discovered Top 40 radio playing their music.

I was ten years old in 1955 when radio's evolution was taking place. With the exception of a few evenings with my parents listening to what remained of the old shows, all my radio experiences were tied up with the new disc jockeys and the new music—rock 'n' roll. I had no understanding of the plight of the radio industry against the threat of television. I assumed that it had been that way all along, one station for me and my friends, another for my parents.

Now that I've studied radio of that time and have had the chance to inter-view people who either made it happen or watched it happen, I find the stories of rejuvenation in stark contrast to what was called radio's "death." There's further contradiction: The number of stations in the U.S. has grown from 700 then to more than 11,000 now! There's no indication that the number of signals is going to decrease. Technology was good to radio in terms of its ability to expand its numbers, just as technology has created radio's challenges.

Futurist Alvin Toffler wrote that "new technologies make diversity as cheap as uniformity." That applies to radio, too. The industry's newly formulated formats fragmented into narrower niches as space was made available on the dial for new facilities.

As radio expanded in the seventies and eighties, so did cable television. Again, radio was challenged by technology.

CNN and MTV were the eighties equivalent of "moving from the radio desk to the TV desk." Both networks borrowed radio's strong, cohesive positioning by using 24-hour vertical programming in the emerging cable medium. A staggering number of video music shows followed in MTV's wake. Record companies transferred promotional dollars from radio play to production of video clips.

What was radio's response? Radio met cable's challenge head on with adaptation. The result was Contemporary Hit Radio in the eighties, then a new breed of Country for the nineties.

Another of radio's Golden Ages is emerging as this book goes to press: the impact of talk radio on national discourse. A cohesive, singular, pop music doesn't exist amid the format fragmentation and fractionalization of the nineties. So words are stimulating Baby Boomers as adults the way rock 'n roll did when they were teenagers.

Rush Limbaugh made his mark by speaking for large numbers of conservative Baby Boom males. Howard Stern expressed in words (often outrageous words) what rock singers have said in music. From politics to sex to comedy, the spoken word creates a new context for radio.

There may be context, but content is often no more than rhetoric. Talk show hosts rely on neatly turned slogans that amount to bumper sticker philosophy. At the root of the rise of talk radio is the glut of information confronting consumers. Talk shows are an attempt to sort it all out, to filter data from information overload.

There's a theme here. In each of radio's "Golden Ages," as I've defined them, technology is the driving force:

In the 1930s and 1940s radio itself was the driving technology. There was total national distribution of a free-entertainment medium.

In the fifties, TV challenged. At first radio worried, then it reinvented itself.

Radio's reinvention of the fifties and sixties took advantage of the phonograph record, especially the new 45 rpm discs.

In the late sixties and early seventies the LP record became the primary medium for popular music. Radio capitalized.

The technology of improved audio married the LP and FM sound to reinvent radio once again in the seventies.

The CD did the same for radio in the eighties.

The cable boom of the eighties caused radio to react, first with new vertical formats, then with music that had been seen on MTV and other music channels.

Instant and constant access to information prompted the rise of radio talk shows to "sort it all out" and to speak for the "average person."

The next Golden Age will be created by users of this book responding to the medium's next challenge: How do we use radio's unique attribute—portability—in a wired world?

Broadcasters line up on either side of that question. Some say on-line services will offer most of what radio can deliver—news, weather, music, odd facts—and that the click of a mouse can kill radio as we know it.

I don't believe it. I'm too optimistic about radio and its ability to adapt. I'm optimistic, too, about your ability to respond to what you'll learn in this book—a real world view of a medium that knows how to reinvent itself and re-energize itself.

See you in the Golden Age—the one you create!

Radio waves
sing, dance,
glide and spin
through the air
like strange magic,
delivering messages
as individual
as the fingerprints
of each listener.

Lindy Bonczek

Preface

It seems remarkable that a decade has passed since the publication of the first edition of this text. Truly gratifying is the fact that during this time it has become the most widely adopted book of its kind at colleges throughout the country and enjoys a sizable following overseas, as well as among industry practitioners. It is what may be called the "standard," and it has become so because of its 3-C formula—comprehensive, concise, and candid.

Since the first outing of this book, the radio industry has witnessed enormous change. It has been a time of significant technical innovation and economic flux. The advent of digital audio promises to revolutionize broadcast signal transmission and reception. Of course, with such a transformation come challenges and concerns, and these will doubtlessly occupy the thoughts of broadcasters well into the next millennium.

When this book was first published in 1986, radio was enjoying unprecedented prosperity. The prices being paid for radio properties were soaring, and station revenues were at exceptional levels. Life was good for almost everyone in the industry, or so it seemed. Many AM station owners were not in on the opulent banquet, and a growing number were pulling the plug on their operations. Yet on the whole, the eighties were auspicious years for the magic medium.

The tide shifted as the final decade of the twentieth century began. The nation had slipped into a nasty recession taking radio with it on its downward slide, but soon the medium's fortunes were on the upswing. The industry has indeed experienced many ups and downs since its inception over 75 years ago, however, and it will doubtlessly know the thrill of ascent and the angst of decline again.

It is impossible to imagine a world without David Sarnoff's radio music box, but that is not necessary for it is safe to assume that radio in some form will continue to be an integral part of our lives for a very long time to come.

The mission of this book has not changed. This edition, like the previous ones, is the result of a desire and effort to provide the student of radio with the most complete account of the medium possible, from the insider's view, if you will. It is presented from the perspective of the radio professional, drawing on the insights and observations of those who make their daily living by working in the industry.

What continues to set this particular text apart from others is that not one or two but literally hundreds of radio people have contributed to this effort to disseminate factual and relevant information about the medium in a way that captures its reality. These professionals represent the top echelons of network and group-owned radio, as well as the rural daytime-only outlets spread across the country.

We have sought to create a truly practical, timely, illustrative (a picture can be worth a thousand words—stations explain and reveal themselves through visuals), and accessible book on commercial radio station operations; a book that reflects through its structure and organization the radio station's own. Therefore, the departments and personnel that comprise a radio station are our principle focus. We begin by examining the role of station management and then move into programming, sales, news, engineering, production, and traffic, as well as other key areas that serve as the vital ingredients of any radio outlet.

Since our strategy was to draw on the experience of countless broadcast and allied professionals, our debt of gratitude is significant. It is to these individuals who contributed most directly to its making that we also dedicate this book.

In addition, we would like to express our sincere appreciation to the many individuals and organizations that assisted us in so many important ways. Foremost among them is Ed Shane, whose contribution to this edition brings new meaning to the term generosity. A profound word of thanks is also due Jay Williams, Jr., Ralph Guild, Bill Siemering, Lee Abrams, Lynn Christian, Chris Sterling, Donna Halper, Bruce DuMont, Paul Fiddick, Norm Feuer, Ward Quaal, Frank Bell, Allen Myers, Ted Bolton, et al.—the list is endless. My hat is off to every individual and station cited in this book.

Countless companies contributed to the body of this work. They include the ABC Radio Networks, Arbitron Ratings Company, Auditronics Inc., The Benchmark Company, Bolton Research, BMI, BPME, Broadcast Electronics, *Broadcasting and Cable*, Burkhart Douglas and Associates, Coleman Research, Communication Graphics, CFM, CRN, Denon, DMR, the FCC, FMR Associates, Halper and Associates, IGM Inc., Interep Radio Store, International Demographics, Jacobs Media, Jefferson Pilot Data Systems, Lund Consultants, Marketron Inc., Mediabase, Metro Traffic Network, MMR, Museum of Broadcast Communications, National Association of Broadcasters, Premiere Radio Networks, Public Radio International, QuikStats, Radio Advertising Bureau, Radio Aahs, *Radio Ink, Radio and Records*, RTNDA, Radio Computer Systems, RCA, Satelillite Music Network, Shane Media, Society of Broadcast Engineers, Jim Steele, Annette Steiner, Superaudio, Tapscan, TM Century, Westinghouse Broadcasting, and Westwood One.

Since the publication of earlier editions, it is safe to assume—in an industry noted for its nomadic nature—that a number of contributors have moved on to positions at other stations. Moreover, it is equally certain that many stations have changed call letters, because that is the name of the game too. Due to the sheer volume of contributors, it would be difficult to establish the current whereabouts or status of each without employing the services of the FBI, CIA, and Secret Service. We have therefore let stand the original addresses and call letters of contributors except when new information has become available; in those cases, changes have been made.

1

State of the Fifth Estate

IN THE AIR—EVERYWHERE

As we approach the new millennium, radio remains ubiquitous. It continues to be the most pervasive medium on earth. There is no patch of land, no piece of ocean surface, untouched by the electromagnetic signals beamed from the more than twenty-seven thousand radio stations worldwide. Over a third of these broadcast outlets transmit in America alone. Today, over eleven thousand stations in this country reach 99 percent of all households, and less than 1 percent have fewer than five receivers. There are nearly a billion working radios in the United States.

Contemporary radio's unique personal approach has resulted in a shift of the audience's application of the medium: from family or group entertainer before 1950, to individual companion in the last half of the century. Although television usurped radio's position as the number one entertainment medium four decades ago, radio's total reach handily exceeds that of the video medium. More people tune in to radio for its multifaceted offerings than to any other medium—print or electronic. Practically every automobile (95 percent) has a radio. "There are twice as many car radios in use (approximately 132 million) as the total circulation (60 million) of all daily newspapers, and four out of five adults are reached by radio each week," contends Kenneth Costa, vice president of marketing for the Radio Advertising Bureau (RAB).

Seven out of ten adults are reached weekly by car radio.

The average adult spends over two-and-a-half hours per day listening to radio. A survey conducted by RADAR (Radio's All Dimensional Audience Research) concluded that 95 percent of all persons over twelve years old tuned in to radio. In the 1990s, this computes to around 200 million Americans, and the figure continues to grow.

The RADAR report also found that working women account for nearly 65 percent of radio listening by women, a statistic that reflects the times.

The number of radio receivers in use in America has risen by more than 50 percent since 1970, when 321 million sets provided listeners a wide range of audio services. In recent years, technological innovations in receiver design alone have contributed to the ever-increasing popularity of the medium. Boxes and walk-alongs, among others, have boosted receiver sales over the three billion dollar mark annually, up 30 percent since 1970. There are 25 million walk-along listeners. Radio's ability to move with its audience has never been greater. Out-of-home listeners account for over 60 percent of the average audience Monday through Friday. In addition, the Radio Advertising Bureau concluded that seven out of ten computer purchasers and wine and beer drinkers tune into the medium daily.

Radio appeals to everyone and is available to all. Its mobility and variety of offerings have made it the most popular medium in history. To most of us, radio is as much a part of our day as morning

FIGURE 1.1
Radio celebrated its
75th birthday in
1995.

coffee and the ride to work or school. It is a companion that keeps us informed about world and local events, gives us sports scores, provides us with the latest weather and school closings, and a host of other information, not to mention our favorite music, and asks for nothing in return. A recent Katz Radio Group study concluded that "only radio adapts to the lifestyle of its audience." The report dispelled the belief that radio listening drops during the summer, as does TV viewing, proving that radio is indeed a friend for all seasons.

FIGURE 1.2
Courtesy RAB.

RADIO AUDIENCES

**Radio Reaches Almost Everyone
...Every Week**

Demo		Weekly Reach
Persons	12+	95.5%
Teens	12-17	98.8%
Adults	18+	95.1%
Men	18+	95.7%
	18-34	97.6%
	25-54	97.5%
	35-64	97.2%
	55+	90.2%
	65+	85.5%
Women	18+	94.6%
	18-34	97.7%
	25-54	97.6%
	35-64	97.0%
	55+	87.3%
	65+	82.9%

*Source: RADAR ® 50, Fall 1994. © Copyright Statistical Research, Inc.
(Monday-Sunday, 6am-Mid.)*

95.5%

It is difficult to imagine a world without such an accommodating and amusing cohort, one that not only has enriched our lives by providing us with a nonstop source of entertainment, but has also kept us abreast of happenings during times of national and global crisis. To most Americans, radio is an integral part of daily life.

A HOUSEHOLD UTILITY

Although radio seems to have been around for centuries, it is a relatively recent invention. Many people alive today lived in a world without radio—hard to imagine, yet true. The world owes a debt of gratitude to several "wireless" technologists who contributed to the development of the medium. A friendly debate continues to be waged today as to just who should rightfully be honored with the title "father of radio." There are numerous candidates who actually take us as far back as the last century. For example, there is physicist James Clerk Maxwell, who theorized the existence of electromagnetic waves, which later in the century were used to carry radio signals. Then there is German scientist Heinrich Hertz, who validated Maxwell's theory by proving that electromagnetic waves do indeed exist.

The first choice of many to be called grand patriarch of radio is Guglielmo Marconi, who is credited with devising a method of transmitting sound without the help of wires—thus "wireless telegraphy." There are a host of other inventors and innovators who can, with some justification, be considered for the title. Nikola Tesla experimented with various forms of wireless transmission, and although he has been largely neglected by historians, today there are Tesla Societies that maintain he is responsible for the invention of wireless transmission and modern radio. Lee De Forest, Ambrose Fleming, Reginald Fessenden, and David Sarnoff are a few others whose names have been associated with the hallowed designation. (A further discussion of radio's preeminent

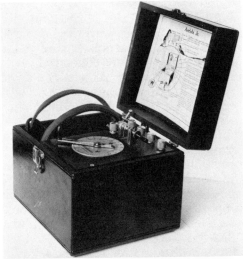

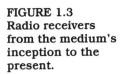

FIGURE 1.3
Radio receivers
from the medium's
inception to the
present.

technologists can be found in Chapter 10.) However, of the aforementioned, perhaps the pioneer with the most substantial claim is Sarnoff. A true visionary, Sarnoff allegedly conceived of the ultimate application of Marconi's device in a now-famous memorandum. In what became known as the "radio music box" memo, Sarnoff supposedly suggested that radio receivers be mass produced for public consumption and that music, news, and information be broadcast to the households that owned the appliance. According to legend, at first his proposal was all but snubbed. Sarnoff's persistence eventually paid off, and in 1919 sets were available for general purchase. Within a very few years, radio's popularity would exceed even Sarnoff's estimations. Recently some scholars have concluded that Sarnoff's memo may have been written several years later, if at all, as a means of securing his status in the history of the radio medium.

A TOLL ON RADIO

Though not yet a household word in 1922, radio was surfacing as a medium to be reckoned with. Hundreds of thousands of Americans were purchasing the crude, battery-operated crystal sets of the day and tuning the two frequencies (750 and 833 kc) set aside by the Department of Commerce for reception of radio broadcasts. The majority of stations in the early 1920s were owned by receiver manufacturers and department stores that sold the apparatus. Newspapers and colleges owned nearly as many. Radio was not yet a commercial enterprise. Those stations not owned by parent companies often depended on public donations and grants. These outlets found it no small task to continue operating. Interestingly, it was not one of these financially pinched stations that conceived of a way to generate income, but rather AT&T-owned WEAF in New York.

The first paid announcement ever broadcast lasted ten minutes and was bought by Hawthorne Court, a Queens-based real estate company. Within a matter of weeks other businesses also paid modest "tolls" to air messages over WEAF. Despite AT&T's attempts to monopolize the pay-for-broadcast concept, a year later in 1923 many stations were actively seeking sponsors to underwrite their expenses as well as to generate profits. Thus, the age of commercial radio was launched. It is impossible to imagine what American broadcasting would be like today had it remained a purely noncommercial medium as it has in many countries.

BIRTH OF THE NETWORKS

The same year that Pittsburgh station KDKA began offering a schedule of daily broadcasts, experimental network operations using telephone lines were inaugurated. As early as 1922, stations were forming chains, thereby enabling programs to be broadcast simultaneously to several different areas. Sports events were among the first programs to be broadcast in network fashion. Stations WJZ (later WABC) in New York and WGY in Schenectady linked for the airing of the 1922 World Series, and early in 1923 WEAF in New York and WNAC in

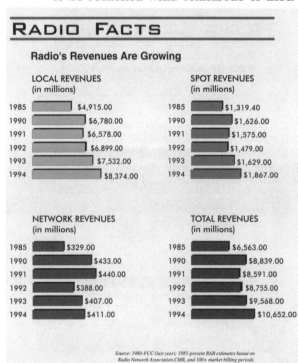

RADIO FACTS

Radio's Revenues Are Growing

LOCAL REVENUES
(in millions)

1985	$4,915.00
1990	$6,780.00
1991	$6,578.00
1992	$6,899.00
1993	$7,532.00
1994	$8,374.00

SPOT REVENUES
(in millions)

1985	$1,319.40
1990	$1,626.00
1991	$1,575.00
1992	$1,479.00
1993	$1,629.00
1994	$1,867.00

NETWORK REVENUES
(in millions)

1985	$329.00
1990	$433.00
1991	$440.00
1992	$388.00
1993	$407.00
1994	$411.00

TOTAL REVENUES
(in millions)

1985	$6,563.00
1990	$8,839.00
1991	$8,591.00
1992	$8,755.00
1993	$9,568.00
1994	$10,652.00

Source: 1980–FCC (last year); 1985-present RAB estimates based on Radio Network Association,CMR, and 100+ market billing periods.

Boston transmitted a football game emanating from Chicago. Later the same year, President Coolidge's message to Congress was aired over six stations. Chain broadcasting, a term used to describe the earliest networking efforts, was off and running.

The first major broadcast network was established in 1926 by the Radio Corporation of America (RCA) and was named the National Broadcasting Company (NBC). The network consisted of two dozen stations—several of which it had acquired from AT&T, which was encouraged by the government to divest itself of its broadcast holdings. Among the outlets RCA purchased was WEAF, which became its flagship station. Rather than form one exclusive radio combine, RCA chose to operate separate Red and Blue networks. The former comprised the bulk of NBC's stations, whereas the Blue network remained relatively small, with fewer than half a dozen outlets. Under the NBC banner, both networks would grow, the Blue network remaining the more modest of the two.

Less than two years after NBC was in operation, the Columbia Broadcasting System (CBS, initially Columbia Phonograph Broadcasting System) began its

FIGURE 1.5
David Sarnoff, the man who helped put radio into the home. Courtesy RCA.

network service with sixteen stations. William S. Paley, who had served as advertising manager of his family's cigar company (Congress Cigar), formed the network in 1928 and would remain its chief executive into the 1980s.

A third network emerged in 1934. The Mutual Broadcasting System went into business with affiliates in only four cities—New York, Chicago, Detroit, and Cincinnati. Unlike NBC and CBS, Mutual did not own any stations. Its primary function was that of program supplier. In 1941, Mutual led its competitors with 160 affiliates.

FIGURE 1.6
Radio fans circa 1921. Courtesy Westinghouse Electric.

Although NBC initially benefited from the government's fear of a potential monopoly of communication services by AT&T, it also was forced to divest itself of a part of its holdings because of similar apprehensions. When the Federal Communications Commission (FCC) implemented more stringent chain broadcasting rules in the early 1940s, which prohibited one organization from operating two separate and distinct networks, RCA sold its Blue network, retaining the more lucrative Red network.

The FCC authorized the sale of the Blue network to Edward J. Noble in 1943. Noble, who had amassed a fortune as owner of the Lifesaver Candy Company, established the American Broadcasting Company (ABC) in 1945. In the years to come, ABC would eventually become the largest and most successful of all the radio networks.

By the end of World War II, the networks accounted for 90 percent of the radio audience and were the greatest source of individual station revenue. Today most of the major networks are under the auspices of megacorporations. In 1995, Disney purchased Cap Cities/ABC and Westinghouse bought CBS. A few years earlier, GE reclaimed NBC.

CONFLICT IN THE AIR

The five years that followed radio's inception saw phenomenal growth. Millions of receivers adorned living rooms throughout the country, and over seven hundred stations were transmitting signals. A lack of sufficient regulations and an inadequate broadcast band contributed to a situation that bordered on catastrophic for the fledgling medium. Radio reception suffered greatly as the result of too many stations broadcasting, almost at will, on the same frequencies. Interference was widespread. Frustration increased among both the listening public and the broadcasters, who feared the strangulation of their industry.

Concerned about the situation, participants of the National Radio Conferences (1923–1925) appealed to the secretary of commerce to impose limitations on station operating hours and power. The bedlam continued, since the head of the Commerce Department lacked the necessary power to implement effective changes. However, in 1926, President Coolidge urged Congress to address the issue. This resulted in the Radio Act of 1927 and the formation of the Federal Radio Commission (FRC). The five-member commission was given authority to issue station licenses, allocate frequency bands to various services, assign frequencies to individual stations, and dictate station power and hours of operation.

Within months of its inception, the FRC established the Standard Broadcast band (500–1500 kc) and pulled the plug on 150 of the existing 732 radio outlets. In less than a year, the medium that had been on the threshold of ruin was thriving. The listening public responded to the clearer reception and the increasing schedule of entertainment programming by purchasing millions of receivers. More people were tuned to their radio music boxes than ever before.

RADIO PROSPERS DURING THE DEPRESSION

The most popular radio show in history, "Amos 'n' Andy," made its debut on NBC in 1929, the same year that the stock market took its traumatic plunge. The show attempted to lessen the despair brought on by the ensuing Depression by addressing it with light-hearted humor. As the Depression deepened, the stars of "Amos 'n' Andy," Freeman Gosden and Charles Correll, sought to assist in the president's recovery plan by helping to restore confidence in the nation's banking system through a series of recurring references and skits. When the "Amos 'n' Andy" show aired, most of the country stopped what it was doing and tuned in. Theater owners complained that on the evening the show was broadcast, ticket sales were down dramatically.

As businesses failed, radio flourished. The abundance of escapist fare that the medium offered, along with the important fact that it was provided free to the listener, enhanced radio's hold on the public.

Not one to overlook an opportunity to give his program for economic recuperation a further boost, President Franklin D. Roosevelt launched a series of broadcasts on March 12, 1933, which became known as the "fireside chats." Although the president had never received formal broadcast training, he was completely at home in front of the microphone. The audience perceived a man of sincerity, intelligence, and determination. His sensitive and astute use of the medium went a long way toward helping in the effort to restore the economy.

In the same year that Roosevelt took to the airwaves to reach the American people, he set the wheels in motion to create an independent government agency whose sole function would be to regulate all electronic forms of communication, including both broadcast and wire. To that end, the Communications Act of 1934 resulted in the establishment of the Federal Communications Commission.

As the Depression's grip on the nation weakened in the late 1930s, another crisis of awesome proportions loomed—World War II. Once again radio would prove an invaluable tool for the national good. Just as the medium completed its second decade of existence, it found itself enlisting in the battle against global tyranny. By 1939, as the great firestorm was nearing American shores, 1,465 stations were authorized to broadcast.

RADIO DURING WORLD WAR II

Before either FM or television had a chance to get off the ground, the FCC saw fit to impose a wartime freeze on the con-

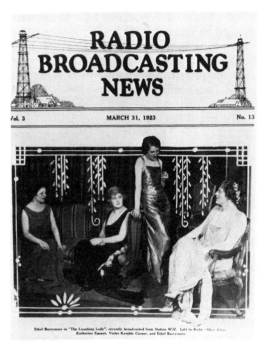

FIGURE 1.7
Radio trade paper in 1923 serving a growing industry. Courtesy Westinghouse Electric.

struction of new broadcast outlets. All materials and manpower were directed at defeating the enemy. Meanwhile, existing AM stations prospered and enjoyed increased stature. Americans turned to their receivers for the latest information on the war's progress. Radio took the concerned listener to the battle fronts with dramatic and timely reports from war correspondents, such as Edward R. Murrow and Eric Sevareid, in Europe and the South Pacific. The immediacy of the news and the gripping reality of the sounds of battle brought the war into stateside homes. This was the war that touched all Americans. Nearly everyone had a relative or knew someone involved in the effort to preserve the American way of life.

FIGURE 1.8
In the 1930s, radio offered an impressive schedule of programs for children.

FIGURE 1.9
Today many of these pre–World War II catalin table model receivers, which originally cost under $20, sell for thousands of dollars as collector's items.

Programs that centered on concerns related to the war were plentiful. Under the auspices of the Defense Communications Board, radio set out to do its part to quash aggression and tyranny. No program of the day failed to address issues confronting the country. In fact, many programs were expressly propagandistic in their attempts to shape and influence listeners' attitudes in favor of the Allied position.

Programs with war themes were popular with sponsors who wanted to project a patriotic image, and most did. Popular commentator Walter Winchell, who was sponsored by Jergen's Lotion, closed his programs with a statement that illustrates the prevailing sentiment of the period: "With lotions of love, I remain your New York correspondent Walter Winchell, who thinks every American has at least one thing to be thankful for on Tuesday next. Thankful that we still salute a flag and not a shirt."

Although no new radio stations were constructed between 1941 and 1945, the industry saw profits double and the listening audience swell. By war's end, 95 percent of homes had at least one radio.

TELEVISION APPEARS

The freeze that prevented the full development and marketing of television was lifted within months of the war's end. Few radio broadcasters anticipated the dilemma that awaited them. In 1946 it was business as usual for the medium, which enjoyed new-found prestige as the consequence of its valuable service during the war. Two years later, however, television was the new celebrity on the block, and radio was about to experience a significant decline in its popularity.

While still an infant in 1950, television succeeded in gaining the distinction of being the number one entertainment medium. Not only did radio's audience begin to migrate to the TV screen, but many of the medium's entertainers and sponsors jumped ship

as well. Profits began to decline, and the radio networks lost their prominence.

In 1952, as television's popularity continued to eclipse radio's, three thousand stations of the faltering medium were authorized to operate. Several media observers of the day predicted that television's effect would be too devastating for the older medium to overcome. Many radio station owners around the country sold their facilities. Some reinvested their money in television, while others left the field of broadcasting entirely.

Recalling this bleak period in radio's history in a 1958 magazine article in *Wisdom,* David Sarnoff wrote:

"In the Spring of 1949, the cry went up that 'radio is doomed.' Some of the prophets of doom predicted that within three years sound broadcasting over national networks would be wiped out, with television taking its place.

"I did not join that gloomy forecast in '49, nor do I now. Years have passed, and radio broadcasting is still with us and rendering nationwide service. It plays too vital a role in the life of this nation to be canceled out by another medium. I have witnessed too many cycles of advance and adaptation to believe that a service so intimately integrated with American life can become extinct.

"We would be closing our eyes to reality, however, if we failed to recognize that radio has been undergoing fundamental changes. To make the most of its great potentials, it must now be operated and used in ways which take cognizance of the fact that it is no longer the only broadcast medium. A process of adjustment is necessary, and it is taking place."

A NEW DIRECTION

A technological breakthrough by Bell Laboratory scientists in 1948 resulted in the creation of the transistor. This innovation provided radio manufacturers the chance to produce miniature portable receivers. The new transistors, as they were popularly called, enhanced radio's mobility. Yet the medium continued to flounder throughout the early 1950s as it attempted to formulate a strategy that would offset the effects of television. Many radio programmers felt that the only way to hold onto their dwindling audience was to offer the same material, almost program for program, aired by television. Ironically, television had appropriated its programming approach from radio, which no longer found the system viable.

By 1955, radio revenues reached an unimpressive 90 million dollars, and it was apparent to all that the medium had to devise another way to attract a more formidable following. Prerecorded music became a mainstay for many stations that had dropped their network affiliations in the face of decreased program schedules. Gradually, music became the primary product of radio stations and the disc jockey (deejay, jock) their new star.

RADIO ROCKS AND ROARS

The mid-fifties saw the birth of the unique cultural phenomenon known as rock 'n' roll, a term invented by deejay Alan Freed to describe a new form of music derived from rhythm and blues. The new sound took hold of the nation's youth and helped return radio to a position of prominence.

In 1955, Bill Haley's recording of "Rock Around the Clock" struck paydirt and sold over a million copies, thus ushering in a new era in contemporary music. The following year Elvis Presley tunes dominated the hit charts. Dozens of stations around the country began to focus their playlists on the newest music innovation. The Top 40 radio format, which was conceived about the time rock made its debut, began to top the ratings charts. In its original form, Top 40 appealed to a much larger cross section of the listening public because of the diversity of its offerings. At first artists such as Perry Como, Les Paul and Mary Ford,

and Doris Day were more common than the rockers. Then the growing penchant of young listeners for the doo-wop sound figured greatly in the narrowing of the Top 40 playlist to mostly rock 'n' roll records. Before long the Top 40 station was synonymous with rock and teens.

A few years passed before stations employing the format generated the kind of profits their ratings seemed to warrant. Many advertisers initially resisted spending money on stations that primarily attracted kids. By 1960, however, rock stations could no longer be denied since they led their competitors in most cities. Rock and radio formed the perfect union.

FM'S ASCENT

Rock eventually triggered the wider acceptance of FM, whose creator, Edwin H. Armstrong, set out to produce a static-free alternative to the AM band. In 1938 he accomplished his objective, and two years later the FCC authorized FM broadcasting. However, World War II and RCA (which had a greater interest in the development of television) thwarted the implementation of Armstrong's innovation. It was not until 1946 that FM stations were under construction. Yet FM's launch was less than dazzling. Television was on the minds of most Americans, and the prevailing attitude was that a new radio band was hardly necessary.

Over 600 FM outlets were on the air in 1950, but by the end of the decade the number had shrunk by one hundred. Throughout the 1950s and early 1960s, FM stations directed their programming to special-interest groups. Classical and soft music were offered by many. This conservative, if not somewhat highbrow, programming helped foster an elitist image. FM became associated with the intellectual or, as it was sometimes referred to, the "egghead" community. Some FM stations purposely expanded on their snob appeal image in an attempt to set themselves apart from popular, mass appeal radio. This, however, did little to fill their coffers.

FM remained the poor second cousin to AM throughout the 1960s, a decade that did, however, prove transitional for FM. Many FM licenses were held by AM station operators who sensed that someday the new medium might take off. An equal number of FM licensees used the unprofitable medium for tax write-off purposes. Although many AM broadcasters possessed FM frequencies, they often did little when it came to programming them. Most chose to simulcast their AM broadcasts. It was more cost effective during a period when FM drew less than 10 percent of the listening audience.

In 1961 the FCC authorized stereo broadcasting on FM. This would prove to be a benchmark in the evolution of the medium. Gradually more and more recording companies were pressing stereo disks. The classical music buff was initially considered the best prospect for the new product. Since fidelity was of prime concern to the classical music devotee, FM stations that could afford to go stereo did so. The "easy listening" stations soon followed suit.

Another benchmark in the development of FM occurred in 1965 when the FCC passed legislation requiring that FM broadcasters in cities whose populations exceeded one hundred thousand break simulcast with their AM counterparts for at least 50 percent of their broadcast day. The commission felt that simply duplicating an AM signal did not constitute efficient use of an FM frequency. The FCC also thought that the move would help foster growth in the medium, which eventually proved to be the case.

The first format to attract sizable audiences to FM was Beautiful Music, a creation of program innovator Gordon McLendon. The execution of the format made it particularly adaptable to automation systems, which many AM/FM combo operations resorted to when the word came down from Washington that simulcast days were over. Automation kept staff size and production expenses to a minimum. Many stations assigned FM operation to their engineers, who kept the system fed with reels of music tapes and cartridges containing commercial material. Initially, the idea was to keep the FM as a form of garnishment for the more lucrative AM operation. In other words, at combo stations the FM was thrown in as a perk to attract advertisers—two stations for the price of one. To the surprise of more than a few station managers, the FM side began to attract impressive numbers. The more-music, less-talk (meaning fewer commercials) stereo operations made money. By the late 1960s, FM claimed a quarter of the radio listening audience, a 120 percent increase in less than five years.

Contributing to this unprecedented rise in popularity was the experimental Progressive format, which sought to provide listeners with an alternative to the frenetic, highly commercial AM sound. Rather than focus on the best-selling songs of the moment, as was the tendency on AM, these stations were more interested in giving airtime to album cuts that normally never touched the felt of studio turntables. The Progressive or album rock format slowly chipped away at Top 40s ratings numbers and eventually earned itself part of the radio audience.

The first major market station to choose a daring path away from the tried and true chart hit format was WOR-FM in New York. On July 30, 1966, the station broke from its AM side and embarked upon a new age in contemporary music programming. Other stations around the country were not long in following their lead.

The FM transformation was to break into full stride in the early and mid-1970s. Stereo component systems were a hot consumer item and the preferred way to listen to music, including rock. However, the notion of Top 40 on FM was still alien to most. FM listeners had long regarded their medium as the alternative to the pulp and punch presentation typical of the Standard Broadcast band. The idea of contemporary hit stations on FM offended the sensibilities of a portion of the listening public. Nonetheless, Top 40 began to make its debut on FM and for many license holders it marked the first time they enjoyed sizable profits. By the end of the decade, FM's profits would triple, as would its share of the audience. After three decades of living in the shadow of AM, FM achieved parity in 1979 when it equaled AM's listenership. The following year it moved ahead. In the late 1980s, studies demonstrated that FM now attracts as much as 85 percent of the radio audience.

With 400 more commercial AM stations (4,985) than FM stations (4,570), and only a quarter of the audience, the older medium is now faced with a unique challenge that could determine its very survival. In an attempt to retain a share of the audience, many AM stations have dropped music in favor of news and talk. WABC-AM's shift from Musicradio to Talkradio in 1982 clearly illustrated the metamorphosis that AM was undergoing. WABC had long been the nation's foremost leader in the pop-rock music format.

In a further effort to avert the FM sweep, hundreds of AM stations have gone stereo and many more plan to do the same. AM broadcasters are hoping this will give them the competitive edge they urgently need. Music may return to the AM side, bringing along with it some unique format approaches. A number of radio consultants believe that the real programming innovations in the next few years will occur at AM stations. "Necessity is the mother of invention," says Dick Ellis, programming consultant and former radio format specialist for Peters Productions in San Diego, California. "Expect some very exciting and interesting things to happen on AM."

AM STEREO

Hoping to help AM radio out of its doldrums, in the early 1980s the FCC authorized stereocasting on the senior band. However, the Commission failed to declare a technical standard, leaving it to the marketplace to do so. This resulted in a very sluggish conversion to the two channel system, and by the 1990s only a few hundred AM outlets offered stereo broadcasting. Those that did were typically the more prosperous metro market stations that ultimately featured talk and information formats.

Eventually, the FCC declared Motorola the industry standard bearer, but by the mid-1990s the hope that stereo would provide a cure for AM's deepening malaise had dimmed considerably. By this time, many AM outlets, which may have benefitted by having a stereo signal, were in a weaker financial state and unable to convert or were less than enthusiastic about any potential payback.

NONCOMMERCIAL RADIO

Over 1500 stations operate without direct advertiser support. Noncommercial stations, as they are called, date back to the medium's heyday, and were primarily run by colleges and universities. The first "noncoms" broadcast on the AM band but moved to the FM side in 1938. After World War II the FCC reconstituted the FM band and reserved the first twenty channels (88 to 92 MHz) for noncommercial facilities. Initially this gave rise to low-power (10 watt) stations known as Class D's. The lower cost of such operations was a prime motivator for schools that wanted to become involved with broadcasting.

In 1967 the Corporation for Public Broadcasting (CPB) was established as the result of the Public Broadcasting Act. Within three years National Public Radio (NPR) was formed. Today over 400 stations are members of NPR, which provides programming. Many NPR affiliates are licensed to colleges and universities, while a substantial number are owned by nonprofit organizations.

Member stations are the primary source of funding for NPR. They contribute 60 percent of its operating bud-

FIGURE 1.10
NPR News reaches millions of listeners daily and is regarded by many as the medium's foremost news service. Courtesy National Public Radio and Corey Flintoff. Photo credit: Shannon Henry.

FIGURE 1.11
A Public Radio International promotional piece. Courtesy PRI.

get. Affiliates in turn are supported by listeners, by community businesses, and by grants from the CPB.

NPR claims that over 12 million Americans tune in to their member stations. Their literature states "NPR's news and performance programming attracts an audience distinguished by its level of education, professionalism, and community involvement. [Programs such as "All Things Considered" and "Morning Edition" have become the industry's premier news and information features, achieving both popular

A   B   O   U   T

" 🌐 "

PRI Public Radio International℠

P ublic Radio International (PRI) is the source of more public radio programming than any other distributor in the United States. Founded in 1983 as American Public Radio, PRI is steadily moving forward with its second decade goal to "provide expanded global perspectives on world news, current events, and culture to public radio audiences."

In addition to acquiring finished programming from station-based and independent producers around the globe, PRI actively shapes and develops new programs and program formats. PRI also seeks to expand the reach, impact, and relevancy of public radio for audiences who have not traditionally been public radio listeners.

The network emphasizes programming in three general areas:
• News and information
• Classical music
• Comedy/Variety and contemporary music

PRI also distributes an average of four special programs per month.

Keep us in mind and on file. We look forward to talking with **YOU.**

" 🌐 "
PRI Public Radio International"

Public Radio International
100 North Sixth Street, Suite 900A
Minneapolis, Minnesota 55403
Telephone: 612.338.5000
Facsimile: 612.330.9222

This recycled paper is made from 50% waste paper and contains 10% postconsumer waste.

and critical acclaim.] Research shows NPR listeners are consumers of information from many sources and are more likely than average Americans to buy books. They are motivated citizens involved in public activities, such as voting and fund raising. They address public meetings, write letters to editors, and lead business and civic groups."

Public Radio International (formerly American Public Radio) debuted in 1983 and operated much like NPR. It provided listeners with additional noncommercial options, airing popular programs like Garrison Keillor's "A Prairie Home Companion" and many others.

Noncommercial stations can be divided into at least three categories: public, college (noncommercial educational), and community. A fourth category, noncommercial religious stations, has emerged during the last couple of decades.

Many public radio stations, especially those affiliated with NPR, choose to air classical music around the clock, while others opt to set aside only a portion of their broadcast day for classical programming.

According to the Intercollegiate Broadcasting System (IBS), more than eight hundred schools and colleges hold noncommercial licenses. The majority of these stations operate at lower power, some with as little as 10 watts. Since the late 1970s, a large percentage of college stations have upgraded from Class D to Class A stations and now radiate hundreds of watts or more. Most college stations serve as training grounds for future broadcasters while providing alternative programming for their listeners.

Community noncoms are usually licensed to civic groups, foundations, school boards, and religious associations. Although the majority of these stations broadcast at low power, they manage to satisfy the programming desires of thousands of listeners.

Only in rare instances do noncoms pose a ratings threat to commercial stations, and then it is usually in the area of classical music programming. Consequently, commercial and noncommercial radio stations manage a

William Siemering

On Public Radio

Although the distance between commercial and public radio has narrowed in recent years they remain notably different. Since the largest single source of income for public radio stations is listener contributions, the programming is listener driven and the program directors read the Arbitron ratings just as their commercial colleagues do. However, their programming must be significantly different in content and quality from commercial programs to elicit free-will contributions. (In many European countries, public service broadcasting is supported by a tax on receivers, not voluntary contributions.) Commercial stations, while often involved in community service, have one goal: make a profit. Public stations have a mission to serve unmet cultural,

information, and community needs.

The most listened-to programs on public radio are news and information programs: "Morning Edition," "All Things Considered," and "Fresh Air." These are characterized by both the thoroughness of their coverage and breadth of subjects. They regard news of the arts/popular culture to be as important to understanding the world as news of politicians. No commercial network comes close to replicating these programs. Local stations frequently sponsor town meetings on important public affairs issues or sponsor concerts in the community to strengthen their local links. Local stations may broadcast jazz, Triple A (Adult Alternative Album), blue grass, acoustic, classical music, or an array of talk programs. Some stations are directed to specific audiences such as Hispanic or Indian.

Public radio listeners tend to be educated and to include decision makers and influentials in their communities so their influence is great. Eight out of ten newspaper editors rely on it as an important source of information as do network television producers and anchors. *Publisher's Weekly* said public radio is the single most important medium for the book business.

FIGURE 1.12

In addition to support from listeners, public radio also receives corporate and business underwriting, and grants from foundations and the Corporation for Public Broadcasting, which was created in 1967 to distribute federal funds and protect stations against political pressure on programming. Managing a public radio station is, therefore, more complex because its funding is so diverse. At the same time it must protect the independence of the programming from funder influence.

Recently, some politicians have questioned continued federal funding for public broadcasting and various alternative systems of this support are being explored.

---

fairly peaceful and congenial coexistence. (See "Suggested Further Readings" for additional information on noncommercial radio.)

PROLIFERATION AND FRAG-OUT

Specialization, or *narrowcasting* as it has come to be called, salvaged the medium in the early 1950s. Before that time, radio bore little resemblance to its sound during the age of television. It was the video medium that copied radio's approach to programming during its golden age. Sightradio, as television was sometimes ironically called, drew from the older electronic medium its programming schematic and left radio hov-

ering on the edge of the abyss. Gradually radio station managers realized they could not combat the dire effects of television by programming in a like manner. To survive they had to change. To attract listeners they had to offer a different type of service. The majority of stations went to spinning records and presenting short newscasts. Sports and weather forecasts became an industry staple.

Initially, most outlets aired broadappeal music. Specialized forms, such as jazz, rhythm and blues, and country, were left off most playlists, except in certain regions of the country. Eventually these all-things-to-all-people stations were challenged by what is considered to be the first popular attempt at format specialization. As legend now has it, radio programmer Todd Storz and his assistant Bill Steward of KOWH-AM in Omaha, Nebraska, decided to limit their station's playlist to only those records that currently enjoyed high sales. The idea for the scheme struck them at a local tavern as they observed people spending money to play mostly the same few songs on the jukebox. Their programming concept became known as "Top 40." Within months of executing their new format, KOWH topped the ratings. Word of their success spread, inspiring other stations around the nation to take the pop record approach. They too found success.

By the early 1960s other formats had evolved, including Beautiful Music, which was introduced over San Francisco station KABL, and All-News, which first aired over XETRA located in Tijuana, Mexico. Both formats were the progeny of Gordon McLendon and were successfully copied across the country.

The diversity of musical styles that evolved in the mid-1960s, with the help of such disparate performers as the Beatles and Glen Campbell, gave rise to myriad format variations. While some stations focused on 1950s rock 'n' roll ("blasts from the past," "oldies but goodies"), others stuck to current hits and still others chose to play more obscure rock album cuts. The 1960s saw the advent of the radio formats of Soft Rock and Acid and Psychedelic hard rock. Meanwhile, Country, whose popularity had been confined mostly to areas of the South and Midwest, experienced a sudden growth in its acceptance through the crossover appeal of artists such as Johnny Hartford, Bobbie Gentry, Bobby Goldsboro, Johnny Cash, and, in particular, Glen Campbell, whose sophisticated country-flavor songs topped both the Top 40 and Country charts.

As types of music continued to become more diffused in the 1970s, a host of new formats came into use. The listening audience became more and more fragmented. *Frag-out*, a term coined by radio consultant Kent Burkhart, posed an ever-increasing challenge to program directors whose job it was to attract a large enough piece of the radio audience to keep their stations profitable.

The late 1970s and early 1980s saw the rise and decline of the Disco format, which eventually evolved into Urban Contemporary, and a wave of interest in synthesizer-based electropop. Formats such as Soft Rock faded from the scene only to be replaced by a narrower form of Top 40 called Contemporary Hit. New formats continue to surface with almost predictable regularity. Among the most recent batch are All-Comedy, Children's Radio, All-Sports, Eclectic-Oriented Rock, All-Weather, New Age, All-Motivation, All-Business, and All-Beatles.

Although specialization saved the industry from an untimely end over three decades ago, the proliferation in the number of radio stations (which nearly quadrupled since 1950) competing for the same audience has brought about the age of hyperspecialization. Today there are more than one hundred format variations in the radio marketplace, as compared to a handful when radio stations first acknowledged the necessity of programming to a preselected segment of the audience as the only means to remain in business. (For a more detailed discussion on radio formats, see Chapter 3.)

PROFITS IN THE AIR

Although radio has been unable to regain the share of the national advertising dollar it attracted before the arrival of television, it does earn far more today than it did during its so-called heyday. About 7 percent of all money spent on advertising goes to radio. This computes to billions of dollars.

Despite the enormous gains since WEAF introduced the concept of broadcast advertising, radio cannot be regarded as a get-rich-quick scheme. Many stations walk a thin line between profit and loss. While some major market radio stations demand and receive over a thousand dollars for a one-minute commercial, an equal number sell time for the proverbial "dollar a holler."

While the medium's earnings have maintained a progressive growth pattern, it also has experienced periods of recession. These financial slumps or dry periods have almost all occurred since 1950. Initially television's effect on radio's revenues was devastating. The medium began to recoup its losses when it shifted from its reliance on the networks and national advertisers to local businesses. Today 70 percent of radio's revenues come from local spot sales as compared to half that figure in 1948.

By targeting specific audience demographics, the industry remained solvent. In the 1980s, a typical radio station earned fifty thousand dollars annually in profits. As the medium regained its footing after the staggering blow administered it by television, it experienced both peaks and valleys financially. In 1961, for example, the FCC reported that more radio stations recorded losses than in any previous period since it began keeping records of such things. Two years later, however, the industry happily recorded its greatest profits ever. In 1963 the medium's revenues exceeded 636 million dollars. In the next few years earnings would be up 60 percent, surpassing the 1.5 billion mark, and would leap another 150 percent between 1970 and 1980. FM profits have tripled since 1970 and have significantly contributed to the overall industry figures.

The segment of the industry that has found it most difficult to stay in the black is the AM daytimer. These radio stations are required by the FCC to cease broadcasting around sunset so as not to interfere with other AM stations. Of the twenty-three hundred daytimers in operation, nearly a third have reported losses at one time or another over the past decade. The unique problem facing daytime-only broadcasters has been further aggravated by FM's dramatic surge in popularity. The nature of their license gives daytimers subordinate status to fulltime AM operations, which have found competing no easy trick, especially in the light of FM's success. Because of the lowly status of the daytimer in a marketplace that has become increasingly thick with rivals, it is extremely difficult for these stations to prosper, although some do very well. Many day-

Acid Rock	Jazz
Adult Contemporary	Lite
Album-Oriented Rock	MAC
Arena Rock	Mellow Rock
Beautiful Music	Middle-of-the-Road
Big Band	Mix
Black	Modern Rock
Bluegrass	Motown
Bubble Gum	News
Children's	News/Talk
Classical	New Wave
Classic Hits	Nostalgia
Classic Rock	Oldies
Contemporary Country	Pop
Contemporary Hits	Progressive
Country and Western	Punk Rock
Chicken Rock	Religious
Dance	Rhythm and Blues
Disco	Soft Rock
Easy Listening	Southern Rock
Eclectic	Standards
English Rock	Talk
Ethnic	Top 40
Folk Rock	Urban Contemporary
Free Form	Urban Country

FIGURE 1.13
Some popular radio program formats from 1960 to the present. New formats are constantly evolving.

timers have opted for specialized forms of programming to attract advertisers. For example, religious and ethnic formats have proven successful.

At this writing a number of proposals to enhance the status of AM stations are being considered by the FCC. One such proposal suggests that the interference problem can be reduced if cer-

FIGURE 1.14
Courtesy NAB.

THE AM RENAISSANCE

The '90s decade is the decade of change in the radio industry.

As the "underdog," AM radio is in a unique position to increase its strength in the marketplace.

AM offers more flexibility and room for niche-programming to expand radio's audience base.

The growth in the News/Talk, Adult Standards, Spanish and Religious formats documents AM's elasticity.

Innovation is the watchword of AM programming. New on the format scene and "enjoying varying amounts of success," according to an Interep Radio Store report, are Sports Radio and Children's programming.

tain stations shift frequencies to the extended portion (1605–1705 KHz) of the AM band. The FCC's Docket 87-267, issued in the latter part of 1991, cited the preceding as a primary step in the improvement of the AM situation. It inspired many skeptics who regard it as nothing more than a Band-Aid™. Other elements of the plan include tax incentives for AM broadcasters who pull the plug on their ailing operations and multiple AM station ownership in the same market.

As a consequence of the formidable obstacles facing the AM daytime operation, many have been put up for sale, and asking prices have been alarmingly low.

Meanwhile, the price for FM stations has skyrocketed since 1970. Many full-time metro market AMs have sold for multimillions, for the simple reason that they continue to appear in the top of their respective ratings surveys.

In general, individual station profits have not kept pace with profits industry-wide due to the rapid growth in the number of outlets over the past two decades. To say the least, competition is keen and in many markets downright fierce. It is common for thirty or more radio stations to vie for the same advertising dollars in large cities, and the introduction of other media in recent years, such as cable, intensifies the skirmish over sponsors.

ECONOMICS AND SURVIVAL

Following the general financial euphoria and binge-buying of the 1980s, the early 1990s experienced a considerable economic downturn, which had a jarring impact on the radio industry. Had the medium become a "top-down" industry, to use the vernacular of the day?

Nineteen ninety-one is considered by many to be one of the worst years ever. "As a result of the proliferation of stations, the excessively high prices paid for them during the deregulatory buying and selling binge of the late 1980s, and the recession, more than half of the

stations in the country ran in the red," observed Rick Sklar of Sklar Communications. (Mr. Sklar passed away shortly after this interview.)

Producer Ty Ford agreed with Sklar, adding "The price fallout of the late 1980s and early 1990s was due in great part to the collapse of the property-value spiral that was started by deregulation and the negative effect that investors had on the broadcasting business. This resulted in depressed or reduced salaries and an inability to make equipment updates due to the need to pay off highly leveraged station loans."

According to leading radio consultant Kent Burkhart, "The recession in the first third of the 1990s crippled financing of radio properties. The banks were under highly leveraged transaction (HLTs) rules regarding radio loans; thus the value of stations dropped by one-third to one-half. The recession created advertising havoc too. Instead of five-deep buys, we were looking for one through three deep. Emotional sales pitches were rejected. Stations streamlined costs due to the economic slump. Air shifts were expanded and promotion budgets slashed. The top ten to fifteen markets did reasonably well in the revenue column, but those markets outside of the top majors went searching for new ad dollars, which were difficult to find."

The total value of radio station sales declined 65 percent in a six-month period between 1990 and 1991, and radio revenues dropped 4 percent during the same period, according to statistics in *Broadcasting* magazine (September 9, 1991). All of this took place during a time when operating expenses rose. These were troubling figures when compared to the salad days of 1988 when $5.8 billion was paid for 955 radio stations.

Cash flow problems were the order of the day, and this resulted (as Ty Ford pointed out) in significant budgetary cuts. This was made very evident by a report in the December 1991 issue of *Broadcast Engineering* magazine that stated, "All radio budgets show a decrease." The survey showed that budgets for equipment purchases were being delayed and that "planned spending for most areas is somewhat below last year's."

In order to counter the sharp reversal of fortunes, many broadcasters formed local marketing (also called *management*) agreements (LMAs), whereby one radio station leases time and/or facilities from another area station. The buzz-term in the early 1990s became *LMAs*.

LMAs allowed radio stations to enter into economically advantageous, joint operating ventures, stated the editors of *Radio World*. They believe that LMAs should remain the province of the local marketplace and not be regulated by the federal government. The publication asserted that LMAs provide broadcasters a means of functioning during tough economic times and in a ferociously competitive marketplace.

Those who opposed LMAs feared that diversity would be lost as stations combined resources (signals, staffs, and facilities). A few years later, the relaxation of the duopoly rules would raise similar concerns. Proponents argued that this was highly improbable given the vast number of frequencies that light up the dial. In other words, there is safety in numbers, and the public will continue to be served.

However, Pat McNally, general manager of KITS-FM, warned, "In the long run LMAs and consolidation may cause a loss of available jobs in our business and help to continue the erosion of creative salesmanship and conceptual selling. Radio station sales staffs will become like small rep firms."

Century 21 Programming's Dave Scott observes that LMAs have inspired some interesting arrangements. "Some of the novel partnerships include a suburban station north of Atlanta that bought a suburban station south of Atlanta and created one studio to feed them both. A similar situation occurred in San Francisco/San Jose, and I believe they're on the same frequency (or maybe a notch apart). Around Los Angeles, someone got two or three stations on the same frequency. Necessity is the mother of invention, they say."

While many industry people were guardedly hopeful about the future of radio, many saw change as inevitable. "Major adjustments are being made as the consequence of the recent minicrash. In the future, we will depend on fewer nonrevenue producers at stations. We'll see less people per facility. The belt will be tightened for good," notes Bill Campbell, vice president and general manager of WSNE-FM in Providence, Rhode Island.

Rick Sklar accurately predicted that satellite-supplied stations will have a role in this. "To save money and stay in business, large numbers of stations are going satellite and others are turning to suppliers of twenty-four-hour formats for their programming because of the economic environment. (See Chapter

---

FCC SAYS RADIO IS IN 'PROFOUND FINANCIAL DISTRESS'

"Small [radio] stations—the bulk of the industry—are in profound financial distress."

That's the first line and bottom line of an internal FCC report on the state of the radio business distributed to Chairman Alfred Sikes and other commissioners last week.

"Radio today is a world of large haves and little have-nots," the report says. "Industry revenue and profits are overwhelmingly concentrated in the small number of large radio stations, while most small stations struggle to remain solvent."

One indicator of the "distress": by the FCC's count, 287 radio stations have gone dark, 53% in just the last 12 months.

Not surprisingly, the findings undergird pending proposals to relax the radio ownership rules, which prohibit a company from owning more than one AM or FM in a market and from owning more than 12 AM's and 12 FM's. (A minority-controlled company may own up to 14 stations of each type.) The FCC may vote on the ownership proposal in March or April.

"The potential economies from consolidation would materially improve industry profitability," the report says. "If a conservative 10% of general and administrative costs could be eliminated, for example, the savings would raise industry profitability by 30%," the report continues. "Alternatively, these savings could immediately boost flat per-station programing outlays by 5% and still raise industry profits by 15%," the report says.

The top-50 large-market stations, just one-half of 1% of the some 10,000 stations now on the air, account for 11% of industry revenue and 50% of industry profit in 1990, the report says. Yet stations with less than $1 million in annual revenues—75% of all stations—on average, lost money in 1990, it says. —*NU*

FIGURE 1.15
In the early 1990s stories like this revealed the downside of one of radio's most turbulent years. Courtesy *Broadcasting*.

11 for more on syndicated programming.)

Fred Jacobs, president of Jacob's Media, assessed the state of the Fifth Estate. "In the first third of the 1990s, everything seems to be converging, and radio's future is up in the air. Many operators are in debt and are feeling the pressures of a long economic recession. Similarly, the FCC is not providing regulatory focus. There's no consensus about the legality of LMAs, multiple station ownership in the same market, and so on. Format fragmentation has made for a more competitive environment in most cities, including medium markets. The available revenue pie is now being split among more players. Like cable television, radio has become very niche-oriented. To make matters even more uncertain, Digital Audio Broadcasting (DAB) is also a murky issue at the moment. It's coming, but we don't really have a fix on when and in what form. Will it create even more outlets? If so, where will existing broadcasters end up and what will be the value of their properties? Will there be enough advertising revenue to go around?"

Despite the medium's shaky start in the 1990s, B. Eric Rhoads possessed a positive outlook. "I believe that radio is poised for a strong future over the next ten years. Cable is floundering in its local sales efforts, and TV is having huge problems. Nationally the networks are still high-priced and have reduced viewing. In the end, radio is the only stable medium. It is targeted and cost effective. This fact will keep the industry alive for a long time. Radio will survive and thrive."

RAB's senior vice president of radio, Lynn Christian, believed that the economic ills of radio had been overstated. "To paraphrase Mark Twain, 'Reports of radio's death have been grossly exaggerated.' At a recent RAB Management gathering, the mood was pretty upbeat. In fact, a number of broadcasters said their profits were up. I think the recession of the early 1990s has been less injurious to the medium than it has been to other industries. Radio's going to be fine."

CONSOLIDATION AND DOWNSIZING

As the medium entered the middle of the decade, it was doing more than just fine. The headlines in the industry trade publications revealed exactly how well the medium had recovered: "Radio Draws Advertisers as Economy Strengthens" (*Broadcasting*, May 1994), "Recovery" (*Radio Ink*, December 1993), "Radio Revenues Hit One Billion in May" (*Radio World*, August 1994), "National Spot Revenue up 38 Percent" (*Broadcasting*, May 1995).

The primary cause of this dramatic upsurge was the relaxation of FCC rules, foremost among them station ownership caps and duopoly. By the mid-1990s, individual companies could own several stations in the same market, and this spurred active trading and mergers of broadcast properties. The idea was to reduce competition and thus overhead. The consolidation and downsizing prompted by LMAs would pick up steam with the elimination of the duopoly rule, which prevented dual station ownership in the same market.

Says Lynn Christian, "Consolidation—market by market—is the word best describing what is happening in commercial radio in the 1990s. The legal authority to own and operate several radio stations in almost every market is rapidly changing radio's landscape. Radio station operators are becoming more like local cable operators, offering a variety of formats on the FM and AM dials."

One alarming effect of downsizing for aspiring broadcasters is the reduction of available jobs. "Individual station staffs get small as companies grow in station holdings. A direct result of duopoly softening and the increase in ownership limits is fewer jobs, more generalization, and less specialization. Jacks and Jills of all trades will be valuable. Group presidents will be taking jobs as station managers, especially in duopoly/multiple station operations," observes Ed Shane, president of Shane Media.

Another concern inspired by consolidation is the potential loss of programming diversity. Says Christian, "While cost savings and profits are central to the concept behind downsizing and multiple ownership, the creative forces in radio are taking a hit. In point of fact, in the past few years no exciting new programming ideas have been developed."

Despite the many concerns, business is good and getting better in the mid-1990s. The dollar volume of station transactions (number of stations changing hands) approved by the FCC is creeping up to the astronomic levels of the pre-1991 slump, and local revenue was up 7 percent in 1995.

BUYING AND SELLING

Today brokerage firms handle the sale of many radio stations. "It's difficult to overlook the importance of Wall Street and the financial community in the future of radio," notes Ed Shane. Bill Campbell, co-owner of Blue River Communications, says the future is now. "Wall Street is where much of the buying and selling of radio outlets occurs nowadays. Things have changed to where stations are sold through lawyers and brokerage houses more than they are from broadcaster to broadcaster. Those are pretty much bygone days, and that is kind of sad. It's the 'three-piece suiters' game in the 1990s. There is little direct negotiating, no bargaining between owners over a drink at the corner pub. Stations are commodities to be bought and sold by people who sometimes have little appreciation or understanding of what radio is really all about. Of course, the economic inertia of the first part of this decade inspired more direct negotiations (strategic alliances) between owners, and I think that is good. I'm also detecting a move to drive the MBAs out of our business. Broadcasters who gain general experience beyond just management are the future."

For their services, brokers receive an average commission of 7 to 8 percent on sales, and in some cases they earn

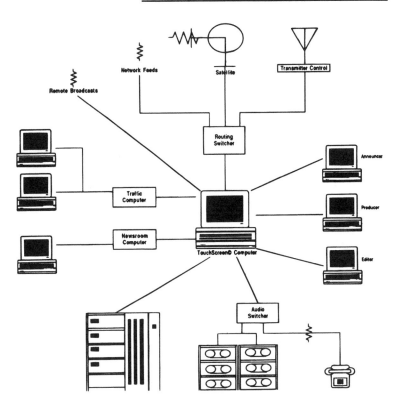

additional incentives based on the size of the transaction. In recent years brokers have been very successful in negotiating large profits for their clients.

Brokerage firms promote the sale of stations through ads in industry trade magazines, direct mailings, and appearances at broadcast conferences. Interested buyers are provided with all the pertinent data concerning a station's geographical location, physical holdings, operating parameters, programming, and income history, as well as economic, competitive, and demographic information about the area within reach of the station's signal.

Another recent approach to the buying and selling of radio properties is the auction method, although this means of selling a station is perceived by some as a kind of last-resort effort at getting rid of profitless stations, most of which are AM. (See Figure 1.18.)

The average price of an AM station in the late 1980s was $450,000, with some selling for as little as a few thousand dollars, while others sold for several million. In 1986 New York station WMCA-AM sold for $11 million, and in

FIGURE 1.16
The fully computerized radio station has arrived. The idea behind it is to enhance the efficiency of station operations. In many cases, the computerization has improved a station's product while reducing expenses, often in the area of personnel.

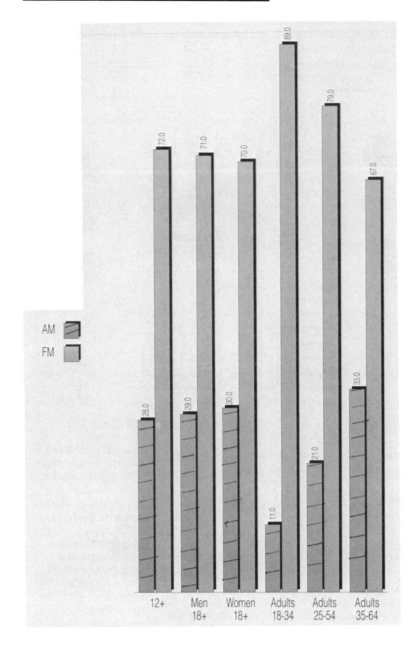

FIGURE 1.17
AM/FM share of
listening. Courtesy
of National
Database.

station sales in the early 1990s as reflected in a broadcast trade paper:

City and station: Port Arthur, TX.,
                     KLTN FM
                     Price: $3.65 million.
City and station: Staunton, VA.,
                     WTON AM/FM
                     Price: $1 million.
City and station: Cabo Roio, Puerto
                     Rico, WEKO AM
                     Price: $1.1 million.
City and station: Hampton, N.H.,
                     WZEA FM
                     Price: $1.1 million.

DAB REVOLUTION

DAB makes analog amplitude modulation (AM) and frequency modulation (FM) outmoded systems. With the great popularity of home digital music equipment (CD and DAT), broadcasters are forced to convert their signals to remain competitive. Thus DAB looms large in the future of radio. The days of analog signal propagation are numbered. (For an explanation of both digital and analog signaling, see Chapters 9 and 10.)

In the mid-1980s, compact disc players were introduced to the consumer market. Today, CD players rank as the top consumer item for home music reproduction. Turntables have gone by the board, and the analog tape cassette market is beginning to downsize. Digital is here to stay, at least until something better comes along.

At first DAB was perceived as a threat by broadcasters. The National Association of Broadcasters (NAB) looked at the new sound technology adversarially. In an interview in the July 23, 1990, issue of *RadioWeek*, John Abel, NAB's executive vice president of operations, stated, "DAB is a threat and anyone who plans to stay in business for a while needs to pay careful attention."

As time went on, DAB was regarded as a *fait accompli*, something that was simply going to happen. Soon broadcasters assumed a more proactive pos-

1992, WFAN-AM received a bid of $70 million, proving that AM stations can still command enormous sums.

The average price for FM stations is higher than it is for AM. In the early 1990s the average price for an FM station exceeded $3 million.

Many AM broadcasters look for full-time status, improved reception, and stereo to increase the value of their properties in the coming years.

Here are some representative radio

ture regarding the technology, and then the concern shifted to where to put the new medium and how to protect existing broadcast operations.

Early on, NAB proposed locating DAB in the L-Band portion of the electromagnetic spectrum. It also argued for in-band placement. Eventually the FCC saw fit to recommend that DAB be allocated room in the S-Band, and it took its proposal to the World Administrative Radio Conference (WARC) held in Spain in February 1992. This spectrum designation is expected to help in-band terrestrial development. In-band digital signaling would permit broadcasters to remain on their existing frequencies. This is something they favor, as satellite DAB signal transmission is regarded as a significant threat to the local nature of U.S. broadcasting.

On the other hand, many countries are fully supportive of a satellite DAB system because they do not have the number of stations the U.S. possesses and thus lack the coverage and financial investment.

As of this writing, many of the DAB issues are still being debated, and this will likely be the case throughout much of the decade. However, it appears a relative certainty that DAB will be available to the U.S. public in the not-too-distant future. Of course, this will require a whole new direction in home/car/walk-along receiver design. Simply put, DAB renders existing radios obsolete. This is cause for some anxiety among broadcasters who wonder how quickly the buying public will convert. However, considerable confidence exists, since the consumer's huge appetite for new and improved sound shows no sign of abating.

America is a nation of audiophiles, demanding high-quality sound. Analog broadcasting cannot compete with the interference-free reception and greater frequency dynamics of digital signals.

Digital signaling heralds a new age in radio broadcasting. Jeff Tellis, president of the Intercollegiate Broadcasting System (IBS), explains why. "The reason for the great interest in DAB is its considerable number of advantages." Among them are:

Significantly improved coverage using significantly less power.

Dramatic improvement in the quality of the signal; compare CD to vinyl.

More precise coverage control using multiple transmitters similar to cellular phone technology.

No adjacent channel reception problems.

On-channel booster capabilities eliminating the need to use separate frequencies to extend the same signal.

Easy transmission of auxiliary services, including format information, traffic, weather, text, and selective messaging services.

Sharing of transmitting facilities—common transmitter and antenna.

Telecommunications professor Ernest Hakanen expands on the cost advantages of digital broadcasting. "DAB also promises to be economically efficient. Since there is no interstation interference between DAB signals and because of the appeal of the spectrum efficiency provided by the interleaved environment, all of the channel operators in an area could utilize the same transmitter. The transmission facilities could be operated by a consortium for the construction, operation, and maintenance of the common transmission plant. Antenna height for DAB systems is also lower than current FM standards. Electrical power conservation and savings are a huge advantage of DAB."

Picking up on Hakanen's point about consolidating broadcast operations, RAB's Lynn Christian says, "In the future, the consortium (radio station malls) approach to maintaining and operating a station will be commonplace for obvious economic reasons. DAB is very conducive to a collaborative relationship among broadcasters."

Prior to the WARC meeting in 1992, NAB's DAB Task Force proposed a set of standards to ensure that the technology would operate effectively. The specifications included:

CD quality sound

Enhanced coverage area

Accommodation of existing AM and FM frequencies

FIGURE 1.18
Radio is going to
bits. Courtesy NAB.

RADIOACTIVE

A 'Bit' About The Future

NAB has devoted a substantial amount of resources to protect and nurture broadcasting's "playing field of the future." The future has some radio broadcasters nervous about their place in the changing electronic environment.

The assignment of spectrum space and the various distribution systems and techniques, have captured trade press headlines. One of the nation's leading futurists, Nicholas Negroponte from the MIT Media Lab in Boston, told a recent NAB Futures Summit that it will

Does radio have a future as an important entertainment and information medium in the 21st Century?

simply boil down to "bits and bytes" (see *RadioWeek* 1/20). According to Negroponte, "The separation of radio and TV will be arcane" because we will all have licenses "to radiate so many bits per second." Also, TV sets will carry names such as IBM or Apple. Both he and NAB Exec VP/Operations John Abel encouraged broadcasters to become active in data broadcasting to protect their future.

As this debate continues, the next generation of radio broadcasters will question their role in radio with these pending technological changes.

Does radio have a future as an important entertainment and information medium in the 21st Century? Will the proliferation of media choices, the aging of America's population, and the changing ethnic structure in the U.S. allow radio an opportunity to maintain its dynamic "voice of the people" role?

Sounds, voices, music and events are all vital modules of the American lifestyle. As the soundtrack for many generations of listeners, radio will remain prosperous and evolve if broadcasters (or, if you will, "bit-casters") continue to be innovative. The single most important ingredient for future success in radio will be creativity. Tell the people entering the field not to worry about radio's distribution system — a good system, or several systems, will be available in the next century.

Now is the time to think about developing our programming product to fill these high-quality stereo pipelines: new concepts, new formats, and new ideas are needed. Programming and marketing offer especially exciting creative challenges on the "new radio" we are investing in today. Encourage everyone to think positive about radio's future — just as we do at NAB.

— Lynn Christian, Senior VP/NAB Radio

Immunity to multipath interference
Immunity to stoplight fades
No interference to existing AM and FM
   broadcasters
DAB system interference immunity
Minimization of transmission costs
Receiver complexity
Additional data capacity
Reception area threshold

After nearly a century of analog signal transmission, radio is about to venture into the digital domain, which will keep it relevant to the demands of a technologically sophisticated listening marketplace as it embarks on its next 100 years.

CABLE AND SATELLITE RADIO

Radio broadcasters have a wary eye on the evolving digital audio services being made available by cable and satellite services. It is the threat of increased competition that inspires contempt for these new audio options. While broadcasters have long employed satellite programming and network services to enhance their over-the-air terrestrial signals, the idea of a direct-to-consumer alternative is not greeted with enthusiasm, especially since these are already available in digital sound, something broadcasters are still investigating. At this writing, the FCC was deciding where to locate satellite radio in the electromagnetic spectrum, while the NAB argued against its introduction into the local marketplace. Despite all the brouhaha, it appears certain that conventional earthbound radio will be joined in its difficult quest for listeners and advertisers by these new delivery systems and that the already fragmented radio audience will be splintered even further.

Subscriber music services, such as Digital Music Express and Music Choice, are growing rapidly with the aid of satellite delivery systems, like DirecTV and Primestar Partners, and their appeal is considerable—no chatter, no commercials, tailor-made formats, and so on.

RADIO ON-LINE

Your favorite radio station on your laptop computer? Why not? Many stations already use e-mail for interactivity with their audiences, and now stations are preparing to download signals to PC's.

The term *network* has existed nearly since the medium's inception, and now the information highway term *Internet*, which embraces this vintage word, is making significant inroads into the industry's vernacular.

While some broadcasters view the computer as just another unwanted competitor, many are happily exploiting

FIGURE 1.19
The local nature of
U.S. radio is
thought to be
threatened by
national DAB
satellite delivery.

it as a new opportunity and potential revenue stream. Internet Talk Radio is already garnering an impressive audience around the country, and a service called RealAudio provides radio shows through on-line service to computer speakers. NPR and ABC both have download services, and the ranks of computer radio programmers is swelling.

A good example is KJHK-FM in Lawrence, Kansas, which claims to be among the first stations broadcasting around the clock on the Internet. A growing number of outlets (over 600 at this writing) are establishing station web sites, whereby listeners can access programming and interact with the station by dialing up through their modems. The brave new world of the cyber-station is here!

Stations are finding the Internet to be an outstanding tool for station promotion. It is quick, easy, cheap, and universal.

Interactive radio is a growing reality, as is the opportunity for everyone with the right computer and software to be a broadcaster or cybercaster. With an Internet encoder, the home user can transmit to an international audience. This prospect prompts a collective sigh from station managers, who are losing track of the new forms of competition.

Notes Lynn Christian, "The major concern regarding the future of radio is centered on new competition from satellite, cable, and on-line sources. Those companies which are planning to partner with these new media choices, and develop data services, will undoubtedly be the big winners in the 21st century. Broadcast radio, as I have known it during the past 40 years, will not be the same in 10 years. But what American business will be the same in the year 2000? These are revolutionary times in radio and in the world."

Professors Jeff Harman and Bruce Klopfenstein have attempted to determine (in a special Delphi Forecast Study for the NAB) the future of new and evolving technologies in the radio industry. Among other things, their research concluded that by the year 2000 most radio stations will be engaged in some form of:

FIGURE 1.20
Cable radio
services rely on
satellite technology
to deliver
programming to
subscriber systems
nationwide.
Courtesy
Superaudio.

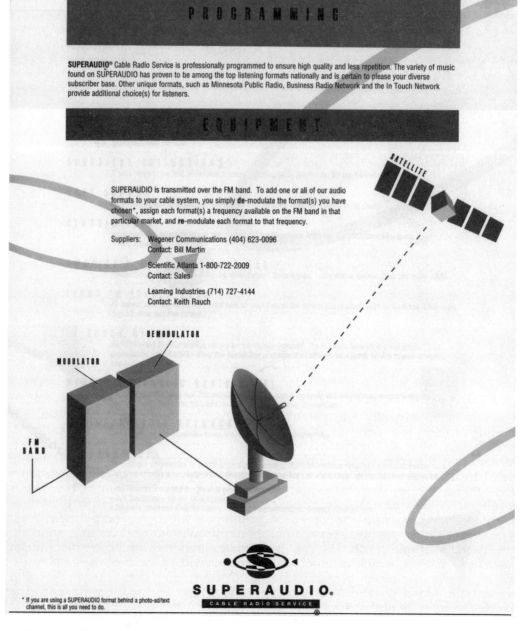

wide-area datacasting, computer paging, digital transmission (DARS), smart receivers (RDS), high-speed data broadcasting, digital production, and Internet services.

Ed Shane adds desktop broadcasting to the list of present and future innovations. "Mainly related to all the downsizing, one station or even seven will be babysat by a computer."

As radio heads warp speed into this "future world," it is obvious that aspiring broadcasters will have to know their way around a computer, because the audio studio will exist within the digital and cyber domains.

RADIO AND GOVERNMENT REGULATIONS

Almost from the start it was recognized that radio could be a unique instrument for the public good. This point was

```
Enter your selection:
Unknown selection - please re-enter

                 Welcome to the FCC Public Access Link
                                "P A L"
             We now support 2400 baud.  Try it, you'll like it!

     1 - Access Equipment Authorization Database
     2 - Definitions - Terms/Codes used in Application Records
     3 - Applying for an Equipment Authorization (1/92)
     4 - Other Commission Activities and Procedures (8/92)
     5 - Laboratory Operational Information
     6 - Public Notices (6/93)
     7 - Bulletins / Measurement Procedures (3/93)
     8 - Rulemakings (8/93)
     9 - Help
     a - Information Hotline (2/93)
     b - Processing Speed of Service (1/94)
     c - Test Sites on File per Sec 2.948 (2/94)
     d - ADVANCED TELEVISION SERVICE SCHEDULE OF MEETINGS

     0 - Exit PAL

Enter your selection:
  ALT-F10  HELP | VT-100    | FDX |  2400 N81 | LOG CLOSED | PRT OFF | CR    | CR
```

FIGURE 1.21
The FCC can be accessed via the Internet, providing stations with needed information and quick answers.

never made more apparent than in 1912, when, according to legend (which was recently debunked), a young wireless operator named David Sarnoff picked up the distress signal from the sinking *Titanic* and relayed the message to ships in the vicinity, which then came to the rescue of those still alive. The survivors were the beneficiaries of the first attempt at regulating the new medium. The Wireless Ship Act of 1910 required that ships carrying fifty or more passengers have wireless equipment on board. The effective use of the medium from an experimental station in New York City's Wanamaker Building resulted in the saving of seven hundred lives.

Radio's first practical application was as a means of communicating from ship to ship and from ship to shore. During the first decade of this century, Marconi's wireless invention was seen primarily as a way of linking the ships at sea with the rest of the world. Until that time, when ships left port they were beyond any conventional mode of communications. The wireless was a boon to the maritime services, including the Navy, which equipped each of its warships with the new device.

Coming on the heels of the *Titanic* disaster, the Radio Act of 1912 sought to expand the general control of radio on the domestic level. The secretary of commerce and labor was appointed to head the implementation and monitoring of the new legislation. The primary function of the act was to license wireless stations and operators. The new regulations empowered the Department of Commerce and Labor to impose fines and revoke the licenses of those who operated outside the parameters set down by the communications law.

Growth of radio on the national level was curtailed by World War I, when the government saw fit to take over the medium for military purposes. However, as the war raged on, the same young wireless operator, David Sarnoff, who supposedly had been instrumental in saving the lives of passengers on the ill-fated *Titanic*, was hard at work on a scheme to drastically modify the scope of the medium, thus converting it from an experimental and maritime communications apparatus to an appliance designed for use by the general public. Less than five years after the war's end, receivers were being bought by the millions, and radio as we know it today was born.

As explained earlier, the lack of regulations dealing with interference nearly resulted in the premature end of radio. By 1926 hundreds of stations clogged the airways, bringing pandemonium to the dial. The Radio Act of 1912 simply did not anticipate radio's new application. It was the Radio Act of 1927 that

FIGURE 1.22
Digital audio is now
available on the
Internet. RealAudio
makes your
computer a stereo
radio. Courtesy
RealAudio.

Audio-on-demand for the Internet...

RealAudio Player and Encoder
Now your browser can play audio on demand --
no more long file transfers!
● Read about RealAudio 2.0
● Download FREE Player
● Download FREE Encoder

RealAudio Server
Make your Web site come alive with sound
effects and dialogue.
● Evaluate Server ver. 1.0
● Beta test the Personal Server
● Purchase the Server.

...Discover RealAudio 2.0 Now!

Sites And Sounds

Hear the most exciting new ways people are using RealAudio:

Now Hear This

The latest news and announcements from Progressive Networks

What's New
Last Update: January 23rd; Next Update: January 26th

WebActive comes alive
Our newest Web publication keeps you up-to-date on the latest in activism
and progressive politics.

RealAudio 2.0 Sites
Hear live and on-demand music at these popular sites.

RealAudio in the News
Progressive Networks announces RealAudio Server Products for the
Macintosh.

• U.S. Internet Broadcast Stations

FIGURE 1.23
Internet links to
station web sites.

o WXYC
   Chapel Hill, North Carolina, USA's 89.3 FM station is also broadcasting live on
   the internet (MBONE)! Check out how **Internet Broadcasting** is done and how
   you can listen-in on your Mac, UNIX or PC box here! Oh yeah, their program
   schedule, rotation list and assorted radio info resides in this sunsite.unc.edu server
   as well (email shoffner@cs.unc.edu).

o KJHK
   KJHK 90.7 FM in Lawrence, KS (KU's student run college radio station) began
   broadcasting its programming 24 hours over the Internet on December 3rd, 1994.

o KUGS
   Bellingham, Washington's KUGS 89.3 FM radio station serving the students of
   Western Washington University is also broadcasting on the Internet.

o Nexus UTV, England
   On the 13th May 1995 Nexus UTV (the student television station of the University
   of East Anglia) began a LIVE 24 hour broadcast down the Internet.

o Radio HK
   Radio HK claims to be the first 24 hour continuous radio station (not broadcasting
   over the airwaves) broadcasting over the Internet.

o NEW WBBR - Bloomberg Information Radio
   Using Xing Technology's Streamworks, we have WBBR broadcasting the news
   and information network of Bloomberg radio.

o NEW KING
   KING-FM is a commercial classical music station in Seattle that in the past 5 years
   has twice won the NAB's Marconi award as best classical music station in the U.S.
   They are the first classical station in the world to send live feeds over the net via
   RealAudio 2. Also in the works is a virtual tour of their studios.

first approached radio as a mass medium. The Federal Regulatory Commission's five commissioners quickly implemented a series of actions that restored the fledgling medium's health.

The Communications Act of 1934 charged a seven-member commission with the responsibility of ensuring the efficient use of the airways—which the government views as a limited resource that belongs to the public and is leased to broadcasters. Over the years the FCC has concentrated its efforts on maximizing the usefulness of radio for the public's benefit. Consequently, broadcasters have been required to devote a portion of their airtime to programs that address important community and national issues. In addition, broadcasters have had to promise to serve as a constant and reliable source of information, while retaining certain limits on the amount of commercial material scheduled.

The FCC has steadfastly sought to keep the medium free of political bias and special-interest groups. In 1949 the commission implemented regulations making it necessary for stations that present a viewpoint to provide an equal amount of airtime to contrasting or opposing viewpoints. The Fairness Doctrine obliged broadcasters "to afford reasonable opportunity for the discussion of conflicting views of public importance." Later it also stipulated that stations notify persons when attacks were made on them over the air.

Although broadcasters generally acknowledge the unique nature of their business, many have felt that the government's involvement has exceeded reasonable limits in a society based upon a free enterprise system. Since it is their money, time, and energy they are investing, broadcasters feel they should be afforded greater opportunity to determine their own programming.

Jay Williams, Jr.

The Future of Radio

FIGURE 1.24

No other industry in the United States is changing and evolving as dramatically as the entire telecommunications industry. Almost as fast, marketing and advertising are also undergoing an unprecedented upheaval. And, in the midst of these changes, regulations governing radio are being relaxed, spurring new development and investment enthusiasm. Radio is in the eye of the hurricane of change.

Here are some of the factors changing radio:

Telecommunications: Digital has revolutionized technology, allowing for capacity and flexibility never before imagined. For example, telephone companies no longer just provide lines for communication, they now also provide the products and services of communication. Increasingly in competition (or alliances) with cable and on-line services, these giant delivery systems provide entertainment, services information, and marketing opportunities. As radio is one of the original seven "media groups" (which include TV, newspapers, books, magazines, etc.), radio will soon be in competition with these new services for consumer/listener time and attention and, not much later, for revenue. Digital itself has literally changed the communications paradigm; in the face of this new technology, communications, computer, and telephone companies will merge, acquire, or form alliances in an attempt to dominate the time

and total involvement of the consumer. Program and content providers will align themselves with these new delivery systems in the new media world.

Technology: Digital and regulatory factors will increasingly add to the more efficient radio operations. Changes in EBS, unattended operations, digital workstations, and the like are but a tip of the iceberg. Expanded band, satellite radio, floating platforms allowing networked radio personalities to interface with locally programmed music, radio programming on the

Internet, digital transmission, and more will change the face of radio.

Regulatory factors: Technology can only expand as rapidly as regulatory constraints allow. With the vast array of choices now available to the listener/viewer from these new media sources, the justification for both content and technological restraints covering radio is evaporating. Clearly, if technological advancements are not soon encouraged, radio will be relegated as obsolete by consumers. Thus, reality dictates the direction of deregulation; politics unfortunately, dictates the pace.

Duopoly and multistation ownership: As the regulatory climate is relaxed, mature market economics will continue to impact the radio industry. The problems of increasingly expensive talent and personnel (and the difficulty in finding qualified people, particularly in smaller markets) are being solved by increasingly sophisticated computer equipment which can run major areas of a station easily and efficiently. Jobs will continue to be combined as computer-networked personnel are able to handle multiple internal tasks for off-site stations as well as those in the next office. Similar equipment allows stations to reduce the once large on-air staffs to a handful of employees.

Types of owners: Although publicly traded companies can own radio stations, it wasn't until the regulatory climate changed that their interest grew in radio. Over the past few years, the elimination of the three year rule, the so-called Fairness Doctrine, and the 7-7-7 ownership rule created a more positive investment climate for these demanding companies. More recent regulatory changes include the creation of LMA's duopolies, and higher limits on station ownership among these well-financed companies, driving up multiples in large and medium-sized markets as they build clusters of four, five, and even more stations in a single market. The result is that radio is moving quickly from a "cottage industry," with many family-owned stations, to a network of fewer but more powerful group owners (a transition that parallels those in retailing and other industries).

Marketing: The degree of intense competitive programming among radio stations will decrease as multiple stations in a market are co-owned. Co-owned stations in direct format competition will be altered by ownership to decrease cannibaliza-tion of listenership and to maximize revenue opportunities. Thus, the internal focus will shift more to sales and, soon after, to nonspot revenue as pressure grows on spot inventory. Advertisers, however, are also changing; increasingly shown specific user information by other media groups, advertisers will force radio to present more than "listener estimates" and generic qualitative information. Pressure will grow on the rating services to provide specific information; yet the solutions ultimately must lie in listener databases created and maintained by individual stations in order to prove specific station listenership by geodemographic and consumer behavioral data. Promotional demands on stations, by advertising agencies and clients alike, will continue to increase as advertisers attempt to create additional reasons for people to become acquainted with or sample their products or services.

The concept of broadcasting is dead. Broadcasting, as with all media, must become interactive to survive. The listener must be allowed to "talk-back" and communicate with the station. The VCR created a whole new paradigm; no longer did the family have to gather for the CBS Monday night movie on Monday night. Real time will become increasingly more obsolete as the listener begins to take more control of when and what he or she hears (creating "my time"). Initially this can be done by allowing listeners to preselect prizes they want to win in a contest or promotion, giving them special listener status with individual incentives, allowing them to respond directly in the medium of their choice (phone, Internet, write-in) and receive personal responses. Radio broadcasters will learn that real time, station time, is no longer as important as "my time," when the listener wants to listen.

Broadcasters who develop the means to segment listeners effectively (both for sales and marketing and to create listener loyalty) and provide ways for those listeners to receive the information/entertainment/contests and prizes when they want them will create the "new era of radio."

In the late 1970s, a strong movement headed by Congressman Lionel Van Deerlin sought to reduce the FCC's role in broadcasting, in order to allow the marketplace to dictate how the industry conducted itself. Van Deerlin actually proposed that the Broadcast Branch of the commission be abolished and a new organization with much less authority created. His bill was defeated, but out of his and others' efforts came a new attitude concerning the government's hold on the electronic media.

President Reagan's antibureaucracy, free-enterprise philosophy gave impe-tus to the deregulation move already under way when he assumed office. The FCC, headed by Chairman Mark Fowler, expanded on the deregulation proposal that had been initiated by his predecessor, Charles Ferris. The dereg-ulation decision eliminated the re-quirement that radio stations devote a portion of their airtime (8 percent for AM and 6 percent for FM) to nonenter-tainment programming of a public affairs nature. In addition, stations no longer had to undergo the lengthy process of ascertainment of commu-nity needs as a condition of license

Air Concert "Picked Up" By Radio Here

Victrola music, played into the air over a wireless telephone, was "picked up" by listeners on the wireless receiving station which was recently installed here for patrons interested in wireless experiments. The concert was heard Thursday night about 10 o'clock, and continued 20 minutes. Two orchestra numbers, a soprano solo—which rang particularly high and clear through the air—and a juvenile "talking piece" constituted the program.

The music was from a Victrola pulled up close to the transmitter of a wireless telephone in the home of Frank Conrad, Peen and Peebles avenues, Wilkinsburg. Mr. Conrad is a wireless enthusiast and "puts on" the wireless concerts periodically for the entertainment of the many people in this district who have wireless sets.

Amateur Wireless Sets, made by the maker of the Set which is in operation in our store, are on sale here $10.00 up.

—*West Basement*

FIGURE 1.25
History-making ad in 1920. Courtesy Westinghouse Electric.

renewal, and guidelines pertaining to the amount of time devoted to commercial announcements were eliminated. The rule requiring stations to maintain detailed program logs was also abolished. A simplified postcard license renewal form was adopted, and license terms were extended from three to seven years. In a further step the commission raised the ceiling on the number of broadcast outlets a company or individual could own from seven AM, seven FM, and seven television stations to twelve each. As of March 1992, the FCC saw fit to raise the caps on ownership again, this time to 30 AM and 30 FM (see Figure 1.22). Later in the year, these were reduced to 18 AM and 18 FM. Three years later, with a Republican Congress in place, a new telecommunications bill proposed to eliminate ownership caps completely. The bill also sought to relax the cross-ownership rule, which kept a single entity from possessing a radio, TV, and newspaper company in the same market. Meanwhile, radio license terms were to be raised to ten years.

On August 4, 1987, the FCC voted to eliminate the thirty-eight-year-old Fairness Doctrine, declaring it unconstitutional and no longer applicable to broadcasters. A month before, President Reagan had vetoed legislation that would have made the policy law.

The extensive updating of FCC rules and policy was based on the belief that the marketplace should serve as the primary regulator. Opponents of the reform feared that with their new-found freedom, radio stations would quickly turn their backs on community concerns and concentrate their full efforts on fattening their pocketbooks.

Those who support the position that broadcasters should first serve the needs of society are concerned that deregulation (unregulation) has further

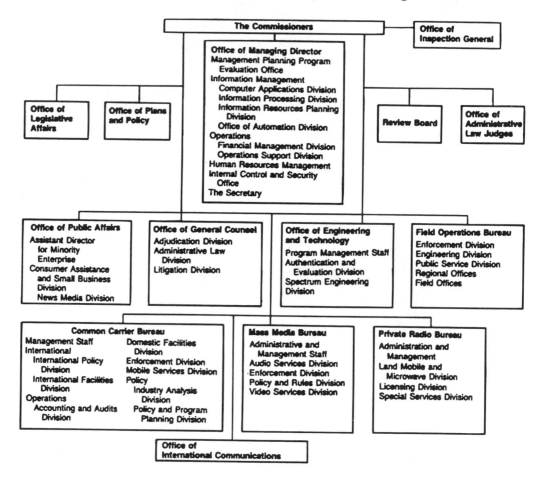

FIGURE 1.26
FCC organizational chart.

reduced the medium's "good-citizen" role. "Radio, especially the commercial sector, has long since fallen down on its 'interest, convenience, and necessity' obligation born of the Radio Act of 1927. While a small segment of the industry does exert an effort to address the considerable problems facing society today, the overwhelming majority continue to be fixated on the financial bottom line. There needs to be more of a balance," observes Robert Hilliard, former FCC Chief of Public and Educational Broadcasting. Proponents of deregulation applauded the FCC's actions, contending that the listening audience would indeed play a vital role in determining the programming of radio stations, since the medium had to meet the needs of the public in order to prosper.

While the government continues to closely scrutinize the actions of the radio industry to ensure that it operates in an efficient and effective manner, it is no longer perceived as the fearsome, omnipresent Big Brother it once was. Today broadcasters more fully enjoy the fruits of a laissez-faire system of economy.

In the spring of 1996, the telecommunications bill (made a Telecommunication Act) mentioned above became a reality and ownership caps were all but eliminated.

JOBS AND EQUALITY IN RADIO

Today the radio industry employs over 100,000 people, but with all the downsizing and consolidation, this figure is likely to shrink some in the coming years. Since 1972 75,000 individuals have found full-time employment in radio. Today, opportunities for women and minorities are greater than ever. Radio has until fairly recently been a male-dominated profession. In 1975 men in the industry outnumbered women nearly four to one. But that has changed. Now women are being hired more than ever before, and not just for office positions. Women have made significant inroads into programming,

---

How FCC Rules Are Made

I. Initiation of Action

Suggestions for changes to the FCC Rules and Regulations can come from sources outside of the Commission either by formal petition, legislation, court decision or informal suggestion. In addition, a Bureau/Office within the FCC can initiate a Rulemaking proceeding on its own.

II. Bureau/Office Evaluation

When a petition for Rulemaking is received, it is sent to the appropriate Bureau(s)/Office(s) for evaluation. If a Bureau/Office decides a particular petition is meritorious, it can request that the Dockets department assign a Rulemaking number to the petition.

A similar request is made when a Bureau/Office decides to initiate a Rulemaking procedure on its own. A weekly notice is issued listing all accepted petitions for Rulemaking. The public has 30 days to submit comments. The Bureau/Office then has the option of generating an agenda item requesting one of four actions by the Commission. If a Notice of Inquiry (NOI) or Notice of Proposed Rulemaking (NPRM) is issued, a docket is instituted and a docket number is assigned.

III. Possible Commission Actions

Major changes to the Rules are presented to the public as either an NOI or NPRM. The Commission will issue an NOI when it is simply asking for information on a broad subject or trying to generate ideas on a given topic. An NPRM is issued when there is a specific change to the Rules being proposed.

If an NOI is issued, it must be followed by ei-ther an NPRM or a Memorandum Opinion and Order (MO&O) concluding the inquiry.

IV. Comments and Replies Evaluated

When an NOI or NPRM has been issued, the public is given the opportunity to comment initially, and then respond to the comments that are made. When the Commission does not receive sufficient comments to make a decision, a further NOI or NPRM may be issued.

It may be determined that an oral argument before the Commission is needed to provide an opportunity for the public to testify before the Commission, as well as for the Bureau(s)/Office(s) to present diverse opinions concerning the proposed Rule change.

V. Report and Order Issued

A Report and Order is issued by the Commission stating the new or amended Rule, or stating that the Rules will not be changed. The proceeding may be terminated in whole or in part.

The Commission may issue additional Report and Orders in the docket.

VI. Reconsideration Given

Petitions for reconsideration may be filed by the public within 30 days. They are reviewed by the appropriate Bureau(s)/Office(s) and/or by the Commission.

VII. Modification Possible

As a result of its review of a petition for reconsideration, the Commission may issue an MO&O modifying its initial decision or denying the petition for reconsideration.

Provided by FCC

FIGURE 1.27 Courtesy of Radio World.

---

sales, and management positions, and there is no reason to think that this trend will not continue. It will take a while, however, before an appropriate proportion of women and minorities are working in the medium. The FCC's insistence on equal opportunity employment (at this writing there was concern that this requirement would be dropped) within the broadcast industry makes prospects good for all who are interested in broadcasting careers.

Still there is room for improvement in the participation levels of different ethnic groups in the radio industry. Says NAB's vice president for human resources, Dwight Ellis, "Current employment trends reported by the FCC reveal very little growth between 1980 and 1992 in Native American broadcast employment, for example. Other groups have experienced slow growth too. The NAB is committed to assisting the growth of employment and station ownership for minorities in the industry. The NAB has been a vanguard for minority progress in broadcasting.

FIGURE 1.28
Native-owned
stations provide
radio jobs for
American Indians.
Courtesy KTNN.

Nevertheless, more must be done."

A common misconception is that a radio station consists primarily of dee-jays with few other job options available. Wrong! Nothing could be further from the truth. Granted, disc jockeys comprise part of a station's staff, but many other employees are necessary to keep the station on the air.

An average-size station in a medium market employs between eighteen and twenty-six people, and on-air personnel comprise about a third of that figure. Stations are usually broken down into three major areas: sales, programming, and engineering. Each area, in particular the first two, requires a variety of people for positions that demand a wide range of skills. Subsequent chapters in this book will bear this out.

Proper training and education are necessary to secure a job at most stations, although many will train people to fill the less demanding positions. Over a thousand schools and colleges offer courses in radio broadcasting, and most award certificates or degrees. As in most other fields today, the more credentials a job candidate possesses, the better he or she looks to a prospective employer.

Perhaps no other profession weighs practical, hands-on experience as heavily as radio does. This is especially true in the on-air area. On the programming side, it is the individual's sound that wins the job, not the degree. However, it is the formal training and education that usually contribute most directly to the quality of the sound that the program director is looking for when hiring. In reality, only a small percentage of radio announcers have college degrees (the number is growing), but

statistics have shown that those who do stand a better chance of moving into managerial positions.

Many station managers look for the college-educated person, particularly for the areas of news and sales. Before 1965 the percentage of radio personnel with college training was relatively low. But the figure has gradually increased as more and more colleges add broadcasting curricula. Thousands of communications degrees are conferred annually, thus providing the radio industry a pool of highly educated job candidates. Today, college training is a plus when one is searching for employment in radio.

The job application or resume that lists practical experience in addition to formal training is most appealing. The majority of colleges with radio curricula have college stations. These small outlets provide the aspiring broadcaster with a golden opportunity to gain some much-needed on-air experience. Some of the nation's foremost broadcasters began their careers at college radio stations. Many of these same schools have internship programs that provide the student with the chance to get important on-the-job training at professional stations. Again, experience is experience, and it does count to the prospective employer. Small commercial stations often are willing to hire broadcast students to fill part-time and vacation slots. This constitutes "professional" experience and is an invaluable addition to the resume.

Entry-level positions in radio seldom pay well. In fact, many small market stations pay near minimum wage. However, the experience gained at these small-budget operations more than makes up for the small salaries. The first year or two in radio constitutes the dues-paying period, a time in which a person learns the ropes. The small radio station provides inexperienced people the chance to become involved in all facets of the business. Rarely does a new employee perform only one function. For example, a person hired to dee-jay will often prepare and deliver newscasts, write and produce commercials, and may even sell airtime.

In order to succeed in a business as

unique as radio, a person must possess many qualities, not the least of which are determination, skill, and the ability to accept and benefit from constructive criticism. A career in radio is like no other, and the rewards, both personal and financial, can be exceptional. "It's a great business," says Lynn Christian, senior vice president of the Radio Advertising Bureau. "No two days are alike. After thirty some years in radio, I still recommend it over other career opportunities."

CHAPTER HIGHLIGHTS

1. The average adult spends two-and-a-half hours per day listening to radio. Radio is the most available source of entertainment, companionship, and information.

2. Guglielmo Marconi is generally considered the father of radio, although David Sarnoff is a likely contender.

3. As early as 1922, the Department of Commerce set aside two frequencies for radio broadcasts. WEAF in New York aired the first commercial.

4. Today, most radio networks have been subsumed by major corporations (Disney, GE, Westinghouse).

5. Station networks, first called chain broadcasting, operated as early as 1922. Radio Corporation of America (RCA) formed the first major network in 1926, the National Broadcasting Company (NBC). Columbia Broadcasting System (CBS) was formed in 1928, and Mutual Broadcasting System (MBS) followed in 1934. American Broadcasting Company (ABC), formed in 1945, became the largest and most successful radio network.

6. Early station proliferation led to overlapping signals. Signal quality decreased, as did listenership. The Radio Act of 1927 formed the Federal Radio Commission (FRC), a five-member board authorized to issue station licenses, allocate frequency bands, assign frequencies to individual stations, and dictate station power and hours of operation. The FRC established the Standard Broadcast band

Words from Larry King

"I started my career in May 1957. I did news and sports. In 1960 I began a talk show in Miami and have been hosting talk ever since. Back then I also did sports for the Miami Dolphins. Sports is a big love of mine. I've never regarded what I do as a 'challenge,' per se. It is a love affair. My desire is to inform and entertain the listener. That gets me going. For those young people out there who want to pursue a career in radio, I'd say never give up. Always be yourself. Don't try to be Larry King. The only secret is there is no secret. Your talent will come out if you are good. As far as the future of radio is concerned, it is hard to predict. Talk is always in good shape. The public tends to grow as part of radio. There will always be a need for what the medium offers. Bottom line: the opportunities will be there for those good enough to take advantage of them."

FIGURE 1.29
The Larry King Show is aired by over 350 stations. Courtesy Larry King.

THE MOST INFLUENTIAL INTERVIEW PROGRAM ON PLANET EARTH

CNN LARRY KING LIVE

SIMULCAST LIVE EACH DAY ON WESTWOOD ONE ENTERTAINMENT

Catch all the biggest stories, monumental events and exciting personalites as Larry King talks with the movers and shakers in politics and entertainment on Westwood One's daily simulcast of *CNN'S LARRY KING LIVE.* Monday through Saturday 9-10 p.m. (ET).

WESTWOOD ONE ENTERTAINMENT

FOR MORE INFORMATION CONTACT YOUR WESTWOOD ONE REPRESENTATIVE AT (703) 413-8550.

(500–1500 kc).

7. Radio prospered during the Depression by providing cost-free entertainment and escape from the harsh financial realities. "Amos 'n' Andy," which made its debut in 1929, was the most popular radio show in history. President Franklin D. Roosevelt's fireside chats began on March 12, 1933. The Communications Act of 1934 established the seven-member Federal Communications Commission (FCC).

FIGURE 1.30
Organizational flow chart for a medium market radio station.

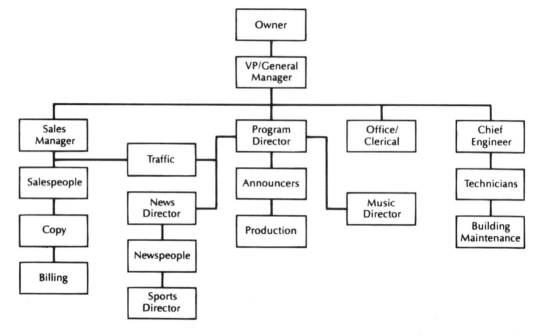

8. World War II led the FCC to freeze construction of new broadcast outlets; therefore, existing AM outlets prospered.

9. Two years after the war's end, television usurped the entertainment leadership.

10. The Bell Lab scientists' invention of the transistor in 1948 helped save radio by providing portability. Music became radio's mainstay.

11. The Top 40 radio format, conceived about the same time as the emergence of rock music, became the most popular format with younger audiences.

Bruce DuMont

On the Museum of Broadcast Communications

The Museum of Broadcast Communications is one of only two broadcast museums in America. It is a museum of popular culture and contemporary American history that preserves and presents historic moments from radio and television. It contains a collection of 75,000 radio and television programs and commercials depicting the historic moments of the past 50 years.

Public programs are presented frequently, featuring many of radio and television's major personalities, as well as those who work behind the scenes. The significant moments of world history in the past fifty years have been captured by microphones and cameras, and the Museum's Archives represent an opportunity for students, scholars, and the general public to tap into that history.

The Archives are fully computerized and are accessible to the general public seven days a week. The Museum of Broadcast Communications contains America's only Radio Hall of Fame, which includes 85 of the major personalities and leaders of the medium. Working radio and television studios provide visitors with an opportunity to experience the thrill of being "on-the-air."

FIGURE 1.31

The Museum of Broadcast Communications at the Chicago Cultural Center is located on Michigan Avenue at Washington Street.

Media Jobs Offered Sorted by Date

FIGURE 1.32
Jobs on the Net.
Provided by
Airwaves Online.

- Post To This Group
- Back To Index Sorted By: [thread][subject][author]

Starting: *Mon 23 Oct 1995 - 21:57:09 CST*
Ending: *Wed 28 Feb 1996 - 20:49:47 CST*
Messages: 44

- HELP WANTED *WRVV@aol.com*
- Help wanted: Morning News/Sidekick *Michael G. Scull*
- Help Wanted: AE for Cape Cod *WXTK*
- Help Wanted: Music Director/AT *MaxPace@aol.com*
- equity position with sales experience *MR STEVE HARTFORD*
- Help Wanted : Entry level news reporter *Kevin McKinnon*
- HELP WANTED: Sales Manager-Birmingham. *BILL THOMAS*
- HELP WANTED: Sales Manager-Birmingham. *BILL THOMAS*
- HELP WANTED: MUSIC DIRECTOR/MID-DAY AT *Rick Daniels*
- Help Wanted: College Students or Graduates for Internships on LI *CleanWilly@aol.com*
- Help Wanted: News Reporter/Anchor, Toledo, OH *BobSeybold@aol.com*
- Help Wanted: Radio Copywriter, UK *Ian Britton*
- Help wanted:OPS MGR/NEWS TALENT *MR DAVID H ISEMAN*
- Help Wanted: Syndication Radio Sales *FMacek@aol.com*
- Help Wanted: Adv.Sales Reps/Illinois *Randy Miller*
- Help Wanted: N/T Air talent *John B. Adams*
- Re: Job opening in Worcester *SidS1045@aol.com*
- Help Wanted: Air Talent/Salesp in North & Cent Illinois *Randy Miller*
- Help Wanted: Radio News Director *Joey Helleny*
- HELP WANTED: Station Manager *T. KentAtkins*
- News/Operations person needed *Mike Petersen*
- Account Manager needed *Mike Petersen*
- Help Wanted: Station Manager (Radio - NW) *doherty@ios.com*
- Help Wanted: Assistant Engineer, Pittsburgh area *CHRISHUDAK@aol.com*
- Help Wanted: Production/Technical coordinator *Richard Paul*
- HELP WANTED TORONTO CAMPUS COMMUNITY STATION NEEDS TECH/PROD. PERSON *Richard Paul*
- help wanted section/engineer *Terry*
- Help Wanted *KGMO*
- Help Wanted: News Anchor/Reporter, Coldwater & Kalamazoo, MI *Sean A. Watson*
- HELP WANTED: P/T Board Operators-Air Talents *WLNO Radio*
- HELP WANTED: KIDS RADIO NEWS ANCHOR *Maura Gallucci*
- Help Wanted: Executive Director, NACB *Michael Black...WEOS-FM*
- Help Wanted: Worcester PM drive opening *SSCHWEIGER@delphi.com*
- Help Wanted: Worcester news opening *SSCHWEIGER@delphi.com*
- Help Wanted: Development Director, Program Director *The_Doge*
- Help Wanted: Chief Engineer, Denver, CO, *Eric Schecter*
- Help Wanted: Radio Transmitter Sales Engineer *Howard M. Ginsberg*
- HELP WANTED PAGE *Myloe@aol.com*
- Help Wanted *John Rosso*
- HELP WANTED: Milwaukee Marketing Director *John Davis*
- HELP WANTED - Affiliate Relations - Syndication - NY/LA *SETHJEN@aol.com*
- Help Wanted: International Producers *Paul Baldwin*
- help wanted *Rbk1959@aol.com*
- HELP WANTED: MILWAUKEE PRODUCTION DIRECTOR *John Davis*

Last message date: *Wed 28 Feb 1996 - 20:49:47 CST*
Archived on: *Fri Mar 01 1996 - 10:41:32 CST*

12. Edwin H. Armstrong developed the static-free FM band in 1938. The FCC authorized stereo FM broadcasting in 1961, and in 1965 it separated FM from AM simulcasts in cities with populations in excess of one hundred thousand. Beautiful Music was the first popular format on FM, which relied heavily on automation. The Progressive format focused on album cuts rather than Top 40. By the mid-1980s FM possessed 70 percent of the listening audience.

13. Today FM commands 80 percent of the audience although some AM stations top their market ratings.

14. The Corporation for Public Broadcasting was established in 1967. Three years later National Public Radio began providing funding and programming to member stations. More than eight hundred schools and colleges hold noncommercial radio licenses.

15. Narrowcasting is specialized programming. Frag-out refers to the fragmentation of audience because of numerous formats.

16. Highly leveraged transactions (HLTs) have created economic problems for many stations. Local Marketing Agreements (LMAs) allow broadcasters to form contracts with one another for mutually beneficial purposes.

17. Consolidation, downsizing, and mergers have been prompted by the FCC's relaxation of the duopoly and ownership rules.

18. Seventy percent of radio's revenues come from local spot sales.

19. Brokerage firms handle the sale of most radio stations. Brokers receive a commission of between 7 and 8 percent on sales.

20. Digital Audio Broadcasting (DAB) is set to replace the conventional analog system of signal transmission and reception.

21. Cable and satellite radio services are providing listeners with options besides traditional terrestrial signal reception.

22. Radio is on the Internet as stations download programming to PCs. The medium uses computer web sites for true interactivity with audiences.

23. In 1949 the FCC formulated the Fairness Doctrine, which obligated broadcasters to present opposing points of view. In 1987 the FCC declared the Doctrine unconstitutional and eliminated it.

24. The Republicans wrote a new telecommunications bill designed to lift significant sanctions from broadcasters. The Telecommunication Act of 1996 all but eliminated radio station ownership ceilings.

25. The radio industry employs over one hundred thousand people, with women and minorities making significant gains in recent years. A combination of practical experience and formal training is the best preparation for a career in broadcasting.

SUGGESTED FURTHER READING

Aitkin, Hugh G.J. *Syntony and Spark.* New York: John Wiley and Sons, 1976.

Archer, G.L. *History of Radio to 1926.* New York: Arno Press, 1971.

Aronoff, Craig E., ed. *Business and the Media.* Santa Monica, Calif.: Goodyear Publishing Company, 1979.

Baker, W. J. *A History of the Marconi Company.* New York: St. Martin's Press, 1971.

Barnouw, Erik. *A Tower of Babel: A History of Broadcasting in the United States to 1933*, vol. 1. New York: Oxford University Press, 1966.

———.*The Golden Web: A History of Broadcasting in the United States 1933 to 1953*, vol. 2. New York: Oxford University Press, 1968.

———. *The Image Empire: A History of Broadcasting in the United States from 1953*, vol. 3. New York: Oxford University Press, 1970.

———. *Media Marathon: A Twentieth-Century Memoir.* Durham: Duke University Press, 1996.

———. *The Sponsor: Notes on a Modern Potentate.* New York: Oxford University Press, 1978.

Bergreen, Laurence. *Look Now, Pay Later: The Rise of Network Broadcast-*

ing. Garden City, N.Y.: Doubleday, 1980.

Bittner, John R. *Broadcasting and Telecommunications*, 2nd ed. Englewood Cliffs, N.J.: Prentice-Hall, 1985.

———. *Professional Broadcasting: A Brief Introduction.* Englewood Cliffs, N.J.: Prentice-Hall, 1981.

Blake, Reed H., and Haroldsen, E. O. *A Taxonomy of Concepts in Communications.* New York: Hastings House, 1975.

Brant, Billy G. *College Radio Handbook.* Blue Ridge Summit, PA: Tab Books, 1981.

Broadcasting Yearbook. Washington, D.C.: Broadcasting Publishing, 1935 to date, annual.

Browne, Bartz, and Coddington (consultants). *Radio Today—And Tomorrow.* Washington, D.C.: National Association of Broadcasters, 1982.

Buono, Thomas J., and Leibowitz, Matthew L. *Radio Acquisition Handbook.* Miami: Broadcasting and the Law, 1988.

Campbell, Robert. *The Golden Years of Broadcasting.* New York: Charles Scribner's Sons, 1976.

Cantril, Hadley. *The Invasion from Mars.* New York: Harper & Row, 1966.

Chapple, Steve, and Garofalo, R. *Rock 'n' Roll Is Here to Pay.* Chicago: Nelson-Hall, 1977.

Cox Looks at FM Radio. Atlanta: Cox Broadcasting Corporation, 1976.

Delong, Thomas A. *The Mighty Music Box.* Los Angeles: Amber Crest Books, 1980.

Ditingo, Vincent M. *The Remaking of Radio.* Boston: Focal Press, 1995.

Douglas, Susan J. *Inventing American Broadcasting, 1899–1922.* Baltimore: Johns Hopkins University Press, 1987.

Dreher, Carl. *Sarnoff: An American Success.* New York: Quadrangle, 1977.

Dunning, John. *Tune In Yesterday.* Englewood Cliffs, N.J.: Prentice-Hall, 1976.

Edmonds, I. G. *Broadcasting for Beginners.* New York: Holt, Rinehart, and Winston, 1980.

Erickson, Don. *Armstrong's Fight for FM Broadcasting.* Birmingham: University of Alabama, 1974.

Fang, Irving E. *Those Radio Commentators.* Ames: Iowa State University Press, 1977.

Fornatale, Peter, and Mills, J.E. *Radio in the Television Age.* New York: Overlook Press, 1980.

Foster, Eugene S. *Understanding Broadcasting.* Reading, Mass.: Addison-Wesley, 1978.

Grant, August E., ed. *Communication Technology Update.* Boston: Focal Press, 1995.

Hall, Claud, and Hall, Barbara. *This Business of Radio Programming.* New York: Hastings House, 1978.

Hasling, John. *Fundamentals of Radio Broadcasting.* New York: McGraw-Hill, 1980.

Head, Sydney W., and Sterling, C.H. *Broadcasting in America: A Survey of Television, Radio, and the New Techologies*, 7th ed. Boston: Houghton Mifflin, 1993.

Hilliard, Robert L. *The Federal Communications Commission: A Primer.* Boston: Focal Press, 1991.

———, ed., *Radio Broadcasting: An Introduction to the Sound Medium*, 3rd ed. New York: Longman, 1985.

———, and Keith, Michael C. *The Broadcast Century: A Biography of American Broadcasting.* Boston: Focal Press, 1992.

———. *Global Broadcasting Systems.* Boston: Focal Press, 1996.

Hunn, Peter. *Starting and Operating Your Own FM Radio Station.* Blue Ridge Summit, PA: Tab Books, 1988.

Inglis, Andrew F. *Behind the Tube.* Boston: Focal Press, 1990.

Keirstead, Phillip O., and Kierstead, S.-K. *The World of Telecommunication.* Stoneham, Mass.: Focal Press, 1990.

Keith, Michael C. *Signals in the Air: Native Broadcasting In America.* Westport, CT.: Praeger Publishing, 1995.

Ladd, Jim. *Radio Waves.* New York: St. Martin's Press, 1991.

Lazarsfeld, Paul F., and Kendall, P.L. *Radio Listening in America.* Englewood Cliffs, N.J.: Prentice-Hall, 1948.

Leinwall, Stanley. *From Spark to*

Satellite. New York: Charles Scribner's Sons, 1979.

Levinson, Richard. *Stay Tuned.* New York: St. Martin's Press, 1982.

Lewis, Peter, ed. *Radio Drama.* New York: Longman, 1981.

Lewis, Tom. *Empire of the Air: The Men Who Made Radio.* New York: Harper-Collins Publishers, 1991.

Lichty, Lawrence W., and Topping, M. C. *American Broadcasting: A Source Book on the History of Radio and Television.* New York: Hastings House, 1976.

Looker, Thomas. *The Sound and the Story.* Boston: Houghton Mifflin, 1995.

MacDonald, J. Fred. *Don't Touch That Dial: Radio Programming in American Life, 1920–1960.* Chicago: Nelson-Hall, 1979.

Matelski, Marilyn. *Vatican Radio.* Westport, CT.: Praeger Publishing, 1995.

McLuhan, Marshall. *Understanding Media: The Extensions of Man.* New York: McGraw-Hill, 1964.

Morrow, Bruce. *Cousin Brucie.* New York: Morrow, 1987.

O'Donnell, Lewis B., et al. *Radio Station Operations: Management and Employee Perspectives.* Belmont, Calif.: Wadsworth Publishing, 1989.

Orlik, Peter B. *Electronic Media Criticism.* Boston: Focal Press, 1994.

Paley, William S. *As It Happened: A Memoir.* Garden City, N.Y.: Doubleday, 1979.

Pease, Edward C., and Dennis, Everette E. *Radio: The Forgotten Medium.* New Brunswick, NJ: Transaction Press, 1995.

Pierce, John R. *Signals.* San Francisco: W. H. Freeman, 1981.

Pusateri, C. Joseph. *Enterprise in Radio.* Washington, D.C.: University Press of America, 1980.

Radio Facts. New York: Radio Advertising Bureau, 1988.

Rhoads, B. Eric. *Blast from the Past.* West Palm Beach, FL: Streamline Press, 1996.

Routt, Ed. *The Business of Radio Broadcasting.* Blue Ridge Summit, Pa.: Tab Books, 1972.

Sarnoff, David. *The World of Television.* Agoura Hills, Calif.: Wisdom, 1958.

Schiffer, Michael B. *The Portable Radio in American Life.* Tucson: University of Arizona Press, 1991.

Seidle, Ronald J. *Air Time.* Boston: Holbrook Press, 1977.

Settle, Irving. *A Pictorial History of Radio.* New York: Grosset and Dunlap, 1967.

Sipemann, Charles A. *Radio's Second Chance.* Boston: Little, Brown, 1946.

Sklar, Rick. *Rocking America: How the All-Hit Radio Stations Took Over.* New York: St. Martin's Press, 1984.

Smith, F. Leslie. *Perspectives on Radio and Television: An Introduction to Broadcasting in the United States.* New York: Harper & Row, 1979.

Sterling, Christopher H. *Electronic Media.* New York: Praeger, 1984.

Wertheim, Arthur F. *Radio Comedy.* New York: Oxford University Press, 1979.

Whetmore, Edward J. *The Magic Medium: An Introduction to Radio in America.* Belmont, Calif.: Wadsworth Publishing, 1981.

———. *MediaAmerica*, 4th ed. Belmont, Calif.: Wadsworth Publishing, 1989.

Woolley, Lynn. *The Last Great Days of Radio.* Dallas: Republic of Texas Press, 1995.

2 Station Management

NATURE OF THE BUSINESS

It has been said that managing a radio station is like running a business that is a combination of theater company and car dealership. The medium's unique character requires that the manager deal with a broad mix of people, from on-air personalities to secretaries and from sales personnel to technicians. Few other businesses can claim such an amalgam of employees. Even the station manager of the smallest outlet with as few as six or eight employees must direct individuals with very diverse backgrounds and goals. For example, radio station WXXX in a small Maine community may employ three to four full-time air people, who likely were brought in from other areas of the country. The deejays have come to WXXX to begin their broadcasting careers with plans to gain some necessary experience and move on to larger markets. Within a matter of a few months the station will probably be looking for replacements.

Frequent turnover of on-air personnel at small stations is a fact of life. As a consequence, members of the air staff often are regarded as transients or passers-through by not only the community but also the other members of the station's staff. Less likely to come and go are a station's administrative and technical staff. Usually they are not looking toward the bright lights of the bigger markets, since the town in which the station is located is often home to them. A small station's sales department may experience some turnover but usually not to the degree that the programming department does. Salespeople also are likely to have been recruited from the ranks of the local citizenry, whereas air personalities more typically come from outside the community.

Running a station in a small market presents its own unique challenges (and it should be noted that half of the nation's radio outlets are located in communities with fewer than twenty-five thousand residents), but stations in larger markets are typically faced with stiffer competition and fates that often are directly tied to ratings. In contrast to station WXXX in the small Maine community, where the closest other station is fifty miles away and therefore no competitive threat, an outlet located in a metropolitan area may share the airwaves with thirty or more other broadcasters. Competition is intense, and radio stations in large urban areas usually succeed or fail based upon their showing in the latest listener surveys. The metro market station manager must pay close attention to what surrounding broadcasters are doing, while striving to maintain the best product possible in order to retain a competitive edge and prosper.

Meanwhile, the government's perception of the radio station's responsibility to its consumers also sets it apart from other industries. Since its inception, radio's business has been Washington's business, too. Station managers, unlike the heads of most other enterprises, have had to conform to the dictates of a federal agency especially conceived for the purpose of overseeing their activities. With the failure to satisfy the FCC's expectations possibly resulting in stiff penalties, such as fines and even the loss of an operating license, radio station managers have been obliged to keep up with a fairly prodigious volume of rules and regulations.

The 1980s and 1990s deregulation actions designed to unburden the broadcaster of what had been regarded by many as unreasonable government intervention have made the life of the station manager somewhat less complicated. Nevertheless, the government continues to play an important role in American radio, and managers who value their license wisely invest energy and effort in fulfilling federal conditions. After all, a radio station without a frequency is just a building with a lot of expensive equipment.

The listener's perception of the radio business, even in this day and age when nearly every community with a small business district has a radio station, is often unrealistic. The portrayal by film and television of the radio station as a hotbed of zany characters and bizarre antics has helped foster a misconception. This is not to suggest that radio stations are the most conventional places to work. Because it is the station's function to provide entertainment to its listeners, it must employ creative people, and where these people are gathered, be it a small town or a large city, the atmosphere is sure to be charged. "The volatility of the air staff's emotions and the oscillating nature of radio itself actually distinguishes our business from others," observes J.G. Salter, general manager, WFKY, Frankfort, Kentucky.

Faced with an audience whose needs and tastes sometimes change overnight, today's radio station has become adept at shifting gears as conditions warrant. What is currently popular in music, fashion, and leisure-time activities will be nudged aside tomorrow by something new. This, says broadcast manager Randy Lane, forces radio stations to stay one step ahead of all trends and fads. "Being on the leading edge of American culture makes it necessary to undergo more changes and updates than is usually the case in other businesses. Not adjusting to what is currently in vogue can put a station at a distinct disadvantage. You have to stay in touch with what is happening in your own community as well as the trends and cultural movements occurring in other parts of the country."

The complex internal and external factors that derive from the unusual nature of the radio business make managing today's station a formidable challenge. Perhaps no other business demands as much from its managers. Conversely, few other businesses provide an individual with as much to feel good about. It takes a special kind of person to run a radio station.

THE MANAGER AS CHIEF COLLABORATOR

There are many schools of thought concerning the approach to managing a radio station. For example, there are the standard X, Y, and Z models or the-

	AM	FM		AM	FM
WABC	770		WLIR		92.7
WADB		95.9	WLIX	540	
WADO	1280		WLTW		106.7
WALK	1370	97.5	WMCA	570	
WAPP		103.5	WMCX		88.1
WAWZ	1380	99.1	WMGQ		98.3
WBAB		102.3	WMJY		107.1
WBAI		99.5	WMTR	1250	
WBAU		90.3	WNBC	660	
WBGO		88.3	WNCN		104.3
WBJB		90.5	WNEW	1130	102.7
WBLI		106.1	WNJR	1430	
WBLS		107.5	WNNJ	1360	
WBRW	1170		WNYC	830	93.9
WCBS	880	101.1	WNYE		91.5
WCTC	1450		WNYG	1440	
WCTO		94.3	WNYM	1330	
WCWP		88.1	WNYU		89.1
WDHA		105.5	WOBM	1170	92.7
WERA	1590		WOR	710	
WEVD		97.9	WPAT	930	93.1
WFAS	1230	103.9	WPIX		101.9
WFDU		89.1	WPLJ		95.5
WFME		94.7	WPOW	1330	
WFMU		91.1	WPRB		103.3
WFUV		90.7	WPST		97.5
WGBB	1240		WPUT	1510	
WGLI	1290		WQXR	1560	96.3
WGRC	1300		WRAN	1510	
WGSM	740		WRCN	1570	103.9
WHBI		105.9	WRFM		105.1
WHLI	1100		WRHU		88.7
WHN	1050		WRKL	910	
WHPC		90.3	WRKS		98.7
WHTG	1410	106.3	WRTN		93.5
WHTZ		100.3	WRVH		105.5
WHUD		100.7	WSBH		95.3
WHWH	1350		WSIA		88.9
WICC	600		WSKQ	620.0	
WINS	1010		WSOU		89.5
WIXL		103.7	WSUS		102.3
WJDM	1530		WTHE	1520	
WJLK	1310	94.3	WUSB		90.1
WKCR		89.9	WVIP	1310	106.3
WKDM	1380		WVOX	1460	
WKJY		98.3	WVRM		89.3
WKMB	1070		WWDJ	970	
WKRB		90.9	WWRL	1600	
WKTU		92.3	WXMC	1310	
WKWZ		88.5	WYNY		97.1
WKXW		101.5	WYRS		96.7
WLIB	1190		WZFM		107.1
WLIM	1580				

FIGURE 2.1
Newspaper listing of radio stations in a major market. Large cities such as New York, Chicago, Los Angeles, and Philadelphia often have over sixty stations.

ories of management (which admittedly oversimplify the subject but give the neophite a basic working model). The first theory embraces the idea that the general manager is the captain of the vessel, the primary authority, with solemn, if not absolute, control of the decision-making process. The second theory casts the manager in the role of collaborator or senior advisor. The third theory forms a hybrid of the preceding two; the manager is both coach and team player, or chief collaborator. Of the three models, broadcast managers tend to favor the third approach.

Lynn Christian, who has served as general manager of several radio stations, preferred working for a manager who used the hybrid model rather than the purely authoritarian model. "Before I entered upper management, I found that I performed best when my boss sought my opinion and delegated responsibility to me. I believe in department head meetings and the full disclosure of projects within the top organization of the station. If you give someone the title, you should be prepared to give that person some authority, too. I respect the integrity of my people, and if I lose it, I replace them quickly. In other words, 'You respect me, and I'll respect you,' is the way I have always managed."

Randy Bongarten, radio network chief, concurs with Christian and adds, "Management styles have to be adaptive to individual situations so as to provide what is needed at the time. In general, the collaborator or team leader approach gets the job done. Of course, I don't think there is any one school of management that is right one hundred percent of the time."

Jim Arcara, another radio network head, also is an advocate of the hybrid management style. "It's a reflection of what is more natural to me as well as my company. Employees are capable of making key decisions, and they should be given the opportunity to do so. An effective manager also delegates responsibility."

KITS-FM general manager Pat McNally finds the collaborative approach suitable to his goals and temperament. "My management style is more collaborative. I believe in hiring qualified professional people, defining what I expect, and allowing them to do their job with input, support, and constructive criticism from me. My door is always open for suggestions, and I am a good listener. I consider this business something special, and I expect an extra special effort."

This also holds true for Steven Woodbury, general manager of KLXK-FM. "I hire the best people as department heads and then work collaboratively with these experts. Department heads are encouraged to run their areas as if they had major ownership in the company. That instills a sense of team spirit too. Their energy level and decision-making efforts reflect this."

The manager/collaborator approach has gained in popularity in the past two decades. Radio functions well in a team-like atmosphere. Since practically every job in the radio station is designed to support and enhance the air product, establishing a connectedness among what is usually a small band of employees tends to yield the best results, contends station manager Jane Duncklee. "I strongly believe that employees must feel that they are a valid part of what is happening and that their input has a direct bearing on those decisions which affect them and the operation as a whole. I try to hire the best people possible and then let them do their jobs with a minimum of interference and a maximum of support."

Marlin R. Taylor, founder of Bonneville Broadcasting System and former manager of several major market radio outlets, including WRFM, New York, and WBCN, Boston, believes that the manager using the collaborative system of management gets the most out of employees. "When a staff member feels that his or her efforts and contributions make a difference and are appreciated, that person will remain motivated. This kind of employee works harder and delivers more. Most people, if they enjoy the job they have and like the organization they work for, are desirous of improving their level of performance and contributing to the health and well-being of the station. I really think that many station managers should devote even more time and energy to people development."

Station general manager Paul Aaron believes that managers must first assert their authority, that is, make it clear to all that they are in charge, before the transition to collaborator can take place. "It's a sort of process or evolution. Actually, when you come right down to it, any effective management approach includes a bit of both the authoritarian and collaborative concepts. The situation at the station will have a direct impact on the management style I personally deem most appropriate. As the saying goes, 'different situations call for different measures.' When assuming the reins at a new station, sometimes it's necessary to take a more dictatorial approach until the organization is where you feel it should be. Often a lot of cleanup and adjustments are necessary before there can be a greater degree of equanimity. Ultimately, however, there should be equanimity."

Surveys have shown that most broadcast executives view the chief collabora-

tor or hybrid management approach as compatible with their needs. "It has pretty much become the standard modus operandi in this industry. A radio manager must direct as well as invite input. To me it makes sense, in a business in which people are the product, to create an atmosphere that encourages self-expression, as well as personal and professional growth. After all, we are in the communications business. Everyone's voice should at least be heard," contends Lynn Christian.

WHAT MAKES A MANAGER

As in any other profession, the road to the top is seldom a short and easy one. It takes years to get there, and dues must be paid along the way. To begin with, without a genuine affection for the business and a strong desire to succeed, it is unlikely that the position can ever be attained. Furthermore, without the

FIGURE 2.2 Keeping the station solvent is the manager's primary task. Companies like this one help the manager do this. Courtesy CFM. do this.

CASH FLOW MANAGEMENT - *THE "KEY" TO SUCCESS IN RADIO*

You know how it goes: Advertising sales made ... a commercial runs ... you bill the client or agency ... and finally, 60 to 90 days later, you receive payment for your services.

It's the nature of the business.

But delayed revenue can keep radio stations painfully short of cash. That restricts sales staff and puts a stranglehold on growth.

CAN YOU AFFORD TO LEAVE YOUR STATION ON HOLD?

FACTORING CLEARS THE WAY

Cash Flow Management has the answer. We specialize in operating capital for the radio industry. We will purchase your accounts receivable and monitor collections.

Factoring relieves cash flow problems on the spot, making cash available to drive sales.

Having **Dollars** available for operating capital is the key to success in the radio industry.

The **Bottom Line**. More profit potential.

EXPERIENCED FINANCIAL SERVICE

Cash Flow Management operates as an extension of your accounting department. You decide what level of your receivables you want to sell. You pay a nominal fee. We take it from there.

Operating in your name, we:
• Provide accounts receivable ledgering
• Review credit on all current and prospective customers
• Monitor invoices
• Submit bi-monthly aging reports

Our long experience in commercial finance can really pay off for you. Already we have established good working relationships in the industry. This means a professional for your team.

THE END OF RESTRICTED CASH FLOW

Why wait three to four months for your much needed revenues? With **Cash Flow Management** there's no debt., no more valuable time spent on hold, and you have immediate cash to apply to your company's growth.

Let us free your capital from the accounts receivable bottleneck today. We offer our services nationwide. For more details, use the reply card or call toll free 1-800-553-5679.

CASH FLOW MANAGEMENT
Reduce The Receivables Cycle

Yes, my working capital has been on hold long enough! Tell me how **Cash Flow Management** can end the bottleneck and get my money moving again.

Name _____

Position _____

Company _____

Address _____

City _____ State ____ Zip _____

Phone _____ Ext. ____

Fax _____

REPLY CARD

proper training and experience, the top job will remain elusive. So then what goes into becoming a radio station manager? First, a good foundation is necessary. Formal education is a good place to start. Hundreds of institutions of higher learning across the country offer programs in broadcast operations. The college degree has achieved greater importance in radio over the last decade or two and, as in most other industries today, it has become a standard credential for those vying for management slots. Anyone entering broadcasting in the 1980s with aspirations to operate a radio station should acquire as much formal training as possible. Station managers with master's degrees are not uncommon. However, a bachelor's degree in communications gives the prospective station manager a good foundation from which to launch a career.

In a business that stresses the value of practical experience, seldom, if ever, does an individual land a management job directly out of college. In fact, most station managers have been in the business at least fifteen years. "Once you get the theory nailed down you have to apply it. Experience is the best teacher. I've spent thirty years working in a variety of areas in the medium. In radio, in particular, hands-on experience is what matters," says Richard Bremkamp, Jr., general manager, KGLD/WKBQ, St. Louis.

To general manager Roger Ingram, experience is what most readily opens the door to management. "While a degree is kind of like a union card in this day and age, a good track record is what wins the management job. You really must possess both."

Jane Duncklee began her ascent to station management by logging commercials for airplay and eventually moved into other areas. "For the past seventeen years I have been employed by Champion Broadcasting Systems. During that time I have worked in every department of the radio station, from traffic, where I started, to sales, programming, engineering, and finally management on both the local and corporate levels."

Many radio station managers are

HELP WANTED—GENERAL MANAGER

Broadcast Group looking for superb talent, not promises. Successful candidate must be college educated, have outstanding references, be self-motivated, and possess leadership qualities. Must also be sales, programming, and bottom line oriented. Send résumé and letter stating your goals, starting salary, management philosophy, and how you can achieve a position of sales and ratings dominance in our growing company. Send response to Box 22.

FIGURE 2.3
Typical classified ad for a station manager in an industry trade magazine.

---

recruited from the sales area rather than programming. Since the general manager's foremost objective is to generate a profit, station owners usually feel more confident in hiring someone with a solid sales or business background. Consequently, three out of four radio managers have made their living at some point selling airtime. It is a widely held belief that this experience best prepares an individual for the realities encountered in the manager's position. "I spent over a decade and a half in media sales before becoming a station manager. In fact, my experience on the radio level was exclusively confined to sales and then for only eight months. After that I moved into station management. Most of my radio-related sales experience took place on the national level with station rep companies," recalls Norm Feuer of Force Two Communications.

Station manager Carl Evans holds that a sales background is especially useful, if not necessary, to general managers. "I spent a dozen years as a station account executive, and prior to entering radio I represented various product lines to retailers. The key to financial success in radio exists in an understanding of retailing."

It is not uncommon for station managers to have backgrounds out of radio, but almost invariably their experience comes out of the areas of sales, marketing, and finance. Paul Aaron of KFBK/KAER worked as a fund raiser for the United Way of America before entering radio, and contends that many managers come from other fields in which they have served in positions allied to sales, if not in sales itself. "Of those managers who have worked in fields other than radio, most have come to radio via the business sector. There are not many former biologists or glass blowers serving as station managers," says Aaron.

While statistics show that the station salesperson has the best chance of being promoted to the station's head position (more general managers have held the sales manager's position than any other), a relatively small percentage of radio's managers come from the programming ranks. "I'm more the exception than the rule. I have spent all of my career in the programming side, first as a deejay at stations in Phoenix, Denver, and Pittsburgh, and then as program director for outlets in Kansas City and Chicago. I'll have to admit, however, that while it certainly is not impossible to become a GM [general manager] by approaching it from the programming side, resistance exists," admits station manager Randy Lane.

The programmer's role is considered by many in the industry to be more an artistic function than one requiring a high degree of business savvy. However accurate or inaccurate this assessment is, the result is that fewer managers are hired with backgrounds exclusively confined to programming duties. Programmers have reason to be encouraged, however, since a trend in favor of hiring program directors (PDs) has surfaced in recent years, and predictions suggest that it will continue.

In its useful text pertaining to how PDs become station managers, *Up the Management Ladder: From Program Director to General Manager*, the NAB states:

The biggest obstacle facing the program director with an eye on management is image.

Programming is an operational expense. To create the "sound" of the station, the program director must spend money to pay staff salaries, to buy programming aids, and to maintain studio equipment. Where does the money come from?

Sales.

And it used to be that only those who were in sales were considered for promotion. The reasoning? Those in sales, through the very nature of their jobs, had a solid understanding of the radio rule: Time = Money.

Those in sales, through the very nature of their jobs, also enjoyed the pursuit of more money.

"I get turned on," said one general sales manager, "when people tell me 'no.' I love getting them to say 'yes.'"

Program directors, it was believed, lacked that basic grounding. Con-cerned with creating the station's image, program directors inadvertently created an image for themselves as "creative" types without concern or respect for the "business" aspects of radio.

So how is the program director to compete?

By presenting skills you already possess in the most positive, business-oriented format and by getting the skills you don't already possess.

In reality, the most attractive candidate for a station management position is the one whose experience has involved both programming and sales responsibilities. No general manager can fully function without an understanding and appreciation of what goes into preparing and presenting the air product, nor can he or she hope for success without a keen sense of business and finance.

Today's highly competitive and complex radio market requires that the person aspiring to management have both formal training, preferably a college degree in broadcasting, and experience in all aspects of radio station operations, in particular sales and programming. Ultimately, the effort and energy an individual invests will bear directly on the dividends he or she earns, and there is not a single successful station manager who has not put in fifteen-hour days. The station manager is expected to know more and do more than anyone else, and rightfully so, since he or she is the person who stands to gain the most.

Group W Radio's president, Dan Mason, relates the qualities he sees in the most successful station managers: "A keen sense of what is 'good business,' humility to take the blame in bad times and to give staff credit in good times, fairness and passion for all, responsiveness to situations (not reactionary), passion for the industry, recognition and knowledge of staff (know by first name), and ability to keep personal problems out of the station."

THE MANAGER'S DUTIES AND RESPONSIBILITIES

A primary objective of the station manager is to operate in a manner that generates the most profit, while maintaining a positive and productive attitude among station employees. This is more of a challenge than at first it may seem, claims KSKU's Cliff Shank. "In order to meet the responsibility that you are faced with daily, you really have to be an expert in so many areas: sales, marketing, finance, legal matters, technical, governmental, and programming. It helps if you're an expert in human nature, too." Jane Duncklee puts it this way: "Managing a radio station requires that you divide yourself equally into at least a dozen parts and be a hundred percent whole in each situation."

Station owner Bill Campbell says that the theme that runs throughout Tom Peters's book *In Search of Excellence* is one that is relevent to the station manager's task. "The idea in Peters's book is that you must make the customer happy, get your people involved, and get rid of departmental waste and unnecessary expenditures. A station should be a lean and healthy organism."

Station managers generally must themselves answer to a higher authority. The majority of radio stations, roughly 90 percent, are owned by companies and corporations which both hire the manager and establish financial goals or projections for the station. It is the station manager's job to see that corporate expectations are met and, ideally, exceeded. Managers who fail to operate a facility in a way that satisfies the corporate hierarchy may soon find themselves looking for another job.

Fewer than 10 percent of the nation's stations are owned by individuals or partnerships. At these radio outlets the manager still must meet the expectations of the station owner(s). In some cases, the manager may be given more latitude or responsibility in determining the station's fate, while in others the owner may play a greater role in the operation of the station.

It is a basic function of the manager's position to formulate station policy and see that it is implemented. To ensure against confusion, misunderstanding, and possible unfair labor practices which typically impede operations, a station policy book is often distributed to employees. The station's position on a host of issues, such as hiring, termination, salaries, raises, promotions, sick leave, vacation, benefits, and so forth usually are contained therein. As standard practice a station may require that each new employee read and become familiar with the contents of the policy manual before actually starting work. Job descriptions, as well as organization flow charts, commonly are outlined to make it abundantly clear to staff members who is responsible for what. A well-conceived policy book may contain a statement of the station's programming philosophy with an explanation of the format it employs. The more comprehensive a policy book, the less likely there will be confusion and disruption.

Hiring and retaining good people are other key managerial functions. "You have some pretty delicate egos to cope with in this business. Radio attracts some very bright and highly talented people, sometimes with erratic temperaments. Keeping harmony and keeping people are among the foremost challenges facing a station manager," claims Norm Feuer.

KLXK's Woodbury agrees with Feuer, adding, "You have to hire the right people and motivate them properly, and that's a challenge. You have to be capable of inspiring people." Actually, if you are unable to motivate your people, the station will fail to reach its potential. Hire the best people you can and nurture them."

As mentioned earlier, managers of

■ tactics: management
A Management Report from Shane Media

THIRTY SECONDS TO HIGHER CUME

Television still commands the largest percentage of radio
ad budgets. <u>Radio Business Report</u> points out that stations
in the top 75 markets spent $141.8 million on TV spots last
year. That was up 8.5% from 1993.

No surprise that second quarter and fourth quarter showed the
greatest expenditures of TV advertising dollars. The second
quarter total was $50.6 million, the fourth quarter total was
$41.8 million.

There are several important points gleaned from the <u>RBR</u>
overview. They apply generally:

- Stations buy TV to maintain top of mind awareness.

- The more mature a station, the more valuable TV is.

- Television helps the audience see the personalities.

- Reach is television's forte. Frequency makes it work.

- Most TV is bought program-by-program, but demos are
  most important to radio. TV can be bought by demo as
  well, looking at indices of demos by program.

This is good time to review the tactics for television spots
that we've discussed before. A refresher never hurts.

1. Have a message. Communicate how the station is
   different from every other station in town.

2. Have only one message. Keep it simple.

3. Use the same message in all promotional efforts.

4. Avoid creativity for its own sake.

5. Get your name or logo on the screen during the first
   few seconds.

6. Keep the logo visible throughout the commercial.

© SHANE MEDIA   HOUSTON, TX   (713) 952-9221

small market radio stations are confronted with a unique set of problems when it comes to hiring and holding onto qualified people, especially on-air personnel. "In our case, finding and keeping a professional-sounding staff with our somewhat limited budget is an ongoing problem. This is true at most small market stations, however," observes station manager J. G. Salter.

The rural station is where the majority of newcomers gain their experience. Because salaries are necessarily low and the fledgling air person's ambitions are usually high, the rate of turnover is significant. Managers of small outlets spend a great deal of time training people. "It is a fact of the business that radio people, particularly deejays, usually learn their trade at the 'out-of-the-way,' low-power outlet. To be a manager at a small station, you have to be a teacher, too. But it can be very rewarding despite the obvious problem of having to rehire to fill positions so often. We deal with many beginners. I find it exciting and gratifying, and no small challenge, to train newcomers in the various aspects of radio broadcasting," says Salter.

Randy Lane enjoys the instructor's role also, but notes that the high turnover rate affects product continuity. "With air people coming and going all the time, it can give the listening public the impression of instability. The last thing a station wants to do is sound schizophrenic. Establishing an image of dependability is crucial to any radio station. Changing air people every other month doesn't help. As a station manager, it is up to you to do the best you can with the resources at hand. In general, I think small market managers do an incredible job with what they have to work with."

While managers of small market stations must wrestle with the problems stemming from diminutive budgets and high employee turnover, those at large stations must grapple with the difficulties inherent in managing larger budgets, bigger staffs, and, of course, stiffer competition. "It's all relative, really. While the small town station gives the manager turnover headaches, the metropolitan station manager usually is caught up in the ratings battle, which consumes vast amounts of time and energy. Of course, even larger stations are not immune to turnover," observes KGLD's Bremkamp.

It is up to the manager to control the station's finances. A knowledge of bookkeeping and accounting procedures is necessary. "You handle the station's purse-strings. An understanding of budgeting is an absolute must. Station economics is the responsibility of the GM. The idea is to control income and expenses in a way that yields a sufficient profit," says Roger Ingram.

The manager allocates and approves spending in each department. Heads of departments must work within the budgets they have helped establish. Budgets generally cover the expenses involved in the operation of a particular area within the station for a specified period of time, such as a six- or twelve-month period. No manager wants to spend more than is absolutely required. A thorough familiarity with what is involved in running the various departments within a station prevents waste and overspending. "A manager has to know what is going on in programming, engineering, sales, actually every little corner of the station, in order to run a tight ship and make the most revenue possible. Of course, you should never cut corners simply for the sake of cutting corners. An operation must spend in order to make. You have to have effective cost control in all departments. That doesn't mean damaging the product through undernourishment either," says Evans.

David Saperstein, president of Metro Networks, observes that "In the early days, radio was a mom and pop type of business. With the huge dollars in radio today, one mistake could cost a station hundreds of thousands or even millions of dollars in revenue."

To ensure that the product the station offers is the best it can be, the station manager must keep in close touch with every department. Since the station's sound is what wins listeners, the manager must work closely with the program director and engineer. Both significantly contribute to the quality of the air product. The program director is responsible for what goes on the air, and the engineer is responsible for the way it sounds.

Meanwhile, the marketing of the product is vital. This falls within the province of the sales department. Traditionally, the general manager works more closely with the station's sales manager than with anyone else.

An excellent air product attracts listeners, and listeners attract sponsors. It is as basic as that. The formula works when all departments in a station work in unison and up to their potential, contends Marlin R. Taylor. "In radio our product is twofold—the programming we send over our frequency and the listening audience we deliver to advertisers. A station's success is linked to customer/listener satisfaction, just like a retail store's. If you don't have what the consumer desires, or the quality doesn't meet his standards, he'll go elsewhere and generally won't return."

In a fast-moving, dynamic industry like radio, where both cultural and technological innovations have an impact on the way a station operates, the manager must stay informed and keep one eye toward the future. Financial projections must be based on data that include the financial implications of prospective and predicted events. An effective manager anticipates change and develops appropriate plans to deal with it. Industry trade journals (*Broadcasting, Radio and Records, The Radio Ink, Radio Business Report*) and conferences conducted by organizations such as the National Association of Broadcasters (NAB), and the Radio Advertising Bureau (RAB) help keep the station manager informed of what tomorrow may bring.

About industry trade journals, Ed Shane observes, "The consultant (not the comedian) George Burns once said 'Never has so much been written by so many about so few.' The differences in today's trades compared to just a few years ago are significant. *Inside Radio, Radio Only, Radio Business Report's* 'Radio News Today,' *R&R's* 'Hot Wax,' and *Alternative Radio Confidential* are all paperless papers. *Broadcasting* is now *Broadcasting and Cable*. What does that say to the student of radio as a career? *R&R [Radio and Records]* claims more company executives subscribe to it than any other trade, which was formerly *Broadcasting*'s forte. Nonetheless, you have to read them to be up-to-date."

Station consultants and "rep companies," which sell local station airtime to national advertising agencies, also support the manager in his efforts to keep on top of things. "A station manager must utilize all that is available to stay in touch with what's out there. Foresight is an essential ingredient for any radio manager. Hindsight is not enough in an industry that operates with one foot in the future," says Lynn Christian, who summarizes the duties and responsibilities of a station manager: "To me the challenges of running today's radio station include building and maintaining audience ratings, attracting and keeping outstanding employees, increasing gross revenues annually, and creating a positive community image for the station, not necessarily in that order." KGLD's Bremkamp is more laconic. "It boils down to one sentence: Protect the license and turn a profit."

ORGANIZATIONAL STRUCTURE

Radio stations come in all sizes and generally are classified as being either small, medium, or large (metro) market outlets. The size of the community that a station serves usually reflects the size of its staff. That is to say, the station in a town of five thousand residents may have as few as six full-time employees. It is a question of economics. However, some small market radio outlets have staffs that rival those of larger market stations because their income warrants it. Few small stations earn enough to have elaborate staffs, however. Out of financial necessity, an employee may serve in several capacities. The station's manager also may assume the duties of sales manager, and announcers often handle news responsibilities. Meanwhile, everyone, including the station's secretary, may write commercial copy. The key word at the small station is flexibility, since each member of the staff is expected to perform numerous tasks.

Medium market stations are located in more densely populated areas. An outlet in a city with a population of between

Volume Two, 1993-1994

FIGURE 2.5
Publications such
as these keep
managers informed
of industry
developments and
issues.

one hundred thousand and a half million may be considered medium sized. Albany, New York; Omaha, Nebraska; and Albuquerque, New Mexico are typical medium markets. Greater competition exists in these markets, more than in the small market where only one or two stations may be vying for the listening audience. Each of the medium markets cited has over a dozen stations.

The medium market radio station averages twelve to twenty employees. While an overlapping of duties does occur even in the larger station, positions usually are more limited to specific areas of responsibility. Seldom do announcers substitute as newspeople. Nor do sales-

people fill airshifts as is frequently the case at small outlets. (Reexamine Fig. 1.31, a medium market flow chart.)

Large (also referred to as "major" or "metro") market radio stations broadcast in the nation's most populated urban centers. New York, Los Angeles, and Chicago rank first, second, and third, respectively. The top twenty radio markets also include cities such as Houston, Philadelphia, Boston, Detroit, and San Francisco. Competition is greatest in the large markets, where as many as seventy stations may be dividing the audience pie.

Major market stations employ as many as fifty to sixty people and as few as twenty, depending on the nature of their

FIGURE 2.6
Organizational
structure of a small
market radio
station.

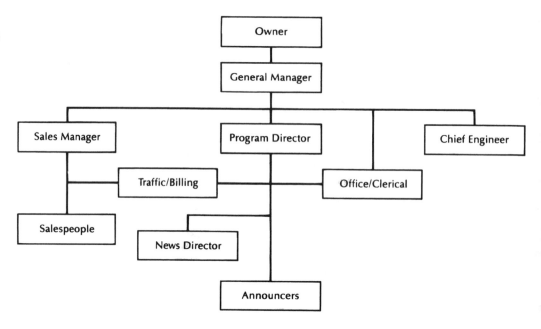

Norman Feuer

The Qualities That Make a Station Manager

1. Smart/intelligent. This is something that I cannot teach or help someone with; they are either smart or they're not.

2. Organized. A GM or GSM [general sales manager] has a lot on their plate, especially with the limited time available to accomplish what has to be done. An unorganized person will waste that time.

3. Ability to communicate. As a group head, I too have a lot on my plate. I must rely on my managers to communicate with me quickly and efficiently. If they can do that, I have the comfort that they are able to communicate effectively with their staff on the station's missions and goals to be accomplished.

4. Strategic thinker. In today's world there's no such thing as a quick fix. Therefore, I need to have someone who can think through the long-term effects of each major decision that he or she makes.

5. Motivated. It is my opinion that you cannot motivate people, they are either self-motivated or not. All I can do is set an environment for them to work in that allows their motivation to work best.

6. Businessperson. I need a person who understands that this is a business, not a hobby, and that every decision that they make has a return on investment and will lead to a successful business conclusion.

7. Leadership qualities. There are a lot of ways to describe this, but I want a person who is a winner, who people want to work for, who will have the ability to read their personnel, hire the best people, and be able to maximize the potential of all their people.

8. Track record. While it is nice to be able to find someone who has a winning track record on all or most of their previous assignments, we also understand that no one is born a GM or GSM. Therefore, it is not always a criterion, and one would look to other agreements to assess future success.

9. Energy level. I've always felt that you can determine a successful person by watching the way they walk down the hallway. I believe a person with a high energy level tends to get his or her people to move at a higher level just because of their own high energy level.

10. Honesty and integrity. It is absolutely critical that you must have the feeling that you can trust your manager and they won't try to make excuses and place blame on other people. This is a very hard ingredient to determine up front and may have to eventually be acquired.

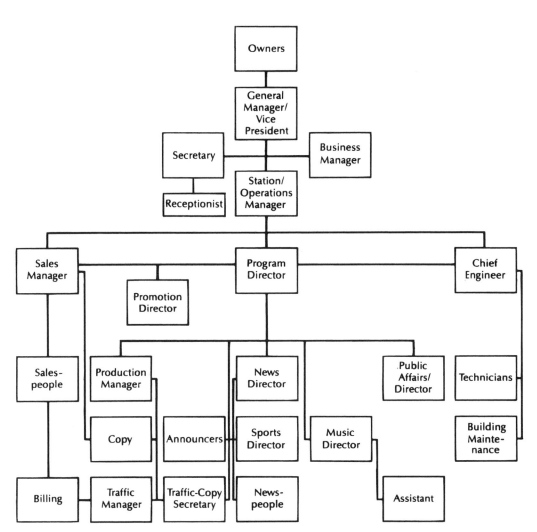

FIGURE 2.7
Organizational
chart of a larger
market radio
station.

format. Stations relying on automation, regardless of the size of their market, usually employ fewer people.

What follows are brief descriptions of department head duties and responsibilities. They are expanded upon in later chapters. These individuals are regarded as station middle management and generally report directly to the station's manager.

Operations Manager

Second only to the general manager in level of authority, the operations manager (not all stations have such a slot) is sometimes considered the station's assistant manager. The operations manager's duties include the following:

Supervising administrative (office) staff

Helping to develop station policies and see that they are implemented

Handling departmental budgeting

Keeping abreast of government rules and regulations pertaining to entire operation

Working as the liaison with the community to ensure that the station provides appropriate service and maintains its "good guy" image

Duties tend to be more skewed toward programming and may even include the job of station programmer. In cases such as this, the operations manager's duties are designed with a primary emphasis on on-air operations. Assistant program directors are commonly appointed to

Ward Quaal

My Management Philosophy

1. To succeed, radio stations must rededicate themselves to "localism" and their involvement in every possible phase of activity within their respective community (this applies regardless of format).

2. Regardless of the size of the market involved, those people who are making the news, whether in the world of academia, government, politics, in general, and civic leadership, must become a part of a radio station's total operation. In short, management must encourage and develop their participation so as to add to the richness and the superior culture of the property.

3. Basic to every radio operation and ultimate success is policy control over commercial material—its length, its acceptability to a station of the type involved, and passing of the "test" as to whether the commercial and the station complement one another.

FIGURE 2.8

work with the operations manager to accomplish programming goals. Not all stations have established this position, preferring the department head reporting to station manager approach.

Program Director

One of the three key department head positions at a radio station, the program director is responsible for the following:

Developing and executing format
Hiring and managing air staff
Establishing the schedule of airshifts
Monitoring the station to ensure consistency and quality of product
Keeping abreast of competition and trends that may affect programming
Maintaining the music library
Complying with FCC rules and regulations
Directing the efforts of the news and public affairs areas

Sales Manager

The sales manager's position is a pivotal one at any radio station, and involves the following:

Generating station income by directing the sale of commercial airtime

Supervising sales staff

Working with the station's rep company to attract national advertisers

Assigning lists of retail accounts and local advertising agencies to sales-people

Establishing sales quotas

Coordinating on-air and in-store sales promotions

Developing sales materials and rate cards

Chief Engineer

The chief engineer's job is a vital station function. Responsibilities include the following:

Operating the station within prescribed technical parameters established by the FCC

Purchasing, repairing, and maintaining equipment

Monitoring signal fidelity

Adapting studios for programming needs

Setting up remote broadcast operations

Working closely with the programming department

WHOM MANAGERS HIRE

Managers want to staff their stations with the most qualified individuals. Their criteria go beyond education and work experience to include various personality characteristics. "A strong resume is important, but the type of person is what really decides it for me. The goal is to hire someone who will fit in nicely with the rest of the station's members. A station is a bit like a family in that it is never too large and people are in fairly close contact with one another. Integrity, imagination, intelligence, and a desire to succeed are the basic qualities that I look for when hiring," says Christopher Spruce, general manager of WZON-AM, Bangor, Maine, a station that is owned by novelist Stephen King. (The "ZON" in the call letters was adopted from the 1960s television program "Twilight Zone.")

Ambition and a positive attitude are attributes that rank high among most managers. "People with a sincere desire to be the best, to win, are the kind who really bolster an operation. Our objective when I headed Viacom Radio was to be number one at what we did, so any amount of negativity or complacency was viewed as counterproductive," says Norm Feuer. Lynn Christian, who once served as Feuer's boss while manager of a Miami radio station, says that competitiveness and determination are among the qualities he, too, looks for, and adds loyalty and dedication to the list. "Commitment to the station's philosophy and goals must exist in every employee or there are soft spots in the operation. In today's marketplace you have to operate from a position of strength, and this takes a staff that is with you all the way."

Stability and reliability are high on any manager's list. Radio is known as a nomadic profession, especially the programming end. Deejays tend to come and go, sometimes disappearing in the night. "A manager strives to staff his station with dependable people. I want an employee who is stable and there when he or she is supposed to be," states Duncklee.

A station can ill afford to shut down because an air person fails to show up for his scheduled shift. When a no-show occurs, the station's continuity is disrupted both internally and externally. Within the station, adjustments must be made to fill the void created by the absent employee. At the same time, a substitute deejay often detracts from efforts to instill in the audience the feeling that it can rely on the station.

Managers are wary of individuals with fragile or oversized egos. "No prima donnas, please! I want a person who is able to accept constructive criticism and use it to his or her advantage without feeling that he is being attacked. The ability to look at oneself objectively and make the adjustments that need to be made is very important," observes KSKU's Cliff Shank. Carl Evans echoes his sentiments. "Open-mindedness is essential. An employee who feels that he can't learn anything from anyone or is never wrong is usually a fly in the ointment."

Honesty and candor are universally desired qualities, says Bremkamp. "Every station manager appreciates an employee who is forthright and direct, one who does not harbor ill feelings or unexpressed opinions and beliefs. An air of openness keeps things from seething and possibly erupting. I prefer an employee to come to me and say what is on his or her mind, rather than keep things sealed up inside. The lid eventually pops and then you have a real problem on your hands."

Self-respect and esteem for the organization are uniquely linked, claims J.G. Salter. "If an individual does not feel good about himself, he cannot feel good about the world around him. This is not a question of ego, but rather one of appropriate self-perception. A healthy self-image and attitude in an employee makes that person easier to manage and work with. The secret is identifying potential problems or weaknesses in prospective employees before you sign them on. It's not easy without a degree in Freud."

Other personal qualities that managers look for in employees are patience, enthusiasm, discipline, creativeness, logic, and compassion. "You look for as much as you can during the interview and, of course, fill in the gaps with phone calls to past employers. After that, you hope that your decision bears fruit," says WZON's Spruce.

THE MANAGER AND THE PROFIT MOTIVE

Earlier we discussed the unique nature of the radio medium and the particular challenges that face station managers as a consequence. Radio, indeed, is a form of show business, and both words of the term are particularly applicable since the medium is at once stage and store. Radio provides entertainment to the public and, in turn, sells the audience it attracts to advertisers.

The general manager is answerable to many: the station's owners, listeners, and sponsors. However, the party the manager must first please in order to keep his or her own job is the station's owner and, more often than not, this person's number one concern is profit. As in any business, the more money the manager generates, the happier the owner.

Critics have chided the medium for what they argue is an obsessive preoccupation with making money that has resulted in a serious shortage of high-quality, innovative programming. They lament too much sameness. Meanwhile, station managers often are content to air material that draws the kind of audience that advertisers want to reach.

Marlin Taylor observes that too many managers overemphasize profit at the expense of the operation. "Not nearly enough of the radio operators in this country are truly committed to running the best possible stations they can, either because that might cost them more money or they simply don't understand or care what it means to be the best. In my opinion, probably no more than 20 percent of the nation's stations are striving to become the IBM of radio, that is, striving for true excellence. Many simply are being milked for what the owners and managers can take out of them."

The pursuit of profits forces the station manager to employ the programming format that will yield the best payday. In several markets certain formats, such as classical and jazz, which tend to attract limited audiences, have been dropped in favor of those which draw greater numbers. In some instances, the actions of stations have caused outcries by unhappy and disenfranchised listeners who feel that their programming needs are being disregarded. Several disgruntled listener groups have gone to court in an attempt to force stations to reinstate abandoned formats. Since the government currently avoids involvement in programming decisions, leaving it up to stations to do as they see fit, little has come of their protests.

The dilemma facing today's radio station manager stems from the complexity of having to please numerous factions while still earning enough money to justify his or her continued existence at the station. Marlin Taylor has suggested that stations reinvest more of their profits as a method of upgrading the overall quality of the medium. "Overcutting can have deleterious effects. A station can be too lean, even anemic. In other words, you have to put something in to get something out. Too much draining leaves the operation arid and subject to criticism by the listening public. It behooves the station manager to keep this thought in mind and, if necessary, impress it upon ownership. The really successful operations know full well that money has to be spent to nurture and develop the kind of product that delivers both impressive financial returns and listener praise."

While it is the manager who must deal with bottom-line expectations, it is also the manager who is expected to maintain product integrity. The effective manager takes pride in the unique role radio plays in society and does not hand it over to advertisers, notes KGLD's Bremkamp. "You have to keep close tabs on your sales department. They are out to sell the station, sometimes one way or the other. Overly zealous salespeople can, on occasion, become insensitive to the station's format in their quest for ad dollars. Violating the format is like mixing fuel oil with water. You may fill your tank for less money, but you're not likely to get very far. The onus is placed on the manager to protect the integrity of the product while making a dollar. Actually, doing the former usually takes care of the latter."

Conscientious station managers are aware of the obligations confronting them, and sensitive to the criticism that crass commercialism can produce a desert or "wasteland" of bland and uninspired programming. They are also aware that while gaps and voids may exist in radio programming and that certain segments of the population may not be getting exactly what they want, it is up to them to produce enough income to pay the bills and meet the ownership's expectations.

THE MANAGER AND THE COMMUNITY

In the early 1980s, the FCC reduced the extent to which radio stations must become involved in community affairs. (This deregulation process continued into the 1990s with the Republican's creation of a sweeping telecommunications bill.) Ascertainment procedures requiring that stations determine and address community issues have all but been eliminated. If a station chooses to do so, it may spin the hits twenty-four hours a day and virtually divorce itself from the concerns of the community. However, a station that opts to function independently of the community to which it is licensed may find itself on the outside looking in. This is especially true of small market stations which, for practical business reasons, traditionally have cultivated a strong connection with the community.

A station manager is aware that it is important to the welfare of his or her organization to behave as a good citizen and neighbor. While the sheer number of stations in vast metropolitan areas makes it less crucial that a station exhibit civic-mindedness, the small market radio outlet often finds that the level of business it generates is relative to its community involvement. Therefore, maintaining a relationship with the town leaders, civic groups, and religious leaders, among others, enhances a station's visibility and status and ultimately affects business. No small market station can hope to operate autonomously and attract the majority of local advertisers. Stations that remain aloof in the community in which they broadcast seldom realize their full revenue potential.

One of the nation's foremost figures in broadcast management, Ward Quaal, president of the Ward L. Quaal Company, observes, "A manager must not only be tied, or perhaps I should say 'married' to a station, but he or she must have total involvement in the community. This is very meaningful, whether the market is Cheyenne, Cincinnati, or Chicago. The community participation

Paul Fiddick

What Makes a Successful Radio Manager?

In reflecting on "what qualities . . . make a radio manager successful," I've considered the dozens of managers who have worked for me in my stints running the Multimedia radio group (when they had one) and the last nine years with Heritage. And I've decided that there are two "indicators" that are essential (among others, I'm sure) for high performance.

The first of these is a bias for action, or what industrial psychologists describe as "task orientation." The concept of entrepreneurism is closely related to this, as is the quality of decisiveness. The radio manager must be innately tuned to the notion of taking the initiative if opportunities are to be exploited or—at a minimum—order is to be maintained.

The other quality is harder to describe. The term I have used is that successful managers take things personally. This may seem counter-intuitive. Shouldn't managers retain a kind of cool objectivity with regard to their work?

What I've found is that the job is too demanding for that—at least at the highest levels. In order to overcome the frustrations that come from so many aspects of your performance being outside of your direct control (i.e., ratings, competitive changes, the fact that a music station's primary product is produced by someone else), the successful manager must be a driven personality, and that drive comes from an inner, personal need. I do not believe it coincidence that success and obsessiveness go hand in hand in the radio business.

FIGURE 2.9

WDMO BROADCASTING CO., INC.   AS OF 1/25/85
INCOME STATEMENT
WDMO-AM
FEB PERIOD, 1985

	***** CURRENT MONTH *****					REVENUE	****** YEAR TO DATE ******				
	BUDGET	%	THIS YEAR	%	LAST YEAR	%	BUDGET	%	THIS YEAR	%	LAST YEAR	%
NATIONAL SALES	34,200	86.4	31,206	92.4	33,789		70,200	91.2	60,684	93.0	65,254	
LOCAL SALES	83,600	80.0	71,505	111.2	64,321		171,600	85.5	137,226	104.1	131,763	
NETWORK SALES	2,375	111.4	2,715	111.4	2,438		4,875	114.3	5,430	116.0	4,683	
PRODUCTION SALES	808	99.5	825	113.8	725		1,658	102.2	1,650	120.0	1,375	
TRADE SALES	475	116.7		.0	475		975	.0	1,138	136.2	835	
GROSS REVENUE	121,458	82.7	106,251	104.4	101,748		249,308	87.5	206,127	101.1	203,910	
AGENCY COMMISSIONS	17,250	78.9	13,352	.0			33,830	77.4	26,705	.2		
TOTAL NET REVENUE	104,208	83.3	92,898	91.3	101,748		215,478	89.1	179,423	88.0	203,910	
OPERATING EXPENSES												
ENGINEERING	10,560	45.8	3,557	61.8	5,751		20,097	33.7	9,200	80.8	11,384	
PROGRAMMING	14,320	102.2	11,362	.0			28,640	79.3	29,279	.2		
NEWS	8,690	28.7	2,379	.0			17,380	27.4	4,993	.0		
SALES	9,355	99.4	5,942	.0			18,710	63.5	18,601	.0		
ADMINISTRATIVE	15,480	56.0	7,183	.0			30,960	46.4	17,352	.0		
TOTAL OPERATING EXPENSES	58,405	68.6	30,421	529.0	5,751		115,787	51.1	79,425	697.7	11,384	
OPERATING INCOME	45,803	100.3	62,477	65.1	95,997		99,691	136.4	99,998	51.9	192,526	
OTHER INCOME		.0	84	.0				.0	251-	.0		
OTHER EXPENSES	9,200	86.8	6,443	.0			18,400	70.0	15,964	.0		
INCOME BEFORE TAXES	36,603	103.1	56,119	58.5	95,997		81,291	153.3	83,783	43.5	192,526	
PROVISION FOR TAXES		.1		.1				.1		.1		
NET INCOME/LOSS	36,603	103.1	56,119	58.5	95,997		81,291	153.3	83,783	43.5	192,526	

builds the proper image for the station and the manager and concurrently aids, dramatically, business development and produces lasting sales strength."

Cognizant of the importance of fostering an image of goodwill and civic-mindedness, the station manager seeks to become a member in good standing in the community. Radio managers often actively participate in groups or associations, such as the local Chamber of Commerce, Jaycees, Kiwanis, Rotary Club, Optimists, and others, and encourage members of their staff to become similarly involved. The station also strives to heighten its status in the community by devoting airtime to issues and events of local importance and by making its microphones available to citizens for discussions of matters pertinent to the area. In so doing, the station becomes regarded as an integral part of the community, and its value grows proportionately.

Surveys have shown that over a third of the managers of small market radio stations are native to the area their signal serves. This gives them a vested interest in the quality of life in their community and motivates them to use the power of their medium to further improve living conditions.

Medium and large market station managers realize, as well, the benefits derived from participating in community activities. "If you don't localize and take part in the affairs of the city or town from which you draw your income, you're operating at a disadvantage. You have to tune in your audience if you expect them to do likewise," says Bremkamp.

The manager has to work to bring the station and the community together. Neglecting this responsibility lessens the station's chance for prosperity, or even survival. Community involvement is a key to success.

FIGURE 2.10
Station's income statement containing budget. The station manager must be adept at directing station finances.

THE MANAGER AND
THE GOVERNMENT

Earlier in this chapter, station manager Richard Bremkamp cited protecting the license as one of the primary functions of the general manager. By "protecting" the license he meant conforming to the rules and regulations established by the FCC for the operation of broadcast facilities. Since failure to fulfill the obligations of a license may result in punitive actions, such as reprimands, fines, and even the revocation of the privilege to broadcast, managers have to be aware of the laws affecting station operations and see to it that they are observed by all concerned.

A recent article in *Radio Ink* summarized the FCC's punitive actions against stations for rules violations. Here are some examples:

- Construction or operation without authorization—$20,000
- Unauthorized transfer of control—$20,000
- Failure to permit FCC inspections—$18,750
- Failure to respond to FCC communications—$17,500
- Exceeding power limits—$12,500
- EBS equipment broken or not installed—$12,500
- Broadcasting indecent/obscene material—$12,500
- Violation of EEO or political broadcast rules—$12,500
- Violation of main studio rule—$10,000
- Public file violations—$7,500
- Sponsor ID or lottery violations—$6,250

The manager delegates responsibilities to department heads who are directly involved in the areas affected by the commission's regulations. For example, the program director will attend to the legal station identification, station logs, program content, and myriad other concerns of interest to the government. Meanwhile, the chief engineer is responsible for meeting technical standards, while the sales manager is held accountable for the observance of certain business and financial practices. Other members of the station also are assigned various responsibilities applicable to the license. Of course, in the end it is the manager who must guarantee that the station's license to broadcast is, indeed, protected.

Contained in Title 47, Part 73 of the *Code of Federal Regulations (CFR)* are all the rules and regulations pertaining to radio broadcast operations. The station manager keeps the annual update of this publication accessible to all those employees involved in maintaining the license. A copy of the *CFR* may be obtained through the Superintendent of Documents, Government Printing Office, Washington, D.C. 20402. A modest fee is charged. Specific inquiries concerning the publication can be addressed to the Director, Office of the Federal Register, National Archives and Records Service, General Services Administration, Washington, D.C. 20408.

To give an idea of the scope of this document, as well as the government's dicta concerning broadcasters, the *CFR*'s index to radio broadcast services is included as an appendix to this chapter. Immediately apparent from a perusal of the index is the preponderance of items that come under the auspices of the engineering department. Obviously, the FCC is concerned with many areas of radio operation, but its focus on the technical aspect is prodigious. Many other items involve programming and sales, and, of course, responsibilities also overlap into other areas of the station.

To reiterate, although the station manager shares the duties involved in complying with the FCC's regulations with other staff members, he holds primary responsibility for keeping the station out of trouble and on the air.

In Chapter 1 it was noted that many of the rules and regulations pertaining to the daily operation of a radio station have undergone revision or have been rescinded. Since the *CFR* is published annually, certain parts may become obsolete during that period. Martha L. Girard, director of the Office of the Federal Register, suggests that the *Federal Register*, from which the *CFR* derives its information, be consulted monthly. "These two publications must be used together to determine the latest version of any given rule," says Byrne. A station may subscribe to the *Federal Register* or visit the local library.

Since the FCC may, at any time during normal business hours, inspect a radio station to see that it is in accord with the rules and regulations, a manager makes certain that everything is in order. An FCC inspection checklist is contained in the *CFR*, and industry organizations, such as the National Association of Broadcasters (NAB), provide member stations with similar checklists. Occasionally, managers run mock inspections in preparation for the real thing. A state of preparedness prevents embarrassment and problems.

THE PUBLIC FILE

Radio stations are required by the FCC to maintain a Public File. This, too, is ultimately the manager's responsibility. However, other members of a station's staff are typically required to update certain elements of the file.

The purpose of the file is to provide the general population access to information pertaining to the way a station has conducted itself during a license period.

Interest in a station's Public File often increases around license renewal periods, when members of a community may choose to challenge a station's right to continue broadcasting.

The FCC expects a station's Public File to contain the following (this is subject to change as rules and regulations are revised):

The Public and Broadcasting—A Procedural Manual (revised edition)
Annual employment reports
Copies of all FCC applications (power increases, original construction permit, facilities changes, license renewals)
Ownership reports
Political file
Letters from the public
Quarterly issues
Local public notices

The Public File must be located in the community in which the station is licensed. Most stations keep the file at their main studio facility rather than at another public location, which is also permissible. The Public File is often the first place FCC agents look when they inspect a station. So it must be readily available. It is imperative that the file be kept up to date. As a general rule, files are retained for a period of seven years.

More detailed information pertaining to a station's Public File may be found in Section 73.3526 of the FCC's *Rules and Regulations*.

THE MANAGER AND UNIONS

The unions most active in radio are the American Federation of Television and Radio Artists (AFTRA), the National Association of Broadcast Employees and Technicians (NABET), and the International Brotherhood of Electrical Workers (IBEW). Major market radio stations are the ones most likely to be unionized. The overwhelming majority of American stations are nonunion, and, in fact, in recent years union memberships have declined.

Dissatisfaction with wages and benefits, coupled with a desire for greater security, often are what motivate station employees to vote for a union. Managers seldom encourage the presence of a union since many believe that unions impede and constrict their ability to control the destiny of their operations. However, a small percentage of managers believe that the existence of a union may actually stabilize the working environment and cut down on personnel turnover.

It is the function of the union to act as a bargaining agent working in "good faith" with station employees and management to upgrade and improve working conditions. The focus of union efforts usually is in the areas of salary, sick leave, vacation, promotion, hiring, termination, working hours, and retirement benefits, among others.

FIGURE 2.11
FCC Ownership and License application forms.

Federal Communications Commission
Washington, D.C. 20554

APPLICATION FOR NEW BROADCAST STATION LICENSE
(Carefully read instructions before filling out Form)
RETURN ONLY FORM TO FCC

Approved by OMB
3060-0029
Expires 9/30/90

For Commission Fee Use Only		For Applicant Fee Use Only
	FEE NO:	Is a fee submitted with this application? ☐ Yes ☐ No
	FEE TYPE:	If No, indicate reason therefor (check one box):
		☐ Nonfeeable application
	FEE AMT:	Fee Exempt (See 47 C.F.R. Section 1.1112)
		☐ Noncommercial educational licensee
	ID SEQ:	☐ Governmental entity

SECTION I — GENERAL DATA

For Commission Use Only

File No.

Legal Name of Applicant	Mailing Address
	City State Zip Code
	Telephone No. (include area code)

1. Facilities authorized by construction permit

   This application is for: ☐ Commercial   ☐ Noncommercial

   ☐ AM Directional   ☐ AM Non-Directional   ☐ FM Directional   ☐ FM Non-Directional   ☐ TV

Call Letters	Community of License	Construction Permit File No.	Modification of Construction Permit File No(s).	Expiration Date of Last Construction Permit

2. Is the station now operating pursuant to automatic program test authority in accordance with 47 C.F.R. Section 73.1620?   ☐ Yes   ☐ No

   If No, explain.

3. Have all the terms, conditions, and obligations set forth in the above described construction permit been fully met?   ☐ Yes   ☐ No

   If No, state exceptions.

4. Apart from the changes already reported, has any cause or circumstance arisen since the grant of the underlying construction permit which would result in any statement or representation contained in the construction permit application to be now incorrect?   ☐ Yes   ☐ No

   If Yes, explain.

5. Has the permittee filed its Ownership Report (FCC Form 323) or ownership certification in accordance with 47 C.F.R. Section 73.3615(b)?   ☐ Yes   ☐ No

   If No, explain.   ☐ Does not apply

FCC 302
June 1988

A unionized station appoints or elects a shop steward who works as a liaison between the union, which represents the employees, and the station's management. Employees may lodge complaints or grievances with the shop steward, who will then review the union's contract with the station and proceed accordingly. Station managers are obliged to work within the agreement that they, along with the union, helped formulate.

FIGURE 2.11
continued

Approved by OMB
3060-0010
Expires 6/30/92

CERTIFICATION

United States of America
Federal Communications Commission
Washington, D. C. 20554

I certify that I am _____
(*Official title, see Instruction 1*)

of _____
(*Exact legal title or name of respondent*)

Ownership Report

NOTE: Before filling out this form, read attached instructions

that I have examined this Report, that to the best of my knowledge and belief, all statements in the Report are true, correct and complete.

Section 310(d) of the Communications Act of 1934 requires that consent of the Commission must be obtained prior to the assignment or transfer of control of a station license or construction permit. This form may **not** be used to report or request an assignment of license/permit or transfer of control (except to report an assignment of license/permit or transfer of control made pursuant to prior Commission consent).

(*Date of certification must be within 60 days of the date shown in Item 1 and in no event prior to Item 1 date*):

_____ , 19 ____
(*Signature*)　　　　(*Date*)

1. All of the information furnished in this Report is accurate as of

_____ , 19 ____ .

(*Date must comply with Section 73.3615(a), i.e., information must be current within 60 days of the filing of this report, when 1(a) below is checked.*)

Telephone No. of respondent (*include area code*):

Any person who willfully makes false statements on this report can be punished by fine or imprisonment. U.S. Code, Title 18, Section 1001.

This report is filed pursuant to Instruction (*check one*)

1(a) ☐ Annual　1(b) ☐ Transfer of Control or Assignment of License　1(c) ☐ Other

Name and Post Office Address of respondent:

for the following stations:

Call Letters	Location	Class of service

4. Name of entity, if other than licensee or permittee, for which report is filed (*see Instruction 3*):

2. Give the name of any corporation or other entity for whom a separate Report is filed due to its interest in the subject licensee (*See Instruction 3*):

5. Respondent is:

☐ Sole Proprietorship

☐ For-profit corporation

☐ Not-for-profit corporation

☐ General Partnership

☐ Limited Partnership

☐ Other: _____

3. Show the attributable interests in any other broadcast station of the respondent. Also, show any interest of the respondent, whether or not attributable, which is 5% or more of the ownership of any other broadcast station or any newspaper or CATV entity in the same market or with overlapping signals in the same broadcast service, as described in Sections 73.3555 and 76.501 of the Commission's Rules.

If a limited partnership, is certification statement included as in Instruction 4?

☐ Yes　☐ No

FCC 323
February 1990

As stated, unions are a fact of life in many major markets. They are far less prevalent elsewhere, although unions do exist in some medium and even small markets. Most small operations would find it impractical, if not untenable, to function under a union contract. Union demands would quite likely cripple most marginal or small profit operations.

Managers who extend employees every possible courtesy and operate in a fair and reasonable manner are rarely

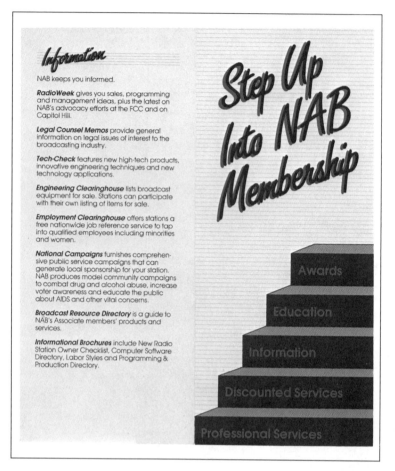

Information

NAB keeps you informed.

RadioWeek gives you sales, programming and management ideas, plus the latest on NAB's advocacy efforts at the FCC and on Capitol Hill.

Legal Counsel Memos provide general information on legal issues of interest to the broadcasting industry.

Tech-Check features new high-tech products, innovative engineering techniques and new technology applications.

Engineering Clearinghouse lists broadcast equipment for sale. Stations can participate with their own listing of items for sale.

Employment Clearinghouse offers stations a free nationwide job reference service to tap into qualified employees including minorities and women.

National Campaigns furnishes comprehensive public service campaigns that can generate local sponsorship for your station. NAB produces model community campaigns to combat drug and alcohol abuse, increase voter awareness and educate the public about AIDS and other vital concerns.

Broadcast Resource Directory is a guide to NAB's Associate members' products and services.

Informational Brochures include New Radio Station Owner Checklist, Computer Software Directory, Labor Styles and Programming & Production Directory.

Step Up Into NAB Membership

Awards
Education
Information
Discounted Services
Professional Services

**FIGURE 2.12
NAB offers
members a variety
of services.
Courtesy NAB.**

affected by unions, whose prime objective is to protect and ensure the rights of station workers.

THE MANAGER AND INDUSTRY ASSOCIATIONS

Every year the NAB and a variety of specialized and regional organizations conduct conferences and seminars intended to generate industry awareness and unity. At these gatherings, held at various locations throughout the country, radio managers and station personnel exchange ideas and share experiences, which they bring back to their stations.

The largest broadcast industry trade organization is the NAB, which was originally conceived out of a need to improve operating conditions in the 1920s.

Initially only a lobbying organization, the NAB has maintained that focus while expanding considerably in scope. The primary objective of the organization is to support and promote the stability and development of the industry.

In the mid-1990s, however, the National Radio Broadcaster's Association, which merged with the NAB in the 1980s, threatened to break from the organization for its alleged overemphasis on nonradio matters. At this writing, the issue was still being debated.

Thousands of radio stations also are members of the Radio Advertising Bureau (RAB), which was founded in 1951, a time when radio's fate was in serious jeopardy due to the rise in television's popularity. "The RAB is designed to serve as the sales and marketing arm of America's commercial radio industry. Members include radio stations, broadcast groups, networks, station representatives, and associated industry organizations in every market, in all fifty states," explains RAB's Kenneth J. Costa.

Dozens of other broadcast trade organizations focus their attention on specific areas within the radio station, and regional and local broadcast organizations are numerous. Listed below is a partial rundown of national organizations that support the efforts of radio broadcasters. A more comprehensive list may be found in *Broadcasting Yearbook*, the definitive industry directory, or by consulting local area directories.

National Association of Broadcasters, 1771 N Street, N.W., Washington, D.C. 20036

Radio Advertising Bureau, 304 Park Avenue, New York, NY 10010

National Association of Farm Broadcasters, 26 E. Exchange Street, St. Paul, MN 55101

American Women in Radio and Television, 1101 Connecticut Avenue, N.W., Suite 700, Washington, D.C. 20036

Broadcast Education Association, 1771 N Street, N.W., Washington, D.C. 20036

Broadcast Pioneers, 320 W. 57th Street, New York, NY 10019

Clear Channel Broadcasting Service, 1776 K Street, N.W., Washington, D.C. 20006

Native American Public Broadcasting Consortium, Box 83111, Lincoln, NE 68501

National Association of Black Owned Broadcasters, 1730 M Street, N.W., Suite 412, Washington, D.C. 20036

National Religious Broadcasters, 299 Webro Road, Parisippany, NJ 07054

Radio Network Association, 1700 Broadway, 3rd floor, New York, NY, 10019

Radio-Television News Directors Association, 1000 Connecticut Avenue, N.W., Washington, D.C. 20006

Society of Broadcast Engineers, Box 20450, Indianapolis, IN 46220

NAB membership dues are based on a voluntary declaration of a station's annual gross revenues. The RAB and others take a similar approach. Some organizations require individual membership fees, which often are absorbed by the radio station as well.

BUYING OR BUILDING A RADIO STATION

The process involved in the purchase of an existing radio facility is fairly complex. There is much to take into consideration. For one thing, it is rarely a quick and easy procedure because the FCC must approve of all station transfers (sales).

The FCC examines the background of any would-be station owner. Licensees must be U.S. citizens and must, among other things:

Not have a criminal record
Be able to prove financial stability
Have a solid personal and professional history

Anyone interested in purchasing a radio property should employ the services of an attorney and/or broker who specializes in this area (consult *Broadcasting and Cable Market Place* for listings or contact the NAB).

The following points should be carefully considered before definitive action is taken to purchase a station:

FIGURE 2.13
The NAB's Crystal Awards are presented to those stations with exemplary community service. Courtesy NAB.

Analyze the market in which the station is located
Evaluate the facilities and assets
Hire a technical consultant
Assess existing contracts, leases, and agreements
Research financial records
Examine the Public File
Probe the FCC's file

According to radio station acquisition expert Erwin G. Krasnow, "The due diligence process involved in the acquisition of a radio station typically includes a review of the general economic and operational conditions, as well as such areas as the financial and accounting systems, programming, technical facilities, legal matters, marketing, employee benefits, and personnel and environmental matters. The objective of due diligence is to obtain information that will (a) influence the decision of whether or not to proceed with the acquisition; and (b) have an effect on the purchase price or working capital adjustment."

Purchasing a broadcast property is unlike any other kind of acquisition because of the unique nature of the business. It is important to keep in mind that in the end a station owner does not own a frequency. An operator is merely granted permission (a license) to propagate a signal for a prescribed period of time (seven years) and then must reapply to continue broadcasting.

If an individual wishes to create a station from scratch, an application for a construction permit (CP) must be filled out and submitted to the FCC. This too is quite involved and requires special expertise as it is necessary to determine whether a new station can be accommodated on the existing broadcast bands in the area of proposed operation.

If it can be proven that an available frequency exists and that no interference will occur, a CP may be granted. The applicant is then given a specified amount of time (usually 18 months) to commence and complete construction.

Before proceeding with a CP request, it is incumbent on the applicant to meet all of the criteria for station ownership set by the FCC.

CHAPTER HIGHLIGHTS

1. Radio's unique character requires that station managers deal with a wide variety of talents and personalities.

2. The authoritarian approach to management implies that the general manager makes all of the policy decisions. The collaborative approach allows the general manager to involve other station staff in the formation of policy. The hybrid or chief-collaborator approach combines elements of both the authoritarian and collaborative management models. The chief-collaborator management approach is most prevalent in radio today.

3. To attain management status, an individual needs a solid formal education and practical experience in many areas of station operation—especially sales.

4. Key managerial functions include: operating in a manner that produces the greatest profit, meeting corporate expectations, formulating station policy and seeing to its implementation, hiring and retaining good people, inspiring staff to do their best, training new employees, maintaining communication with all departments to assure an excellent air product, and keeping an eye toward the future.

5. The operations manager, second only to the general manager at those stations that have established this position, supervises administrative staff, helps develop and implement station policy, handles departmental budgeting, functions as regulatory watchdog, and works as liaison with the community. The program director is responsible for format, hires and manages air staff, schedules airshifts, monitors air product quality, keeps abreast of competition, maintains the music library, complies with FCC rules, and directs the efforts of news and public affairs. The sales manager heads the sales staff, works with the station's rep company, assigns account lists to salespeople, establishes sales quotas, coordinates sales promotions, and develops sales materials and rate cards. The chief engineer operates within the FCC technical parameters; purchases, repairs, and maintains equipment; monitors signal fidelity; adapts studios for programming needs; sets up remote broadcasts; and works closely with programming.

6. Managers hire individuals who possess: a solid formal education, strong professional experience, ambition, a positive attitude, reliability, humility, honesty, self-respect, patience, enthusiasm, discipline, creativity, logic, and compassion.

7. Radio provides entertainment to the public and, in turn, sells the audience it attracts to advertisers. It is the station manager who must ensure a profit, but he or she must also maintain product integrity.

8. To foster a positive community image, the station manager becomes actively involved in the community and devotes airtime to community concerns—even though the FCC recently reduced a station's obligation to do so.

9. Although the station manager delegates responsibility for compliance with FCC regulations to appropriate department heads, the manager is ultimately responsible for "protecting" the license. Title 47, Part 73 of the *Code of Federal Regulations* contains the rules pertaining to radio broadcast operations. Updates of regulations are listed monthly in the *Federal Register*.

10. Radio stations are required to maintain a Public File and to make it available to the public during normal business hours.

11. The American Federation of Television and Radio Artists (AFTRA), the National Association of Broadcast Employees and Technicians (NABET), and the International Brotherhood of Electrical Workers (IBEW) are the unions most active in radio.

12. The National Association of Broadcasters (NAB) and the Radio Advertising Bureau (RAB) are among the largest radio trade industry organizations.

13. A person must be a U.S. citizen to hold a broadcast license. The FCC investigates all would-be station owners. To put a new station on the air, a construction permit (CP) application must be submitted to the FCC.

APPENDIX: *CODE OF FEDERAL REGULATIONS* INDEX

The index to the *Code of Federal Regulations*, Title 47, Part 73, Subparts A, B, G, and H, shown here, deals specifically with commercial radio operations. The department heads most affected by the sections listed in the index are noted in parenthesis after each entry. GM = General Manager, PD = Program Director, E = Engineer, SM = Sales Manager.

Tests

73.961 Tests of the Emergency Broadcast System procedures. (E, PD)
73.962 Closed Circuit Tests of approved national level interconnecting systems and facilities of the Emergency Broadcast System. (E)

SUBPART H—RULES APPLICABLE TO ALL BROADCAST STATIONS

73.1001 Scope. (All dept. heads)
73.1010 Cross reference to rules in other parts.
73.1015 Truthful written statements and responses to Commission inquiries and correspondence. (GM)
73.1020 Station license period. (GM, PD, E)
73.1030 Notifications concerning interference to radio astronomy, research and receiving installations. (E)
73.1120 Station location. (GM, E)
73.1125 Station main studio location. (E)
73.1130 Station program origination. (E)
73.1150 Transferring a station. (GM)
73.1201 Station identification. (PD)
73.1202 Retention of letters received from the public. (GM, PD)
73.1206 Broadcast of telephone conversations. (PD, E)
73.1207 Rebroadcasts. (PD)
73.1208 Broadcast of taped, filmed or recorded material. (PD)
73.1209 References to time. (PD)
73.1210 TV/FM dual-language broadcasting in Puerto Rico. (GM, PD)
73.1211 Broadcast of lottery information. (PD, GM)
73.1212 Sponsorship identification; list retention; related requirements. (GM, PD)
73.1213 Antenna structure, marking and lighting. (E)
73.1215 Specifications for indicating instruments. (E)
73.1216 Licensee-conducted contests. (GM, PD)
73.1217 Broadcast hoaxes. (GM, PD, SM)
73.1225 Station inspection by FCC. (GM, PD, E)
73.1226 Availability to FCC of station logs and records. (GM, PD, E)
73.1230 Posting of station and operator licenses. (PD, E)
73.1250 Broadcasting emergency information. (PD, E)
73.1400 Remote control authorizations. (E)
73.1410 Remote control operation. (E)
73.1500 Automatic transmission system (ATS). (E)
73.1510 Experimental authorizations. (E)
73.1515 Special field test authorizations. (E)
73.1520 Operation for tests and maintenance. (E)
73.1530 Portable test stations. [Definition] (E)
73.1540 Carrier frequency measurements. (E)
73.1545 Carrier frequency departure tolerances. (E)
73.1550 Extension meters. (E)
73.1560 Operating power tolerance. (E)
73.1570 Modulation levels: AM, FM, and TV aural. (E)

73.1580 Transmission system inspections. (E)
73.1590 Equipment performance measurements. (E)
73.1610 Equipment tests. (E)
73.1615 Operation during modification of facilities. (GM, E)
73.1620 Program tests. (E)
73.1635 Special temporary authorizations (STA). (GM, E)
73.1650 International broadcasting agreements. (GM, E)
73.1660 Type acceptance of broadcast transmitters. (E)
73.1665 Main transmitters. (E)
73.1670 Auxiliary transmitters. (E)
73.1675 Auxiliary antennas. (E)
73.1680 Emergency antennas. (E)
73.1690 Modification of transmission systems. (E)
73.1695 Changes in transmission standards. (E)
73.1700 Broadcast day. (GM, PD, E)
73.1705 Time of operation. (PD, E, GM)
73.1710 Unlimited time. (GM, PD, E)
73.1715 Share time. (GM, PD, E)
73.1720 Daytime. (GM, PD, E)
73.1725 Limited time. (GM, PD, E)
73.1730 Specified hours. (GM, PD, E)
73.1735 AM station operation presunrise and postsunset. (GM, PD, E)
73.1740 Minimum operating schedule. (GM, PD, E)
73.1745 Unauthorized operation. (GM, PD, E)
73.1750 Discontinuance of operation. (GM, PD, E)
73.1800 General requirements relating to the station log. (GM, PD)
73.1820 Station log. (PD, E)
73.1835 Special technical records. (E)
73.1840 Retention of logs. (PD, E)
73.1860 Transmitter duty operators. (E)
73.1870 Chief operators. (E)
73.1910 Fairness Doctrine. (GM, PD)
73.1920 Personal attacks. (GM, PD)
73.1930 Political editorials. (GM, PD)
73.1940 Broadcasts by candidates for public office. (GM, PD)
73.1941 Equal opportunities. (GM, PD, SM)
73.1942 Candidate rates. (GM, SM)
73.1943 Political file. (GM, SM)
73.1944 Reasonable access. (GM, PD)
73.2080 Equal employment opportunities. (GM)
73.3500 Application and report forms. (GM, E)
73.3511 Applications required. (GM, E)
73.3512 Where to file; number of copies. (GM, E)
73.3513 Signing of applications. (GM)
73.3514 Content of applications. (GM)
73.3516 Specification of facilities. (GM)
73.3517 Contingent applications. (GM)
73.3518 Inconsistent or conflicting applications. (GM)
73.3519 Repetitious applications. (GM)
73.3520 Multiple applications. (GM)
73.3522 Amendment of applications. (GM)
73.3523 Dismissal of applications in renewal proceedings. (GM)
73.3525 Agreements for removing application conflicts. (GM)
73.3526 Local public inspection file of commercial stations. (GM, PD, E)

73.3533 Application for construction permit or modification of construction permit. (GM, E)

73.3534 Application for extension of construction permit or for construction permit to replace expired construction permit. (GM, E)

73.3535 Application to modify authorized but unbuilt facilities, or to assign or transfer control of an unbuilt facility. (GM, E)

73.3536 Application for license to cover construction permit. (GM, E)

73.3537 Application for license to use former main antenna as an auxiliary. (E)

73.3538 Application to make changes in an existing station. (GM, E)

73.3539 Application for renewal of license. (GM, E)

73.3540 Application for voluntary assignment or transfer of control. (GM)

73.3541 Application for involuntary assignment of license or transfer of control. (GM)

73.3542 Application for temporary or emergency authorization. (GM, E)

73.3543 Application for renewal or modification of special service authorization. (GM, E)

73.3544 Application to obtain a modified station license. (GM, E)

73.3545 Application for permit to deliver programs to foreign stations. (GM, PD)

73.3549 Requests for extension of authority to operate without required monitors, indicating instruments and EBS attention signal devices. (GM, E)

73.3550 Requests for new or modified call sign assignments. (GM, PD)

73.3555 Multiple ownership. (GM)

73.3556 Duplication of programming on commonly owned or time brokered stations. (GM, PD)

73.3561 Staff consideration of applications requiring Commission action. (GM)

73.3562 Staff consideration of applications not requiring action by the Commission. (GM)

73.3564 Acceptance of applications. (GM)

73.3566 Defective applications. (GM)

73.3568 Dismissal of applications. (GM)

73.3570 AM broadcast station applications involving other North American countries. (GM)

73.3571 Processing of AM broadcast station applications. (GM)

73.3573 Processing FM broadcast and FM translator station applications. (GM)

73.3578 Amendments to applications for renewal, assignment or transfer of control. (GM)

73.3580 Local public notice of filing of broadcast applications. (GM)

73.3584 Petitions to deny. (GM)

73.3587 Procedure for filing informal objections. (GM)

73.3591 Grants without hearing. (GM)

73.3592 Conditional grant. (GM)

73.3593 Designation for hearing. (GM)

73.3594 Local public notice of designation for hearing. (GM)

73.3597 Procedures on transfer and assignment applications. (GM)

73.3598 Period of construction. (GM, E)

73.3599 Forfeiture of construction permit. (GM, E)

73.3601 Simultaneous modification and renewal of license. (GM, E)

73.3603 Special waiver procedure relative to applications. (GM)

73.3605 Retention of applications in hearing status after designation for hearing. (GM)

73.3612 Annual employment report. (GM)

73.3613 Filing of contracts. (GM, PD)

73.3615 Ownership reports. (GM)

73.3999 Enforcement of 18 U.S.C. 1464 (restrictions on the transmissions of obscene and indecent material). (GM. PD)

73.4000 Listing of FCC policies. (GM, E)

73.4005 Advertising—Refusal to sell. (GM, SM)

73.4015 Alcoholic beverage advertising. (GM, SM, PD)

73.4017 Application processing: Commercial FM stations. (GM, E)

73.4045 Barter agreements. (GM, SM)

73.4055 Cigarette advertising. (GM, SM)

73.4060 Citizens agreements. (GM, PD)

73.4075 Commercials, loud. (PD, SM, E)

73.4082 Comparative broadcast hearings and specialized programming formats. (PD)

73.4091 Direct broadcast satellites. (PD, E)

73.4094 Dolby encoder. (E)

73.4095 Drug lyrics. (PD)

73.4099 Financial qualifications, certification of. (GM)

73.4100 Financial qualifications; new AM and FM stations. (GM)

73.4102 FAA communications, broadcast of. (E)

73.4104 FM assignment policies and procedures. (GM, PD)

73.4107 FM broadcast assignments, increasing availability of. (GM, E)

73.4108 FM transmitter site map submissions. (E)

73.4110 Format changes of stations. (PD, GM)

73.4135 Interference to TV reception by FM stations. (E)

73.4140 Minority ownership; tax certificates and distress sales. (GM)

73.4154 Network/AM, FM station affiliation agreements. (GM, PD)

73.4157 Network signals which adversely affect affiliate broadcast service. (GM, PD, E)

73.4165 Obscene language. (GM, PD)

73.4170 Obscene lyrics. (GM, PD)

73.4180 Payment disclosure: Payola, plugola, kickbacks. (GM, PD)

73.4185 Political broadcasting, the law of. (GM, PD)

73.4190 Political candidate authorization notice and sponsorship identification. (GM, PD)

73.4210 Procedure Manual: "The Public and Broadcasting." (GM)

73.4215 Program matter: Supplier identification. (GM, PD)

73.4235 Short spacing assignments: FM stations. (E)

73.4240 Sirens and like emergency sound effects in announcements. (PD)

73.4242 Sponsorship identification rules, applicability of. (SM, PD)

73.4246 Stereophonic pilot subcarrier use during monophonic programming. (E)

73.4250 Subliminal perception. (GM, PD)

73.4255 Tax certificates: Issuance of. (GM)

73.4260 Teaser announcements. (GM, PD)

73.4265 Telephone conservation broadcasts (network and like sources). (GM, PD)
73.4266 Tender offer and proxy statements. (GM, SM)
73.4267 Time brokerage. (GM)
73.4275 Tone clusters; audio attention-getting devices. (GM, PD, E)
73.4280 Violation of laws of U.S.A. by station applicants; Commission policy. (GM, PD)

SUGGESTED FURTHER READING

Agor, Weston H. *Intuitive Management.* Englewood Cliffs, N.J.: Prentice-Hall, 1984.

Aronoff, Craig E., ed. *Business and the Media.* Santa Monica, Calif.: Goodyear Publishing Company, 1979.

Albrecht, Karl G. *Successful Management Objectives.* Englewood Cliffs, N.J.: Prentice-Hall, 1979.

Appleby, Robert C. *The Essential Guide to Management.* Englewood Cliffs, N.J.: Prentice-Hall, 1981.

Boyatzis, Richard E. *The Competent Manager.* New York: John Wiley and Sons, 1983.

Brown, Arnold. *Supermanaging.* New York: McGraw-Hill, 1984.

Cole, Barry, and Oettinger, M. *Reluctant Regulators: The FCC and the Broadcast Audience.* Reading, Mass.: Addison-Wesley, 1978.

Coleman, Howard W. *Case Studies in Broadcast Management.* New York: Hastings House, 1978.

Creech, Kenneth C. *Electronic Media Law and Regulation,* 2e. Boston: Focal Press, 1996.

Czech-Beckerman, Elizabeth Shimer. *Managing Electronic Media.* Mass.: Focal Press, 1991.

Ellmore, R. Terry. *Broadcasting Law and Regulation.* Blue Ridge Summit, Pa.: Tab Books, 1982.

Goodworth, Clive T. *How to Be a Super-Effective Manager: A Guide to People Management.* London: Business Books, 1984.

Kahn, Frank J., ed. *Documents of American Broadcasting,* 4th ed. Englewood Cliffs, N.J.: Prentice-Hall, 1984.

Kobert, Norman. *The Aggressive Management Style.* Englewood Cliffs, N.J.: Prentice-Hall, 1981.

Krasnow, Erwin G. *Insider's Guide to Radio Acquisition Contracts.* Springfield, Va.: Radio Business Report, 1992.

Lacy, Stephen, et.al., *Media Management: A Casebook Approach.* Hillsdale, NJ.: Lawrence Erlbaum Associates, 1993.

McCormack, Mark H. *What They Don't Teach You at Harvard Business School.* New York: Bantam, 1984.

Miner, John B. *The Management Process: Theory, Research, and Practice.* New York: Macmillan, 1978.

National Association of Broadcasters. *Up the Management Ladder: From Program Director to General Manager.* Washington, D.C.: NAB, 1991.

Pember, Don R. *Mass Media in America.* Chicago: Science Research Association, 1981.

Pringle, Peter K., Starr, Michael F., and McCavitt, William E. *Electronic Media Management,* 3rd ed. Boston: Focal Press, 1995.

Quaal, Ward L., and Brown, James A. *Broadcast Management,* 2nd ed. New York: Hastings House, 1976.

Rhoads, B. Eric, et.al., eds. *Management and Sales Management.* West Palm Beach, FL.: Streamline Press, 1995.

Routt, Ed. *The Business of Radio Broadcasting.* Blue Ridge Summit, Pa.: Tab Books, 1972.

Schwartz, Tony. *Media, The Second God.* New York: Praeger, 1984.

Shane, Ed. *Cutting Through: Strategies and Tactics for Radio.* Houston, Tex.: Shane Media, 1990.

Townsend, Robert. *Further up the Organization.* New York: Knopf, 1984.

3 Programming

PROGRAM FORMATS

Programming a radio station has become an increasingly complex task. There are twice the number of stations today competing for the audience's attention than existed in the 1960s, and more enter the fray almost daily. Other media have proliferated as well, resulting in a further distraction of radio's customary audience. The government's laissez-faire, "let the marketplace dictate," philosophy concerning commercial radio programming gives the station great freedom in deciding the nature of its air product, but determining what to offer the listener, who is often presented with dozens of alternatives, involves intricate planning.

The basic idea, of course, is to air the type of format that will attract a sizable enough piece of the audience demographic to satisfy the advertiser. Once a station decides upon the format it will program, it then must know how to effectively execute it.

What follows are brief descriptions of the most frequently employed formats in radio today. There are a host of other formats, or subformats—over one hundred, in fact. Many are variations of those listed.

Adult Contemporary (A/C)

In terms of the number of listeners, Adult Contemporary (also referred to as The Mix, Hot AC, Triple A, Urban AC and Lite) was the most popular format in the 1980s and continues to draw impressive audiences in the 1990s, although it has lost ground since the last edition of this book. Says consultant Ed Shane, "Because the A/C target audience is so diverse, the format has been most prone to fragmentation and competition."

A/C is very strong in the 25 to 49 age group, which makes it particularly appealing to advertisers, since this demographic group has significant disposable income. Also, some advertisers spend money on A/C stations simply because they like the format themselves. The Adult Contemporary format is also one of the most effective in attracting women listeners.

A/C outlets emphasize current (since 1970s) pop standards, sans raucous or pronounced beats—in other words, no hard rock. Some A/C stations could be described as soft rockers. However, the majority mix in enough ballads and easy listening sounds to justify their title. The main thrust of this format's programming is the music. More music is aired by deemphasizing chatter. Music is commonly presented in uninterrupted sweeps or blocks, perhaps ten to twelve minutes in duration, followed by a brief recap of artists and song titles. High-profile morning talent or teams became popular at A/C stations in the 1980s and remain so today. Commercials generally are clustered at predetermined times, and midday and evening deejay talk often is limited to brief informational announcements. News and sports are secondary to the music.

Contemporary Hit Radio (CHR)

Also known as Top 40, CHR stations play only those records which currently are the fastest selling. CHR's narrow playlists are designed to draw teens and young adults. The heart of this format's demographic is the twelve- to eighteen-year-old, although in the mid-1980s it enjoyed a broadening of its core audience. Like A/C, it too has experienced erosion in its numbers in recent years. In the *Journal of Radio Studies* (1995-1996), Ed Shane observed that the format "was a statistical loser in the 1990s. What futurist Alvin Toffler called 'the demassification

FIGURE 3.1
Format chart shows the popularity of formats around the country. The chart changes with each ratings sweep. Reprinted with permission from *Radio and Records.*

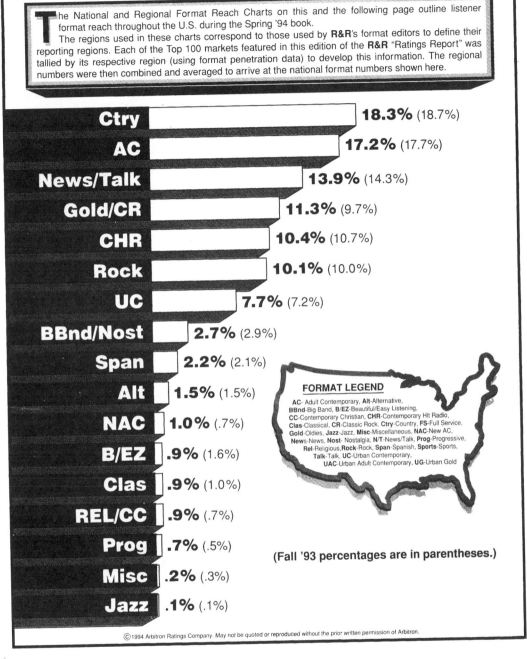

The National and Regional Format Reach Charts on this and the following page outline listener format reach throughout the U.S. during the Spring '94 book.

The regions used in these charts correspond to those used by **R&R**'s format editors to define their reporting regions. Each of the Top 100 markets featured in this edition of the **R&R** "Ratings Report" was tallied by its respective region (using format penetration data) to develop this information. The regional numbers were then combined and averaged to arrive at the national format numbers shown here.

Ctry	**18.3%**	(18.7%)
AC	**17.2%**	(17.7%)
News/Talk	**13.9%**	(14.3%)
Gold/CR	**11.3%**	(9.7%)
CHR	**10.4%**	(10.7%)
Rock	**10.1%**	(10.0%)
UC	**7.7%**	(7.2%)
BBnd/Nost	**2.7%**	(2.9%)
Span	**2.2%**	(2.1%)
Alt	**1.5%**	(1.5%)
NAC	**1.0%**	(.7%)
B/EZ	**.9%**	(1.6%)
Clas	**.9%**	(1.0%)
REL/CC	**.9%**	(.7%)
Prog	**.7%**	(.5%)
Misc	**.2%**	(.3%)
Jazz	**.1%**	(.1%)

FORMAT LEGEND

AC- Adult Contemporary, **Alt**-Alternative, **BBnd**-Big Band, **B/EZ**-Beautiful/Easy Listening, **CC**-Contemporary Christian, **CHR**-Contemporary Hit Radio, **Clas**-Classical, **CR**-Classic Rock, **Ctry**-Country, **FS**-Full Service, **Gold**-Oldies, **Jazz**-Jazz, **Misc**-Miscellaneous, **NAC**-New AC, **News**-News, **Nost**- Nostalgia, **N/T**-News/Talk, **Prog**-Progressive, **Rel**-Religious, **Rock**-Rock, **Span**-Spanish, **Sports**-Sports, **Talk**-Talk, **UC**-Urban Contemporary, **UAC**-Urban Adult Contemporary, **UG**-Urban Gold

(Fall '93 percentages are in parentheses.)

of media' affected CHR the most. . . There were too many types of music to play. No one radio station could create a format with elements as diverse as rapper Ice T, rockers like Nirvana, country artists like George Strait and Randy Travis, or jazz musicians like Kenny G or David Benoit. Each of those performers fits someone's definition of 'contemporary hit radio.' CHR lost its focus."

Consultant Jeff Pollack contends that CHR has lost ground because it is not in tune with what he calls the "streets," and predicts the format will embrace a more dance-rap sound as well as developing more appreciation for alternative rock hits. The format is characterized

FIGURE 3.2
A growing number of stations around the country program exclusively for children. Courtesy Radio Aahs.

Radio Aahs™
America's Only 24 Hour Radio Network For Kids...And Their Parents, Too!

CHILDREN'S RADIO WHAT A CONCEPT!

From our fun-filled wake-up morning show to the story time Evening Theatre, Radio Aahs is 24-hour fun for the whole family -- with music, personalities, kid DJ's, stories, features, games, quizzes, contests, kid-news, entertainment, information, education, laughs and lots, lots more! Actually, we're better described as *family radio*, because when a child tunes in to Radio Aahs, moms and dads are automatically a part of our captive audience!

KIDS HAVE PHENOMENAL BUYING POWER!

* Kids spent $8 Billion in 1991!
* Kids directly influenced another $132 Billion in household purchases!
* Kids (12 and under) combined with their parents represent a whopping 47% of the marketplace!
* 20% of your market is under 12 years old.
* Kids represent 20% of the U.S. population!
* Between 1990 and 2000, the kid population will increase by 3.2 million!

WHO LISTENS TO RADIO AAHS?

* The average child is 7.8 years old and listens to over 30 hours a week.
* Moms listening average 33.6 years old, Dads average 34.7 years. There are also grandparents and primary care provides listening, too.
* 50% of all listening takes place in the car, and since kids under 12 can't drive, the adults with them hear the same programs and advertising.
* On average, every child in America has access to 5.6 radios.

ARBITRON SAYS YES TO KIDS!

Arbitron, the nation's foremost ratings measurement service has finally agreed to measure children's radio, the fastest growing demographic in America!

It's Time Now For Radio To Say "Yes" To Family Spenders And The Advertisers Who Need To Reach Them!

Don't Let America's Fastest Growing Audience Pass You By!

by its swift, often unrelenting pace. Silence, known as "dead air," is the enemy. The idea is to keep the sound hot and tight to keep the kids from station hopping, which is no small task since many markets have at least two hit-oriented stations.

In a 1995 interview in *Radio Ink*, programmer Bill Richards predicted that the high-intensity jock approach would give way to a more laid back, natural sound. "The days of the 'move over and let the big dog eat' sweepers are over. Top 40 will look for more jocks who sound like real people and shy away from the hyped deejay approach." CHR deejays have undergone several shifts in status since the inception of the chart music format in the 1950s. Initially, pop deejay personalities played an integral role in the air sound. However, in the mid-1960s the format underwent a major change when deejay presence was significantly reduced. Programming innovator Bill Drake decided that the Top 40 sound needed to be refurbished and tightened. Thus deejay talk and even the number of commercials scheduled each hour were cut back in order to improve flow. Despite criticism that the new sound was too mechanical, Drake's technique succeeded at strengthening the format's hold on the listening audience.

In the mid- and late 1970s the deejay's role on hit stations began to regain its former prominence, but in the 1980s the format underwent a further renovation (initiated by legendary consultant Mike Joseph) that resulted in a narrowing of its playlist and a decrease in deejay presence. Super or *Hot* hit stations, as they also are called, were among the most popular in the country and could be found either near or at the top of the rating charts in their markets.

At the moment, at least, CHR has a bit less of a frenetic quality to it, and perhaps (it could be said) a more mature sound. Undergoing an image adjustment, the format is keying in on improving overall flow while pulling back on jumping aboard the fad bandwagon. The continued preening of the playlist will keep the format viable, say the experts.

News is of secondary importance on CHR stations. In fact, news programming is considered by many program directors to be a tune-out factor. "Kids don't like news," they claim. However, despite deregulation, which has freed stations of nonentertainment program requirements, most retain at least a modicum of news out of a sense of obligation. CHR stations are very promotion minded and contest oriented.

In the mid-1990s, fewer than five hundred stations (nearly all FM) called themselves CHR. Many of these stations preferred to be called CHURBAN, which combines urban and rock hits, or Modern Hits, a narrower-based version of Top 40 that draws its playlist from MTV and college stations.

Country

Since the 1970s the Country format has been adopted by more stations than any other. Although seldom a leader in the ratings race, its appeal is exceptionally broad. An indication of country music's rising popularity is the fact that there are over ten times as many full-time Country stations today than there were twenty years ago. Although the format is far more prevalent in the South and Midwest, most medium and large markets in the North have Country stations. Due to the diversity of approaches within the format (there are more subformats in Country than in any other, for example, traditional, middle-of-the-road, contemporary hit, and so on), the Country format attracts a broad age group, appealing to young as well as older adults. "The Country format has scored very high among 25- to 54-year-olds," adds Burkhart.

Country radio has always been particularly popular among blue-collar workers. According to the Country Music Association and the Organization of Country Radio Broadcasters, the Country music format is drawing a more upscale audience today than it has in the past. As many FM as AM stations are programming the Country sound in the 1990s, which wasn't the case just a few years ago. Until the 1980s, Country was predominantly an AM offering. Depending on the approach they employ, Country outlets may emphasize or deemphasize air personalities, include news and public affairs features, or confine their programming almost exclusively to music.

According to some programming experts, the Country format peaked in the mid-1990s, and there will be little further growth, at least for the time being. A recent *M Street Journal* survey concluded that over 2,600 stations air some form of country music.

Easy Listening

The Beautiful Music station of the 1960s has become the Easy Listening station of the 1990s. Playlists in this format have been carefully updated in an attempt to attract a somewhat younger audience. The term *Beautiful Music* was exchanged for *Easy Listening* in an effort to dispel the geriatric image the former

FIGURE 3.3
A regional
breakdown of the
effectiveness of
various formats.
Courtesy *Radio
and Records.*

FORMAT REACH CHARTS

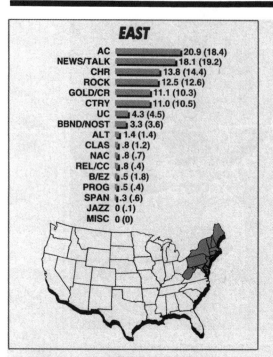

EAST

Format	Value
AC	20.9 (18.4)
NEWS/TALK	18.1 (19.2)
CHR	13.8 (14.4)
ROCK	12.5 (12.6)
GOLD/CR	11.1 (10.3)
CTRY	11.0 (10.5)
UC	4.3 (4.5)
BBND/NOST	3.3 (3.6)
ALT	1.4 (1.4)
CLAS	.8 (1.2)
NAC	.8 (.7)
REL/CC	.8 (.4)
B/EZ	.5 (1.8)
PROG	.5 (.4)
SPAN	.3 (.6)
JAZZ	0 (.1)
MISC	0 (0)

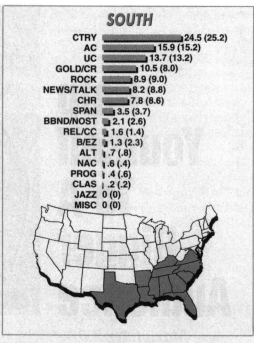

SOUTH

Format	Value
CTRY	24.5 (25.2)
AC	15.9 (15.2)
UC	13.7 (13.2)
GOLD/CR	10.5 (8.0)
ROCK	8.9 (9.0)
NEWS/TALK	8.2 (8.8)
CHR	7.8 (8.6)
SPAN	3.5 (3.7)
BBND/NOST	2.1 (2.6)
REL/CC	1.6 (1.4)
B/EZ	1.3 (2.3)
ALT	.7 (.8)
NAC	.6 (.4)
PROG	.4 (.6)
CLAS	.2 (.2)
JAZZ	0 (0)
MISC	0 (0)

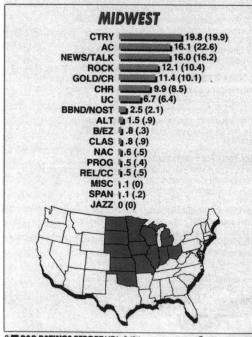

MIDWEST

Format	Value
CTRY	19.8 (19.9)
AC	16.1 (22.6)
NEWS/TALK	16.0 (16.2)
ROCK	12.1 (10.4)
GOLD/CR	11.4 (10.1)
CHR	9.9 (8.5)
UC	6.7 (6.4)
BBND/NOST	2.5 (2.1)
ALT	1.5 (.9)
B/EZ	.8 (.3)
CLAS	.8 (.9)
NAC	.6 (.5)
PROG	.5 (.4)
REL/CC	.5 (.5)
MISC	.1 (0)
SPAN	.1 (.2)
JAZZ	0 (0)

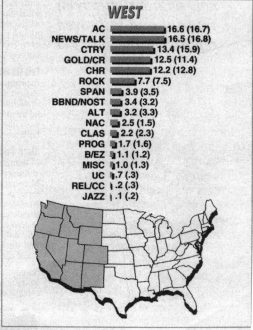

WEST

Format	Value
AC	16.6 (16.7)
NEWS/TALK	16.5 (16.8)
CTRY	13.4 (15.9)
GOLD/CR	12.5 (11.4)
CHR	12.2 (12.8)
ROCK	7.7 (7.5)
SPAN	3.9 (3.5)
BBND/NOST	3.4 (3.2)
ALT	3.2 (3.3)
NAC	2.5 (1.5)
CLAS	2.2 (2.3)
PROG	1.7 (1.6)
B/EZ	1.1 (1.2)
MISC	1.0 (1.3)
UC	.7 (.3)
REL/CC	.2 (.3)
JAZZ	.1 (.2)

term seemed to convey. Easy Listening is the ultimate "wall-to-wall" music format. Talk of any type is kept minimal, although many stations in this format concentrate on news and information during morning drivetime.

Instrumentals and soft vocals of established songs are a mainstay at Easy Listening stations, which also share a penchant for lush orchestrations featuring plenty of strings. These stations boast a devoted audience.

Station hopping is uncommon. Efforts to draw younger listeners into the Easy Listening fold have been moderately successful, but most of the format's primary adherents are over fifty. Music syndicators provide prepackaged (canned) programming to approximately half of the nation's Easy Listening stations, and over three-quarters of the outlets within this format utilize automation systems to varying degrees. Easy Listening has held strong in several markets; WGAY FM/AM in Washington, D.C., and WSSH-FM in Boston have topped the ratings for years, although the format lost some ground in the 1980s to Adult Contemporary and other adult-appeal formats such as New Age, which has been referred to by some media critics as Easy Listening for Yuppies.

Soft Adult and Lite and Easy have become replacement names for Easy Listening, which, like its predecessor, Beautiful Music, began to assume a geriatric connotation.

Album-Oriented Rock (AOR)

The birth of the AOR format in the mid-1960s was the result of a basic disdain for the highly formulaic Top 40 sound that prevailed at the time. In the summer of 1966, WOR-FM, New York, introduced Progressive radio, the forerunner of AOR. As an alternative to the superhyped, ultracommercial sound of the hit song station, WOR-FM programmed an unorthodox combination of nonchart rock, blues, folk, and jazz. In the 1970s the format concentrated its attention almost exclusively on album rock, while becoming less free-form and more formulaic and systematic in its programming approach.

While AOR has done well in garnering the 18- to 34-year-old male, this format has done poorly in winning female listeners, especially when it emphasizes a heavy- or hard-rock playlist. This has proven to be a sore spot with certain advertisers. In the 1980s, the format lost its prominence due, in part, to the meteoric rebirth of hit radio. However, as the decade came to an end, AOR had regained a chunk of its numbers.

Generally, AOR stations broadcast their music in sweeps, or at least segue two or three songs. A large airplay library is typical, in which three to seven hundred cuts may be active. Depending on the outlet, the deejay may or may not have "personality" status. In fact, the more-music/less-talk approach particularly common at Easy Listening stations is emulated by many album rockers. Consequently, news plays a very minor part in the station's programming efforts.

AOR stations are very lifestyle oriented and invest great time and energy developing promotions relevant to the interests and attitudes of their listeners.

In the uncertain 1990s, this format's audience has been further eroded by variations on its patented theme. For instance, Modern Rock/Alternative outlets have snatched away a prominent segment of its younger demographics, while Classic Rock stations have likewise eaten into AOR's more mature listener cells. Reflecting the considerable drift away from the AOR nomenclature and model, WBCN's program director, Oedipus, declared to the world in 1995, "We're Modern Rock!"

News and/or Talk

There are News, News/Talk, News Sports, News Plus, and Talk formats, and each is distinct and unique unto itself. News stations differ from the others in that they devote their entire air schedule to the presentation of news and news-related stories and features. The All-News format was introduced by Gordon McLendon at XETRA (known as "XTRA") in Tijuana, Mexico, in the early 1960s. Its success soon inspired the spread of the format in the United States. Due to the enormous expense involved in the presentation of a purely News format, requiring three to four times the staff and budget of most music operations, the format has been confined to larger markets able to support the endeavor.

The News/Talk format is a hybrid. It combines extensive news coverage with blocks of two-way telephone conversations. These stations commonly "day-part" or segmentalize their programming by presenting lengthened newscasts during morning and afternoon drivetime hours and conversation in the midday and evening periods. The News/Talk combo was conceived by KGO in San Francisco in the 1960s and has gradually gained in popularity so that it now leads both the strictly News and the Talk formats.

Talk radio began at KABC-AM, Los Angeles, in 1960. However, talk shows were familiar to listeners in the 1950s, since a number of adult music stations devoted a few hours during evenings or overnight to call-in programs. The motivation behind most early Talk programming stemmed from a desire to strengthen weak time slots while satisfying public affairs programming requirements. Like its nonmusic siblings, Talk became a viable format in the 1960s and does well today, although it too has suffered due to greater competition. In contrast to All-News, which attracts a slightly younger and more upscale audience, All-Talk amasses a large following among blue-collar workers and retirees.

One of the recent news and information-oriented formats calls itself News Plus. While its emphasis is on news, it fills periods with music, often Adult Contemporary in flavor. News Plus stations also may carry a heavy schedule of sporting events. This combination has done well in several medium and large markets, but has apparently plateaued as this book goes to press.

News and/or Talk formats are primarily located on the AM band, where they have become increasingly prevalent since FM has captured all but a few of radio's music listeners. Meanwhile, the number of nonmusic formats is significantly increasing on FM.

In the middle 1990s, over 1,000 stations offered the information and/or news format. This was up nearly 300 percent since the late 1980s. Over 100 stations alone concentrated on sports exclusively, while dozens of others were beginning to splinter and compartmentalize into news/info niches, such as auto, health, computer, food, business, tourism, entertainment, and so on.

According to a recent survey conducted by *Radio Ink*, News/Talk stations were the second most tuned after Country. National talk networks and syndicated talk shows were proliferating as well in the latter part of the 1990s, as more and more baby-boomers became engaged in the political and social dialogues of the day.

An indication that the News format is achieving broader appeal among younger listeners is the recent emergence on the FM dial of a new hybrid called Talk 'n' Rock.

Classic/Oldies/Nostalgia

While these formats are not identical, they derive the music they play from years gone by. Whereas the Nostalgia station, sometimes referred to as *Big Band*, constructs its playlist around tunes popular as far back as the 1940s and 1950s, the Oldies outlet focuses its attention on the pop tunes of the late 1950s and early 1960s. A typical oldies quarter-hour might consist of songs by Elvis Presley, the Everly Brothers, Brian Hyland, and the Ronettes. In contrast, a Nostalgia quarter-hour might consist of tunes from the pre-rock era performed by artists like Frankie Lane, Les Baxter, the Mills Brothers, and Tommy Dorsey, to name only a few.

Nostalgia radio caught on in the late 1970s, the concept of programmer Al Ham. Nostalgia is a highly syndicated format, and most stations go out-of-house for programming material. Because much of the music predates stereo processing, AM outlets are most apt to carry the Nostalgia sound. Music is invariably presented in sweeps and, for the most part, deejays maintain a low profile. Similar to Easy Listening, Nostalgia pushes its music to the forefront and keeps other program elements at an unobtrusive distance. In the 1980s, Easy Listening/Beautiful Music stations lost some listeners to this format, which claimed a viable share of the radio audience.

IS 93X A MAINSTREAM FORMAT?

- **Today's New Rock**™ has gained enormous mass appeal in recent years and has emerged as America's dominant form of rock music.

- Five of the top ten grossing concerts in the U.S. in December of 1991 featured core artists on **93X**. One of those groups, Guns N' Roses, played to sell out crowds in January at Minneapolis' Target Center.

- America's #1 selling album the week of 1/11/92 was Nirvana's *Nevermind* which surpassed Michael Jackson's *Dangerous*, the previous #1 album. Nirvana is a core artist on 93X but receives little airplay on other Twin Cities FM rock stations. (Radio and Records).

- In a recent *Radio and Records* weekly survey of the most popular rock songs in America, (see chart) approximately 3/4 of the top 40 tracks are from **93X's** artists. Only a small handful of the 40 will receive little more than token airplay on any of the Twin Cities other three FM rock stations.

- The bottom line is that **93X is THE Twin Cities mainstream rock station.**

AOR TRACKS

5 WKS	4 WKS	LW	TW	Artist/Song
1	1	1	1	U2/Mysterious Ways (Island/PLG)
2	2	2	2	VAN HALEN/Right Now (WB)
5	4	3	3	BRYAN ADAMS/There Will Never Be Another... (A&M)
18	10	4	4	GENESIS/I Can't Dance (Atlantic)
19	12	6	5	TOM PETTY & THE.../King's Highway (MCA)
37	27	12	6	RUSH/Ghost Of A Chance (Atlantic)
7	7	7	7	NIRVANA/Smells Like Teen Spirit (DGC)
8	6	8	8	GUNS N' ROSES/November Rain (Geffen)
10	9	9	9	EDDIE MONEY/She Takes My Breath Away (Columbia)
12	8	10	10	METALLICA/The Unforgiven (Elektra)
49	39	15	11	OZZY OSBOURNE/Mama, I'm Coming... (Epic Associated)
BREAKER			12	ERIC CLAPTON/Tears in Heaven (Reprise)
15	11	11	13	BOB SEGER & THE SILVER.../Take A Chance (Capitol)
4	3	5	14	JOHN MELLENCAMP/Love And Happiness (Mercury)
16	13	13	15	TESLA/Call It What You Want (Geffen)
26	22	17	16	TALL STORIES/Wild On The Run (Epic)
BREAKER			17	STEVIE RAY VAUGHAN &.../Empty Arms (Epic)
--	--	58	18	JOHN MELLENCAMP/Again Tonight (Mercury)
38	32	22	19	PEARL JAM/Alive (Epic Associated)
BREAKER			20	JON BON JOVI/Levon (Polydor/PLG)
BREAKER			21	SCORPIONS/Hit Between The... (Mercury/Morgan Creek)
--	--	34	22	STORM/Show Me The Way (Interscope)
35	31	23	23	SOUNDGARDEN/Outshined (A&M)
32	29	25	24	BABY ANIMALS/Painless (Imago)
24	21	18	25	RICHIE SAMBORA/Stranger In This Town (Mercury)
40	35	29	26	CULT/Heart Of Soul (Sire/Reprise)
--	--	50	27	DIRE STRAITS/The Bug (WB)
3	5	14	28	STEVIE RAY VAUGHAN &.../The Sky Is Crying (Epic)
28	26	24	29	SKID ROW/Wasted Time (Atlantic)
23	20	19	30	FOUR HORSEMEN/Rockin' Is... (Def American/Reprise)
44	38	33	31	THUNDER/Love Walked In (Geffen)
6	14	20	32	QUEENSRYCHE/Another Rainy Night (EMI)
53	43	35	33	QUEEN/The Show Must Go On (Hollywood)
--	--	51	34	UGLY KID JOE/Everything About You (Stardog Mercury)
9	15	26	35	RUSH/Roll The Bones (Atlantic)
--	60	47	36	RTZ/Until Your Love Comes Back Around (Giant/Reprise)
--	--	48	37	ROBBIE ROBERTSON/Go Back To Your Woods (Geffen)
54	47	40	38	BODEANS/Good Things (Slash/Reprise)
58	51	41	39	WEBB WILDER/Tough It Out (Praxis/Zoo)
14	8	32	40	STORM/I've Got A Lot To Learn About Love (Interscope)

January 17, 1992

FIGURE 3.4
Media kit items explain a station's market positioning approach. Courtesy KRXX-FM.

The Oldies format was first introduced in the 1960s by programmers Bill Drake and Chuck Blore. While Nostalgia's audience tends to be over the age of fifty, most Oldies listeners are somewhat younger. Unlike Nostalgia, most Oldies outlets originate their own programming, and very few are automated. In contrast with its vintage music cousin, Oldies allows greater deejay presence. At many Oldies stations, air personalities play a key role. Music is rarely broadcast in sweeps, and commercials, rather than being clustered, are inserted in a random fashion between songs.

Consultant Kent Burkhart noted that in the early 1990s, "Oldies stations are scoring very big in a nice broad demographic. These stations are doing quite well today, and this should hold for a while." At the same time, Nostalgia has not been shown as having much growth but remains fairly solid in some markets.

In the middle of the 1990s, Oldies outlets have lost audience ground. However, over 700 stations still call themselves Oldies or Nostalgia.

Meanwhile Classic Rock and Classic Hit stations have emerged as the biggest winners in the late 1980s and early 1990s. These vintage music stations draw their playlists from the chart toppers (primarily in the rock music area) of the 1970s and 1980s (and early 1990s) and often appear in the top ten ratings.

While Classic Rock concentrates on tunes essentially featured by AOR stations over the past two decades, Classic Hit stations fill the gap between Oldies and CHR outlets with playlists that draw from 1970s and 1980s top 40 charts.

Urban Contemporary (UC)

The "melting pot" format, Urban Contemporary, attracts large numbers of Hispanic and Black listeners, as well as White. As the term suggests, stations employing this format are usually located in metropolitan areas with large, heterogeneous populations. UC was born in the early 1980s, the off-spring of the short-lived Disco format, which burst onto the scene in 1978. What characterizes UC the most is its upbeat, danceable sound and deejays who are hip, friendly, and energetic. Although UC outlets stress danceable tunes, their playlists generally are anything but narrow. However, a particular sound may be given preference over another, depending on the demographic composition of the population in the area that the station serves. For example, UC outlets may play greater amounts of music with a Latin or rhythm-and-blues flavor, while others may air larger proportions of light jazz, reggae, or new rock. Several AM stations around the country have adopted the UC format; however, it is more likely to be found on the FM side, where it has taken numerous stations to the forefront of their market's ratings.

UC has had an impact on Black stations, which have experienced erosion in their youth numbers. Many Black stations have countered by broadening their playlists to include artists not traditionally programmed. Because of their high-intensity, fast-paced sound, UC outlets can give a Top 40 impression, but in contrast they commonly segue songs or present music in sweeps and give airplay to lengthy cuts, sometimes six to eight minutes long. While Top 40 or CHR stations seldom program cuts lasting more than four minutes, UC outlets find long cuts or remixes compatible with their programming approach. Remember, UCs are very dance oriented. Newscasts play a minor role in this format, which caters to a target audience aged eighteen to thirty-four. Contests and promotions are important program elements.

As noted earlier, several CHR stations have adopted urban artists in order to offer the hybrid CHURBAN sound. Likewise, many Urban outlets have drawn from the more mainstream CHR playlist in an attempt to expand their listener base.

Classical

Although there are fewer than three dozen full-time commercial Classical radio stations in the country, no other format can claim a more loyal following. Despite small numbers and soft ratings, most Classical stations do manage to generate a modest to good income. Over the years, profits have remained relatively minute in comparison to other formats. However, member stations of the Concert Music Broadcasters Association reported ad revenue increases of up to 40 percent in the 1980s with more growth in the 1990s. Due to its upscale audience, blue-chip accounts find the format an effective buy. This is first and foremost an FM format, and it has broadcast over the megahertz band for almost as long as it has existed.

In many markets, commercial Classical stations have been affected by public radio outlets programming classical music. Since commercial Classical stations must break to air the sponsor messages that keep them operating, they must adjust their playlists accordingly. This may mean shorter cuts of music during particular dayparts—in other words, less music. The noncommercial Classical outlet is relatively free of such constraints and thus benefits as a result. A case in point is WCRB-FM in Boston, the city's only full-time Classical station. While it attracts most of the area's Classical listeners throughout the afternoon and evening hours, it loses many to public radio WGBH's classical segments. At least in part, public radio's success consists in hav-

Roy Acuff • Jerry Lee Lewis • Air Supply** • Paul Anka
Chet Atkins • John Barry • Stephen Bishop • Don Black*
The Black Crowes • Eric Carmen • Alice Cooper • Robert Cray
Creedence Clearwater • Bo Diddley • David Crosby
Crowded House** • Boudleaux & Felice Bryant
Bobby Darin • Dino • Steve Dorff • Fred Ebb
Bee Gees • George Benson • Randy Edelman • Eric Clapton*
Danny Elfman • Sam Cooke
Everly Brothers • Dino • George Fenton • Jude Cole
David Foster • Gerry Goffin • Al Green • Hammer
Isaac Hayes • Holland-Dozier-Holland • Phil Collins*
Buddy Holly • Larnelle Harris • INXS** • Midnight Oil**
Curtis Mayfield • Aretha Franklin
George Jones • Wynonna • Kenny G • B.B. King
John Hiatt • Gloria Estefan • Harlan Howard • Dave Koz
Sting* • Clint Black • Kris Kristofferson • Patti LaBelle
Oleta Adams • Little Richard • Little River Band**
Lynyrd Skynyrd • Loretta Lynn • Melissa Manchester
Michael Jackson • Christine McVie • Janet Jackson
Joni Mitchell • The Beatles* • Mariah Carey • Willie Nelson
Stevie Nicks • Nirvana • Roy Orbison • R.E.M.
Johnny Pacheco • Go West* • Dwight Yoakam • Dolly Parton
Billy Ray Cyrus • Miles Davis • Lionel Hampton
Mick Fleetwood • Sandi Patti • Minnie Pearl • Carl Perkins
Chynna Phillips • Wilson Pickett • The Platters
Mike Post • Elvis Presley • Vince Gill • Otis Redding
Ray Charles • Babyface • The Rolling Stones*
Linda Ronstadt • Diana Ross • Brenda Russell • Tina Turner
Hank Williams, Sr. • Chuck Berry • David Sanborn
James Brown • Dave Brubeck • Lefty Frizzell
George Shearing • Paul Simon • The Beach Boys
Simon & Garfunkel • Brooks & Dunn • Hans Zimmer*
Joe South • Ringo Starr* • Merle Haggard • John Stewart
Ray Stevens • Steven Curtis Chapman • James Taylor
Eddie Money • Allen Toussaint • Carole King
Arrested Development • Travis Tritt • Alabama • Ike Turner
Steve Cropper • Sam & Dave • Muddy Waters
Jody Watley • Andrew Lloyd Webber* • Kitty Wells
Willie Dixon • Allman Brothers • Barry White • Karyn White
Fats Domino • James Brown • Hank Williams, Jr.
Vanessa Williams • Brian Wilson • Carnie Wilson • Bill Monroe
Wendy Wilson • Bebe & Cece Winans • John Williams

Look Who's Got the Hitmakers and the Legends!

Just A Small Sample of the BMI Songwriters Delivering America's Favorite Music!

BMI

America's Hit Music℠

*PRS songwriters represented in U.S. by BMI   **APRA songwriters represented in U.S. by BMI

FIGURE 3.5
Radio stations pay an annual fee to music licensing services such as BMI and ASCAP.

ing fewer interruptions in programming.

Classical stations target the twenty-five- to forty-nine-year-old, higher-income, college-educated listener. News is typically presented at sixty- to ninety-minute intervals and generally runs from five to ten minutes. The format is characterized by a conser-vative, straightforward air sound. Sensationalism and hype are avoided, and on-air contests and promotions are as rare as announcer chatter.

Religious

Live broadcasts of religious programs began while the medium was still in its

FIGURE 3.6
Dozens of other
formats (often
variations on those
listed) have
emerged since the
start of program
specialization in
the 1950s.

Some Format Debuts					
Prior to 1950	1950s	1960s	1970s	1980s	1990s
Classical Country Hit Parade Religious Black Hispanic	Middle-of-the- Road Top 40 Beautiful Music	News Talk News/Talk Progressive Acid/ Psychedelic Jazz All Request Oldies Diversified	Adult Contemporary Album Oriented Rock Easy Listening Contemporary Country Urban Country Mellow Rock Disco Nostalgia British Rock New Wave Public Radio	Arena Rock Hot Adult Contemporary Hit/Radio Urban Contemporary New Age Eclectic Oriented/ Rock All Sports All Motivation All Comedy All Beatles Classic Hits Classic Rock Male Adult Contemporary	Mix All-Children Arrow Rap/Hip Hop Digital Hits Gen X Tourist Radio Triple A NAC Modern Rock Churban Boomer Rock Coupon Radio

experimental stage. In 1919 the U.S. Army Signal Corps aired a service from a chapel in Washington, D.C. Not long after that, KFSG in Los Angeles and WMBI in Chicago began to devote themselves to religious programming. Soon dozens of other radio outlets were broadcasting the message of God. In the 1980s, over six hundred stations broadcast religious formats on a full-time basis, and another fifteen hundred air at least six hours of religious features on a weekly basis. *M Street Journal* reports that over 900 stations air the Religious format today.

Religious broadcasters typically follow one of two programming approaches. One includes music as part of its presentation, while the other does not. The Religious station that features music often programs contemporary tunes containing a Christian or life-affirming perspective. This approach also includes the scheduling of blocks of religious features and programs. Nonmusic Religious outlets concentrate on inspirational features and complementary talk and informational shows.

Religious broadcasters claim that their spiritual messages reach nearly half of the nation's radio audience, and the American Research Corporation in Irvine, California, contends that over 25 percent of those tuned to Religious stations attend church more frequently. Two-thirds of the country's Religious radio stations broadcast over AM frequencies.

An indication of the continued popularity of Religious radio in the latter part of the 1990s is the launch of ChristianNet, a network that offers talk shows from some of the biggest names in conservative chatter.

Ethnic

African-Americans constitute the largest minority in the nation, thus making Black one of the most prominent ethnic formats. Over three hundred radio stations gear themselves to the needs and desires of Black listeners. WDIA-AM in Memphis claims to be the first station to program exclusively to a nonwhite audience. It introduced the format in 1947. Initially growth was gradual in this format, but in the 1960s, as the Motown sound took hold of the hit charts and the Black Pride movement got under way, more Black stations entered the airways.

At its inception, the Disco craze in the 1970s brought new listeners to the Black stations, which shortly saw their fortunes change when all-Disco stations began to surface. Many Black outlets witnessed an exodus of their younger listeners to the Disco stations. This prompted a number of Black stations to abandon their more traditional playlists, which consisted of rhythm and blues, gospel, and soul tunes, for exclusively Disco music. When Disco perished in the early 1980s, the Urban Contemporary format took its place. Today, Progressive Black stations, such as WBLS-FM, New York, combine dance music with soulful rock and contemporary jazz, and many have transcended the color barrier by including certain white artists on their playlists. In fact, many Black stations employ white air personnel in efforts to broaden their demographic base. WILD-AM in Boston, long considered the city's "Black" station, is an example of this trend. "We have become more of a general appeal station than a purely ethnic one. We've had to in order to prosper. We strive for a distinct, yet neuter or deethnicized, sound on the air. The Black format has changed considerably over the years," notes WILD program director Elroy Smith. The old-line R & B and gospel stations still exist and can be found mostly in the South.

Hispanic or Spanish-language stations constitute another large ethnic format. KCOR-AM, San Antonio, became the first all-Spanish station in 1947, just a matter of months after WDIA-AM in Memphis put the Black format on the air. Cities with large Latin populations are able to support the format, and in some metropolitan areas with vast numbers of Spanish-speaking residents, such as New York, Los Angeles, and Miami, several radio outlets are exclusively devoted to Hispanic programming.

Programming approaches within the format are not unlike those prevalent at Anglo stations. That is to say, Spanish-language radio stations also modify their sound to draw a specific demographic. For example, many offer contemporary music for younger listeners and more traditional music for older listeners.

Ed Shane views Hispanic radio as very diverse and vibrant. "An impressive multiplicity of programming styles and approaches are found in this format. Here in Houston, for example, we have two brands of Tejano, one of Exitos (hits), a lot of Ranchera, and a couple of Talk stations. In the Rio Grande Valley

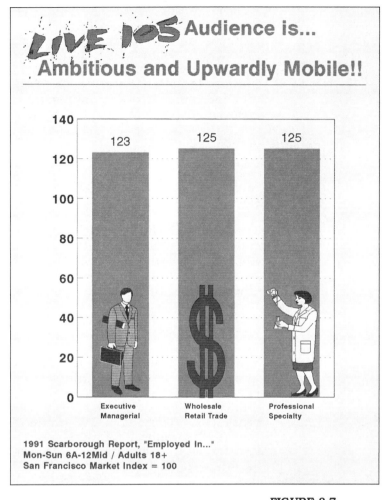

of Texas, there's a lush, instrumental-and-vocal 'Easy Listening' station in Spanish. The L.A. dial is full of Hispanic nuance. Miami, likewise, and it has a totally different slant."

Many Hispanic females are drawn to an approach called The Groove, which mixes Motown and Latin pop artists. The format is marketed by Interstar Programming.

Spanish media experts predict that there will be a significant increase in the number of Hispanic stations through the 1990s. They anticipate that most of this growth will take place on the AM band.

Hundreds of other radio stations countrywide apportion a significant piece of their air schedules (over twenty hours weekly), if not all, to foreign-language programs in Portuguese, German, Polish, Greek, Japanese, and so on.

FIGURE 3.7 Defining the lifestyle of a station's audience plays a crucial role in the molding of a format. Courtesy LIVE 105.

Over twenty-five stations broadcast exclusively to American Indians and Eskimos and are licensed to Native Americans. Today these stations are being fed programming from American Indian Radio on Satellite (AIROS), and the Indigenous Communications Association predicts over 100 Indian-operated stations by the end of the next decade.

Middle-of-the-Road (MOR)/Full-Service/Variety

The Middle-of-the-Road format (also called Full-Service, Variety, General Appeal, Diversified) is ambiguously referred to because it defies more specific labeling. Succinctly, this is the "not too anything" format, meaning that the music it airs is not too current, not too old, not too loud, and not too soft.

MOR has been called the *bridge* format because of its "all things to all people" programming. However, its audience has decreased over the years, particularly since 1980, due to the rise in popularity of more specialized formats. According to radio program specialist Dick Ellis, MOR now has a predominantly over-forty age demographic, several years older than just a decade ago. In some major markets the format continues to do well in the ratings mainly because of strong on-air personalities. But this is not the format that it once was. Since its inception in the 1950s, up through the 1970s, stations working the MOR sound have often dominated their markets. Yet the Soft Rock and Oldies formats in the 1970s, the updating of Easy Listening, and particularly the ascendancy of Adult Contemporary and Talk formats have conspired to significantly erode MOR's numbers. For instance, *M Street Journal* cites fewer than one hundred of these stations today.

MOR is the home of the on-air personality. Perhaps no other format gives its air people as much latitude and freedom. This is not to suggest that MOR announcers may do as they please. They, like any other, must abide by format and programming policy, but MOR personalities often serve as the cornerstone of their station's air product. Some of the best-known deejays in the country have come from the MOR milieu. It would then follow that the music is rarely, if ever, presented in sweeps or even segued. Deejay patter occurs between each cut of music, and announcements are inserted in the same way. News and sports play another vital function at these stations. During drive periods, MOR often presents lengthened blocks of news, replete with frequent traffic reports, weather updates, and the latest sports information. Many MOR outlets are affiliated with major-league teams. With few exceptions, MOR is an AM format. Although it has noted slippage in recent years, it will likely continue to bridge whatever gaps may exist in a highly specialized radio marketplace.

Niche Formats

Experts say that the Alternative formats, with their narrower focus on specific demographic segments, will enjoy greater success in the coming years. In an interview in *Broadcasting and Cable*, Jeff Pollack predicted formats offering more nontraditional approaches to mainstream music (Modern Rock, for example) would be the ratings winners of the next few years. The intense fragmentation of the listening pool means that the big umbrella formats are going to lose out to the ultra-specific ones.

Alternative Rock, which has never fully enjoyed start status, is expected to move up towards the front of the pack in the second half of the 1990s. Of course, when it comes to format prognostication the term *unpredictable* takes on a whole new meaning. Indeed, there will be a rash of successful niche formats in the coming years, due to the ever increasing fragmentation of the radio audience, but exactly what they will be is anyone's guess.

A few years back, no one thought that All-Children's Radio (Radio Aahs) would draw an economically attractive segment of the listening public, but today it is one of the more successful niches on the dial (mostly on AM at this writing), and there are many other wanabees entering the cluttered airwaves.

At best, the preceding is an incomplete list of myriad radio formats that serve the listening public. The program formats mentioned constitute the majority of the basic format categories prevalent today. Tomorrow? Who knows. Radio is hardly a static industry, but one subject to the whims of popular taste. When something new captures the imagination of the American public, radio responds, and often a new format is conceived.

THE PROGRAMMER

Program directors (PDs) are radiophiles. They live the medium. Most admit to being smitten by radio at an early age. "It's something that is in your blood and grows to consuming proportions," admits programmer Peter Falconi. Long-time PD Brian Mitchell recalls an interest in the medium as a small child and for good reason. "I was born into a broadcasting family. My father is a station owner and builder. During my childhood, radio was the primary topic at the dinner table. It fed the flame that I believe was already ignited anyway. Radio fascinated me from the start."

The customary route to the programmer's job involves deejaying and participation in other on-air-related areas, such as copywriting, production, music, and news. It would be difficult to state exactly how long it takes to become a program director. It largely depends on the individual and where he or she happens to be. There are instances where newcomers have gone into programming within their first year in the business. When this happens it is likeliest to occur in a small market where turnover may be high. On the other hand, it is far more common to spend years working toward this goal, even in the best of situations. "Although my father owned the station, I spent a long

time in a series of jobs before my appointment to programmer. Along the way, I worked as station janitor, and then got into announcing, production, and eventually programming," recounts Mitchell.

One of the nation's foremost air personalities, Dick Fatherly, whom *Billboard Magazine* has described as a "long-time legend," spent years as a deejay before making the transition. "In the twenty-five years that I've been in this business, I have worked as a jock, newsman, production director, and even sales rep. Eventually I ended up in program management. During my career I have worked at WABC, WICC, WFUN, WHB, to mention a few. Plenty of experience, you might say," comments Fatherly.

Experience contributes most toward the making of the station's programmer. However, individuals entering the field with hopes of becoming a PD do well to acquire as much formal training as possible. The programmer's job has become an increasingly demanding one as the result of expanding competition. "A good knowledge of research methodology, analysis, and application is crucial. Programming is both an art and a science today," observes Jim Murphy, general manager, WCGY-FM, Lawrence, Massachusetts.

KLSX-FM program director Andy Bloom concurs with Murphy, adding, "A would-be PD needs to school him- or herself in marketing research particularly. Little is done any more that is not based on careful analysis."

Publisher B. Eric Rhoads echoes this stance. "The role has changed. The PD used to be a glorified music director with some background in talent development. Today the PD must be a marketing expert. Radio marketing has become very complex, what with telemarketing, data base marketing, direct mail, interactive communication (fax, computer bulletin boards), and so forth. Radio is changing, and the PD must adapt. No longer will records and deejays make the big difference. Stations are at parity in music, so better ways must be found to set stations apart."

Cognizant of this, schools with programs in radio broadcasting have begun to emphasize courses in audience research, as well as other programming-related areas. An important fact for the aspiring PD to keep in mind is that more people entering broadcasting today have college backgrounds than ever before. While a college degree is not necessarily a prerequisite for the position of PD, it is clearly regarded as an asset by upper management. "It used to be that a college degree didn't mean so much. A PD came up through the ranks of programming, proved his ability, and was hired. Not that that doesn't still happen. It does. But more and more the new PD has a degree or, at the very least, several years of college," contends Joe Cortese, air personality, WBMX-FM, Boston. "I majored in Communication Arts at a junior college and then transferred to a four-year school. There are many colleges offering communications courses here in the Boston area, so I'll probably take some more as a way of further preparing for the day when I'll be programming or managing a station. That's what I eventually want to do," says Cortese, adding

that experience in the trenches is also vital to success.

His point is well taken. Work experience does head the list on which a station manager bases his or her selection for program director. Meanwhile, college training, at the very least, has become a criterion to the extent that if an applicant does not have any, the prospective employer takes notice.

Beyond formal training and experience, Chuck Ducoty, station manager of WIYY-FM, Baltimore, says a PD must possess certain innate qualities. "Common sense and a good sense of humor are necessary attributes and are in rather short supply, I think." Dick Fatherly adds sensitivity, patience, compassion, and drive to the list.

THE PROGRAM DIRECTOR'S DUTIES AND RESPONSIBILITIES

Where to begin this discussion poses no simple problem, because the PD's

Lee Abrams

On Station Programming

Today, the simple heritage or niche programming formulas run into problems too. For example, a station that owns a format in a given market but lets its dominance slip away to one or more stations is what I call a misevolved heritage station. Typically, this situation is caused by greed (leading to over-clutter) and ignoring the audience's needs. Heritage stations that are music driven usually lose their grip during the periods when music goes through significant change, such as in 1954, 1964, 1970, 1980, and 1994. In those years. new musical and

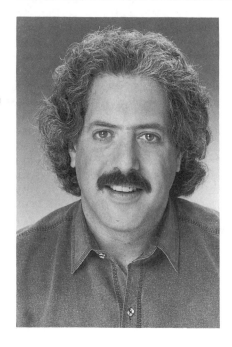

FIGURE 3.8

cultural attitudes evolved, which were not addressed by these stations.

As far as niche stations are concerned, many lack staying power. A good illustration of this point is the 1979 Disco format. WKTU deservedly went to number 1 in just 90 days in New York, prompting scores of other stations to leap to the Disco thing. While the Disco concept helped bring Urban formats into the mainstream, the format was implemented in markets that just weren't right. As a result, many early Disco stations failed to evolve once the "Saturday Night Fever" craze subsided.

The heritage/niche programming scenario seems to be one that hits different format genres at different times, and most stations succumb to it over the long haul. Rock was the first format to experience heritage/niche syndrome. Now we're seeing Urban, and even the highly specialized Ethnic formats, affected by this.

responsibilities and duties are so numerous and wide-ranging. Second in responsibility to the general manager, the program director is the individual responsible for everything that goes over the air. This involves working with the manager in establishing programming and format policy and overseeing their effective execution. In addition, he or she hires and supervises on-air music and production personnel, plans various schedules, handles the programming budget, develops promotions, monitors the station and its competition, assesses research, and may even pull a daily airshift. The PD also is accountable for the presentation of news, public affairs, and sports features, although a news director often is appointed to help oversee these areas.

The program director alone does not determine a station's format. This is an upper management decision. The PD may be involved in the selection process but, more often than not, the format has been decided upon before the programmer has been hired. For example, WYYY decides it must switch from MOR to CHR in order to attract a more marketable demographic. After an in-depth examination of its own market, research on the effectiveness of CHR nationally, and advice from a program consultant and rep company, the format change is deemed appropriate. Reluctantly the station manager concludes that he must bring in a CHR specialist, which means that he must terminate the services of his present programmer, whose experience is limited to MOR. The station manager places an ad in an industry trade magazine, interviews several candidates, and hires the person he feels will take the station to the top of the ratings. When the new program director arrives, he or she is charged with the task of preparing the CHR format for its debut. Among other things, this may involve hiring new air people, the acquisition of a new music library or the updating of the existing one, preparing promos and purchasing jingles, and working in league with the sales, traffic, and engineering departments for maximum results.

On the above points, Corinne Baldasano, vice president of SW Programming, observes, "First of all, of course, you must be sure that the station you are programming fills a market void, i.e., that there is an opportunity for you to succeed in your geographic area with the format you are programming. For example, a young adult alternative rock station may not have much chance for success in an area that is mostly populated by retirees. Once you have determined that the format fills an audience need, you need to focus on building your station. The basic ingredients are making sure your music mix is correct (if you are programming a music station) and that you've hired the on-air talent that conveys the attitude and image of the station you wish to build. At this stage, it's far more important to focus inward than outward. Many stations have failed because they've paid

FIGURE 3.9 Comparing station playlists within the same format is a source of information for programmers. Courtesy of MMR.

MONDAY MORNING REPLAY/ISSUE 204/JANUARY 27, 1992

14

HIT RADIO SAMPLE HOURS

USERS GUIDE: A random sampling of selected radio stations.
This week, sample hours were extracted between 7PM and Midnight.

MARKET COMPARISON — DETROIT

WDFX-FM	WHYT-FM
9PM	**9PM**
FRONT 242/Welcome To Paradise	AUTOGRAPH/Turn Up The Radio
EARL, STACY/Love Me All Up	PUBLIC ENEMY/911 Is A Joke
SWEAT, KEITH/Keep It Comin'	AC/DC/Back In Black
GUNS N' ROSES/Live And Let Die	BEASTIE BOYS/(You Gotta) Fight For Your...
BOYZ II MEN/Uhh Ahh	LL COOL J/I'm Bad
MARKY MARK/Good Vibrations	BOYZ II MEN/Uhh Ahh
BLUR/There's No Other Way	NIRVANA/Smells Like Teen Spirit
WATLEY, JODY/I Want You	HAMMER/Turn This Mutha Out
HAMMER/Addams Groove	ABDUL, PAULA/Vibeology
2 LIVE CREW/Me So Horny	HEAVY D. & THE BOYZ/Now That We Found Love
PARTY/In My Dreams	SALT-N-PEPA/You Showed Me
PENISTON, CECE/We Got A Love Thang	
CAUSE AND EFFECT/You Think You Know Her	

more attention to the competition's product than they have their own."

Once the format is implemented, the program director must work at refining and maintaining the sound. After a short time, the programmer may feel compelled to modify air schedules either by shifting deejays around or replacing those who do not enhance the format. Says Metro Networks president David Saperstein, "You've got to continually fine-tune the station's sound. You must remove any and all negatives, like excessive talk, annoying commercials, technical weaknesses, and so forth. The most critical rule of thumb is that stations should always concentrate on bringing listeners to the station, keeping them tuned in, and providing the right balance of music, personalities, talk, information, and commercials so listeners do not have any reason to tune elsewhere."

The PD prepares weekend and holiday schedules as well, and this generally requires the hiring of part-time announcers. A station may employ as few as one or two part-timers or fill-in people, or as many as eight to ten. This largely depends on whether deejays are on a five- or six-day schedule. At most stations, air people are hired to work a six-day week. The objective of scheduling is not merely to fill slots but to maintain continuity and consistency of sound. A PD prefers to tamper with shifts as little as possible and fervently hopes that he has filled weekend slots with people who are reliable. "The importance of dependable, trustworthy air people cannot be overemphasized. It's great to have talented deejays, but if they don't show up when they are supposed to because of one reason or another, they don't do you a lot of good. You need people who are cooperative. I have no patience with individuals who try to deceive me or fail to live up to their responsibilities," says Brian Mitchell. A station that is constantly introducing new air personnel has a difficult time establishing listener habit. The PD knows that in order to succeed he or she must present a stable and dependable sound, and this is a significant programming challenge.

Production schedules also are prepared by the programmer. Deejays are usually tapped for production duties before or after their airshifts. For example, the morning person who is on the air from six to ten may be assigned production and copy chores from ten until noon. Meanwhile, the midday deejay who is on the air from ten until three is given production assignments from three to five, and so on. Large radio stations frequently employ a full-time production person. If so, this individual handles all production responsibilities and is supervised by the program director.

A program director traditionally handles the department's budget, which generally constitutes 30 to 40 percent of the station's operating budget. Working with the station manager, the PD ascertains the financial needs of the programming area. The size and scope of the budget varies from station to station. Most programming budgets include funds for the acquisition of program materials, such as albums, features, and contest paraphernalia. A separate promotional budget usually exists, and it too may be managed by the PD. The programmer's budgetary responsibilities range from monumental at some outlets to minuscule at others. Personnel salaries and even equipment purchases may fall within the province of the program department's budget. Brian Mitchell believes that "an understanding of the total financial structure of the company or corporation and how programming fits into the scheme of things is a real asset to a programmer."

Devising station promotions and contests also places demands on the PD's time. Large stations often appoint a promotion director. When this is the case, the PD and promotion director work together in the planning, development, and execution of the promotional campaign. The PD, however, retains final veto power should he or she feel that the promotion or contest fails to complement the station's format. When the PD alone handles promotions and contests, he or she may involve other members of the programming or sales department in brainstorming sessions designed to come up with original and interesting concepts. The programmer is aware that the right promotion or contest can have a major impact on ratings. Thus he or she is constantly on the lookout for an appropriate vehicle. In the quest to find the promotion that will launch the station on the path to a larger audience, the PD may seek assistance from one of dozens of companies that offer promotional services.

The program director's major objective is to program for results. If the station's programming fails to attract a sufficient following, the ratings will reflect that unhappy fact. All medium and larger markets are surveyed by ratings companies, primarily Arbitron. Very few small rural markets, with perhaps one or two stations, are surveyed. If a small market station is poorly programmed, the results will be apparent in the negative reactions of the local retailers. Simply put, the station will not be bought by enough advertisers to make the operation a profitable venture. In the bigger markets, where several stations compete for advertising dollars, the ratings are used to determine which is the most effective or cost-efficient station to buy. "In order to make it to the top of the ratings in your particular market, you have to be doing the best job around. It's the PD who is going to get the station the numbers it needs to make a buck. If he doesn't turn the trick, he's back in the job market," observes Dick Fatherly.

Program directors constantly monitor the competition by analyzing the ratings and by listening. A radio station's programming is often constructed in reaction to a direct competitor's. For example, rock stations in the same market often counterprogram newscasts by airing them at different times in order to grab up their competitor's tune-outs. However, rather than contrast with each other, pop stations tend to reflect one another. This, in fact, has been the basis of arguments by critics who object to the so-called mirroring effect. What happens is easily understood. If a station does well by presenting a particular format, other stations are going to exploit the sound in the hopes of doing well also. WYYY promotes commercial-free sweeps of music and captures big ratings, and soon its competitor programs likewise. "Program directors use what has proven to be effective. It is more a matter of survival than anything. I think most of us try to be original to the degree that we can be, but there is very little new under the sun. Programming moves cyclically. Today we're all doing this. Tomorrow we'll all be doing that. The medium reacts to trends or fads. It's the nature of the beast," notes WQIK's Mitchell. Keeping in step with, or rather one step ahead of, the competition requires that the PD know what is happening around him or her at all times.

Nearly 70 percent of the nation's PDs pull an airshift (go on the air themselves) on either a full-time or part-time basis. A difference of opinion exists among programmers concerning their on-air participation. Many feel that being on the air gives them a true sense of the station's sound, which aids them in their programming efforts. Others contend that the three or four hours that they spend on the air take them away from important programming duties. Major market PDs are less likely to be heard on the air than their peers in smaller markets because of addi-

---

SMALL MARKETS
- Program Director
- Airstaff

MEDIUM MARKETS
- Program Director
- Music Director
- Production Director
- News Director
- Engineering
- Promotion
- Airstaff

LARGE MARKETS
- Program Director
- Assistant Program Director
- Music Director
- Music Librarian
- Programming Secretary
- Research Director
- Production Director
- Promotion Director
- News Director
- Tech Assistant
- Marketing Director
- Airstaff

FIGURE 3.10 Personnel structure of the programming department in different size markets.

FIGURE 3.11
Computer software that schedules a station's music is very popular. Courtesy Halper and Associates.

tional duties created by the size and status of the station. Meanwhile, small and medium market stations often expect their PDs to be seasoned air people capable of filling a key shift. "It has been my experience when applying for programming jobs that managers are looking for PDs with excellent announcing skills. It is pretty rare to find a small market PD who does not have a daily airshift. It comes with the territory," says Gary Begin.

Whether PDs are involved in actual airshifts or not, almost all participate in the production of commercials, public service announcements, and promos. In lieu of an airshift, a PD may spend several hours each day in the station's production facilities. The programmer may, in fact, serve as the primary copywriter and spot producer. This is especially true at nonmajor market outlets that do not employ a full-time production person.

In summation, the program director must possess an imposing list of skills to effectively perform the countless tasks confronting him or her daily. There is no one person, other than the general manager, whose responsibilities outweigh the programmer's. The program director can either make or break the radio station.

ELEMENTS OF PROGRAMMING

Few programmers entrust the selection and scheduling of music and other sound elements to deejays and announcers. There is too much at stake and too many variables, both internal and external, which must be considered in order to achieve maximum results within a chosen format. It has become a very complex undertaking, observes KLSX's Bloom. "For instance, all of our music is tested via call-out. At least one or two perceptual studies are done every year, depending on what questions we need answered. Usually a couple of sets of focus groups per year, too. Everything is researched, and nothing is left to chance." In most cases, the PD determines how much music is to be programmed hourly and in what rotation, and when news, public affairs features, and commercials are to be slotted. Program wheels, also variously known as sound hours, hot clocks, and format disks, are carefully designed by the PD to ensure the effective presentation of on-air ingredients. Program wheels are posted in the control studio to inform and guide air people as to what is to be broadcast and at what point in the hour. Although not every station provides deejays with such specific programming schemata, today very few stations leave things up to chance since the inappropriate scheduling and sequencing of sound elements may drive listeners to a competitor. Radio

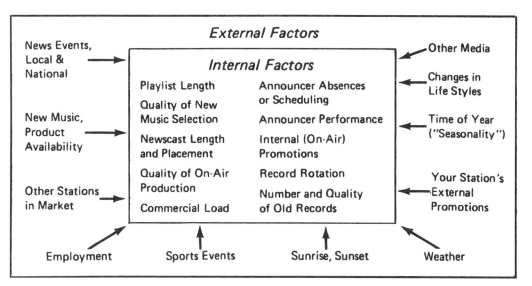

FIGURE 3.12
A model of a
station's
competitive
environment as
conceived by
Arbitron. Courtesy
Arbitron.

programming has be-come that much of an exacting science. With few exceptions, stations use some kind of formula in the conveyance of their programming material.

At one time, Top 40 stations were the unrivaled leaders of formula programming. Today, however, even MOR and AOR outlets, which once were the least formulaic, have become more sensitive to form. The age of free-form commercial radio has long since passed, and it is doubtful, given the state of the marketplace, that it will return. Of course, stranger things have happened in radio. Depending on the extent to which a PD prescribes the content of a sound hour, programming clocks may be elaborate in their detail or quite rudimentary. Music clocks are used to plot out elements. Clocks reflect the minutes of the standard hour, and the PD places elements where they actually are to occur during the hour. Many programmers use a set of clocks, or clocks that change with each hour.

Program clocks are set up with the competition and market factors in mind. For example, programmers will devise a clock that reflects morning and afternoon drive periods in their market. Not all markets have identical commuter hours. In some cities morning drive may start as early as 5:30 a.m., while in others it may begin at 7:00 a.m. The programmer sets up clocks accordingly. A clock parallels the activities of the community in which the station operates.

Music stations are not the only ones that use program wheels. News and Talk stations do so as well. News stations, like music outlets, use key format elements in order to maintain ratings through the hour. Many news stations work their clocks in twenty-minute cycles. During this segment news is arranged according to its degree of importance and geographic relevance, such as local, regional, national, and international. Most news stations lead with their top local stories. News stories of particular interest are repeated during the segment. Sports, weather, and other news-related information, such as traffic and stock market reports, constitute a part of the segment. Elements may be juggled around or different ones inserted during successive twenty-minute blocks to keep things from sounding repetitious.

In the Talk format, two-way conversation and interviews fill the space generally allotted to songs in the music format. Therefore, Talk wheels often resemble music wheels in their structure. For example, news is offered at the top of the hour followed by a talk sweep that precedes a spot set. This is done in a fashion that is reminiscent of Easy Listening.

Of course, not all stations arrange their sound hours as depicted in these pages. Many variations exist, most inspired by computers, but these examples are fairly representative of some of

FIGURE 3.13
Music testing companies provide stations with outside expertise. Courtesy The Benchmark Company.

THE FREQUENCY BASED MUSIC TEST™

Exclusively From The Benchmark Company

WHO DO YOU LIKE?

That's long been the dominant issue in auditorium music testing. Most tests are based on perceived *liking* of a "hook" of a record.

But that causes a problem! For most listeners, there's quite a gap between how much they *"like"* something and how frequently they want to hear it on the radio. Case in point: you play a "hook" of the *"Wind Cries Mary"* for a listener. He gives it the highest score for *"liking."*

But did he give it that score because he hasn't heard it in ten years, or because it's truly one of his favorites? Moreover, that kind of rating doesn't tell you how frequently he wants to hear it on the radio.

FREQUENCY OF AIR PLAY THE TRUE MEASURE

The Benchmark Company has developed a new frequency based music test system based not solely on *"liking"* of a record, but on the more reliable measure of "how often would you like to hear this record." Listeners can respond much more accurately to this type of question—hence, program directors have more reliable data with which to construct play lists and rotations.

The FBMT provides realistic answers for burnout too. While often difficult to measure in the common like-dislike test, burnout evaluation is built right into the FBMT.

CLUSTER ANALYSIS

There continues to be a lot of confusion with some broadcasters about multivariate techniques for analyzing music data. At Benchmark we make it simple to understand and use—after all, we pioneered the technique *ten* years ago!

Simply put, *cluster analysis* is a tool for data reduction. It allows you to identify groups of listeners who "cluster" together in their musical preferences. *Factor analysis* allows you to see how individual songs and artists can be grouped to improve rotations. The FBMT provides the perfect approach to use for cluster and factor analysis. Your FBMT results are turned around quickly and the reports (including cluster and factor analysis) are broadcaster friendly.

The Benchmark Company

*"The FBMT is the **best** new approach I've seen in auditorium research."*
Robb Stewart, PD
WMXC, Charlotte
Format: AC

"The FBMT represents a significant step forward in auditorium music research."
Chris Brodie, PD
KTWV, Los Angeles
Format: NAC

"The FBMT is light years ahead of the field. It really delivered for us at KVIL."
Tom Watson
Adult Contemporary Concepts
(Former PD at KVIL, Dallas)

For More Information on the Frequency Based Music Test™, Contact:
Paige Blount, Director of Research at 1-800-274-5164
1705 S. Capital of Texas Hwy. • Suite 305 • Austin, Texas 78746

ples are fairly representative of some of the program schematics used in today's radio marketplace.

Program wheels keep a station on a preordained path and prevent wandering. As stated, each programming element—commercial, news, promo, weather, and so on—is strategically located in the sound hour to enhance flow and optimize impact. Balance is imperative. Too much deejay patter on a station promoting more music and less talk and listeners become disenchanted. Too little news and information on a station targeting the over-thirty male commuter and the competition benefits. "When constructing or arranging the program clock, you have to work forward

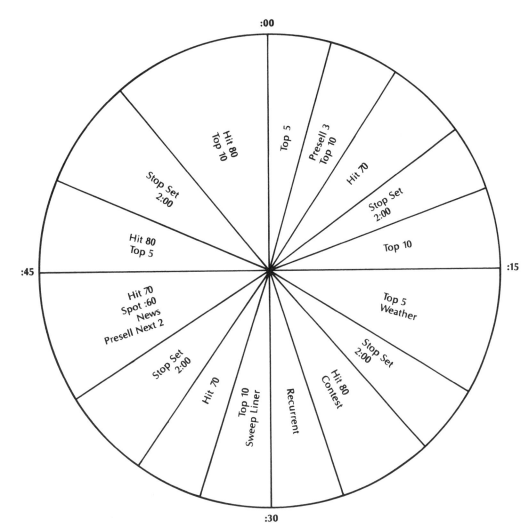

:00

Top 5

Presell 3
Top 10

Hit 70

Stop Set
2:00

Top 10

:15

Top 5
Weather

Stop Set
2:00

Hit 80
Contest

Recurrent

Top 10
Sweep Liner

Hit 70

Stop Set
2:00

Hit 70
Spot :60
News
Presell Next 2

:45

Hit 80
Top 5

Stop Set
2:00

Hit 80
Top 10

:30

FIGURE 3.14
This is a typical morning drivetime CHR clock. It reflects a nine-minute commercial load maximum per hour. Notice the way the stop sets occur away from the quarter-hour. A Top 5 record is often aired on the quarter-hour to give the station the best ratings advantage. This is called "quarter-hour maintenance."

and backward to make sure that everything fits and is positioned correctly. One element out of place can become that proverbial hole in the dam. Spots, jock breaks, music—it all must be weighed before clocking. A lot of experimentation, not to mention research, goes into this," observes radio executive Lorna Ozman.

It was previously pointed out that a station with a more-music slant will limit announcer discourse in order to schedule additional tunes. Some formats, in particular Easy Listening, have reduced the role of the announcer to not much more than occasional live promos and IDs, which are written on liner or flip cards. Nothing is left to chance. This also is true of stations airing the supertight hit music format. Deejays say what is

written and move the music. At stations where deejays are given more control, wheels play a less crucial function. Outlets where a particular personality has ruled the ratings for years often let that person have more input as to what music is aired. However, even in these cases, playlists generally are provided and followed.

The radio personality has enjoyed varying degrees of popularity since the 1950s. Over the years, Top 40, more than any other format, has toyed with the extent of deejay involvement on the air. The pendulum has swung from heavy personality presence in the 1950s and early 1960s to a drastically reduced role in the mid- and later 1960s. This dramatic shift came as the result of programmer Bill Drake's

FIGURE 3.15
A public radio station's program schedule employing the block approach. Courtesy KNTU.

University of North Texas
KNTU
88.1 FM

	SUNDAY	MONDAY	TUESDAY	WEDNESDAY	THURSDAY	FRIDAY	SATURDAY	
6 am	KNTU CLASSICAL							6 am
7 am								7 am
8 am	CLEVELAND ORCHESTRA						La Onda Tejana	8 am
9 am		UNT TALK	In Black America	DIALOGUE	Denton Weekly	Vocal Point		9 am
10 am	DETROIT SYMPHONY ORCHESTRA							10 am
11 am	WEEKEND RADIO							11 am
12 n		Soundings	Vocal Point	UNT TALK	DIALOGUE	Denton Weekly	Jazz Revisited	12 n
1 pm	SAN FRANCISCO ORCHESTRA	SIDE ORDERS FOR LUNCH				Fives in Jazz*		1 pm
2 pm	KNTU CLASSICAL							2 pm
3 pm		Jazz						3 pm
4 pm	AND SELDOM IS HEARD							4 pm
5 pm								5 pm
6 pm		LATE EDITION						6 pm
	20th CENTURY ROMANTICS	In Black America	Denton Weekly	Soundings	Vocal Point	UNT TALK	Jazz Revisited	
7 pm								7 pm
8 pm	KNTU CLASSICAL							8 pm
	TO THE BEST OF OUR KNOWLEDGE							
9 pm	TYPE "A" RADIO							9 pm
10 pm	KNTU BLUES REVIEW	10 PM NEWS UPDATE					Global Mix	10 pm
11 pm		LATE NIGHT SNACK		Piano Jazz	Riverwalk Live from the Landing			11 pm
12 m	SEEDS							12 m

*Jazz South airs the first Friday of every month.

attempt to streamline Top 40. In the 1970s, the air personality regained some of his status, but in the 1980s the narrowing of hit station playlists brought about a new leanness and austerity which again diminished the jock's presence. In the mid-1980s, some pop music stations began to give the deejay more to do. "There's sort of a pattern to it all. For a while, deejays are the gems in the crown, and then they're just the metal holding the precious stones in place for another period of time. What went on in the mid-1970s with personality began to recur in the latter part of the 1980s. Of course, there are a few new twists in the tiara, but what it comes down to is the temporary restoration of the hit radio personality. It's a back and forth movement, kind of like a tide. It comes in, then retreats, but each time something new washes up. Deejays screamed at the teens in the 1950s, mellowed out some in the 1960s and 1970s, went hyper again in the 1980s, but conservatism seems to be gaining favor once again," observes WQIK's Brian Mitchell.

In addition to concentrating on the role deejays play in the sound hour, the PD pays careful attention to the general nature and quality of other ingredients. Music is, of course, of paramount importance. Songs must fit the format to begin with, but beyond the obvious, the quality of the artistry and the audio mix must meet certain criteria. A substandard musical arrangement or a disk with poor fidelity detracts from the station's sound. Jingles and promos must effectively establish the tone and tenor of the format, or they have the reverse effect of their intended purpose, which is to attract and hold listeners. Commercials, too, must be compatible with the program elements that surround them.

In all, the program director scrutinizes every component of the program wheel to keep the station true to form. The wheel helps maintain consistency, without which a station cannot hope to cultivate a following. Erratic programming in today's highly competitive marketplace is tantamount to directing listeners to other stations.

THE PROGRAM DIRECTOR AND THE AUDIENCE

The programmer must possess a clear perception of the type of listener the station management wants to attract. Initially, a station decides on a given format because it is convinced that it will make money with the new-found audience, meaning that the people who tune in to the station will look good to prospective advertisers. The purpose of any format is to win a desirable segment of the radio audience. Just who these people are and what makes them tick are questions that the PD must constantly address in order to achieve reach and retention. An informed programmer is aware that different types of music appeal to different types of people. For example, surveys have long concluded that heavy rock appeals more to men than it does to women, and that rock music, in general, is more popular among teens and young adults than it is with individuals over forty. This is no guarded secret, and certainly the programmer who is out to gain the over-forty crowd is doing himself and his station a disservice by programming even an occasional hard rock tune. This should be obvious.

A station's demographics refer to the characteristics of those who tune in— sex, age, income, and so forth. Within its demographics a station may exhibit particular strength in specific areas, or *cells* as they have come to be termed. For example, an Adult Contemporary station targeting the twenty-four- to thirty-nine-year-old group may have a prominent cell in females over thirty. The general information provided by the major ratings surveys indicate to the station the age and sex of those listening, but little beyond that. To find out more, the PD may conduct an in-house survey or employ the services of a research firm.

Since radio accompanies listeners practically everywhere, broadcasters pay particular attention to the lifestyle activities of their target audience. A sta-

FIGURE 3.16
Program clock for a
network syndicated
format. Courtesy
ABC.

STARDUST

FLEX CLOCK 1
5AM to 10PM (CT)

FUNCTIONS:
1. Local Service Liners
2. 02:00 Net News/Local Option
3. 3:00 Local Break
4. 3:00 Minute Local Option
5. :06 Legal ID
6. Magic Calls (:02, Dedicated Cart)**

Sample Hour: 5:00am - 10:00pm (Central Time)
Commercial Time
Minimum: 3:00 Local/2:00 NET/5:00 Total
Maximum: 16:00 Local/2:00 NET/18:00 Total

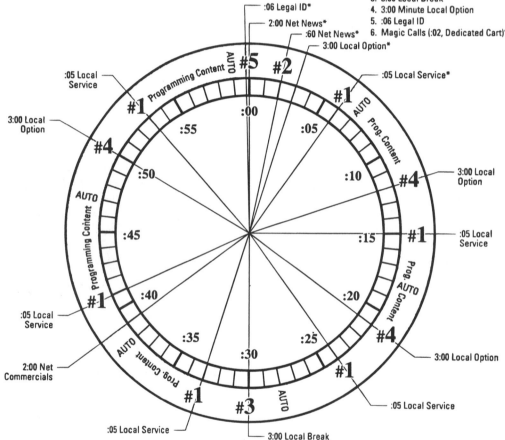

ALL TIMES ARE APPROXIMATE
* =EXACT TIMES
** CALL LETTERS, FREQUENCY OR
SLOGAN THAT RUNS VARIABLY WITHIN
THE HOUR PRECEDING LIVE ANNOUNCERS
AUTO= AUTOMATION UPDATE TIMES

© 1995 ABC Radio Networks

RECOMMENDED SPOT
PLACEMENT:
1st: 32:00 (3:00) (Must be filled)
2nd: 50:00 (3:00)
3rd: 21:00 (3:00)
4th: 12:00 (3:00)
5th: 03:00 (3:00)

ABC RADIO NETWORKS

tion's geographic locale often dictates its program offerings. For example, hoping to capture the attention of the thirty-five-year-old male, a radio outlet located in a small coastal city along the Gulf of Mexico might decide to air a series of one-minute informational tips on outdoor activities, such as tennis, golf, and deep-sea fishing, that are exceptionally popular in the area. Stations have always catered to the interests of their listeners, but in the 1970s audience research became much more lifestyle oriented.

In the 1990s, broadcasters delve further into audience behavior through psychographic research, which, by examining motivational factors, provides programmers with information beyond the purely quantitative. Perhaps one of the best examples of a station's efforts to conform to the lifestyle of its listeners is "dayparting," a topic briefly touched upon in the discussion of program wheels. For the sake of illustration, let us discuss how an AC station may daypart (segmentalize) its broadcast day. To begin with, the station is targeting an over-forty audience, somewhat skewed toward males. The PD concludes that the station's biggest listening hours are mornings between seven and nine and afternoons between four and six, and that most of those tuned in during these periods are in their cars commuting to or from work. It is evident to the programmer that the station's programming approach must be modified during drivetime to reflect the needs of the audience. Obviously traffic reports, news and sports updates, weather forecasts, and frequent time checks are suitable fare for the station's morning audience. The interests of homebound commuters contrast slightly with those of workbound commuters. Weather and time are less important, and most sports information from the previous night is old hat by the time the listener heads for home. Stock-market reports and information about upcoming games and activities pick up the slack. Midday hours call for further modification, since the lifestyle of the station's audience is different. Aware that the

majority of those listening are homemakers (in a less enlightened age this daypart was referred to as "housewife" time), the PD reduces the amount of news and information, replacing them with music and deejay conversation designed specifically to complement the activities of those tuned. In the evening, the station redirects its programming and schedules sports and talk features, going exclusively talk after midnight. All of these adjustments are made to attract and retain audience interest.

The program director relies on survey information and research data to better gauge and understand the station's audience. However, as a member of the community that the station

FIGURE 3.17
Many stations have replaced local deejays with national personalities. Tom Joyner is among the most popular. Courtesy ABC.

```
Page No.     1              C1TISTAT - Chicago
                              WKQX-FM/AC
91.01.18
                 TUNED IN-ALPHABETICAL LIST BY ARTIST

Artist                  Song Title                        Time
-------------------     -------------------------------   ----


**  1
10CC                    The Things We Do For Love         12m

**  A
ADAMS, OLETA            Get Here                          02a
ADAMS, OLETA            Get Here                          06a
ADAMS, OLETA            Get Here                          10a
ADAMS, OLETA            Get Here                          02p
ADAMS, OLETA            Get Here                          07p
ADAMS, OLETA            Get Here                          11p
ALIAS                   More Than Words Can Say           01a
ALIAS                   More Than Words Can Say           12n
ALIAS                   More Than Words Can Say           10p
ASTLEY, RICK           It Would Take A Strong Strong     01p
AUSTIN/INGRAM          Baby, Come To Me                  02a

**  B
BAD ENGLISH             When I See You Smile              01a
BANGLES                 Eternal Flame                     01a
BANGLES                 Manic Monday                      04p
BEACH BOYS              Kokomo                            07p
BERLIN                  Take My Breath Away               01a
BERLIN                  Take My Breath Away               08p
BOLTON, MICHAEL        (Sittin' On) The Dock Of The..    08p
BRANIGAN, LAURA        Self Control                      01a
BROWNE, JACKSON        Doctor My Eyes                    01p
BROWNE, JACKSON        Running On Empty                  09p
BROWNE, JACKSON        Somebody's Baby                   09a

**  C
CAFFERTY, JOHN & BBB   On The Dark Side                  06a
CARA, IRENE            Flashdance (What A Feeling)       01p
CAREY, MARIAH          Love Takes Time                   01a
CAREY, MARIAH          Love Takes Time                   05a
CAREY, MARIAH          Love Takes Time                   11a
CAREY, MARIAH          Love Takes Time                   04p
CAREY, MARIAH          Love Takes Time                   10p
CAREY, MARIAH          Vision Of Love                    10p
CARLISLE, BELINDA      Heaven Is A Place On Earth        02a
CARLISLE, BELINDA      Heaven Is A Place On Earth        11p
CARLISLE, BELINDA      I Get Weak                        06p
CARMEN, ERIC           Hungry Eyes                       09p
CHER                   If I Could Turn Back Time         12m
CHER                   The SI
CHER                   The SI
CHER                   The SI    | Music is broken our alphabetically by artists with the
CHICAGO                Feelir    | exact hour of air noted
CHICAGO                Hard
CHICAGO                Look Away                         11p
CHICAGO                Old Days                          08p
```

FIGURE 3.18
Keeping track of a song's performance is a vital element in retaining the edge in music programming. Courtesy Mediabase.

serves, the programmer knows that not everything is contained in formal documentation. He or she gains unique insight into the mood and mentality of the area within the station's signal simply by taking part in the activities of day-to-day life. A programmer with a real feel for the area in which the station is located, as well as a fundamental grasp of research methodology and its application, is in the best possible position to direct the on-air efforts of a radio station. Concerning the role of audience research, Peter Falconi says, "You can't run a station on research alone. Yes, research helps to an extent, but it can't replace your own observations and instincts." Brian Mitchell

agrees with Falconi. "I feel research is important, but how you react to research is more important. A PD also has to heed his gut feelings. Gaps exist in research, too. If I can't figure out what to do without data to point the way every time I make a move, I should get out of radio. Success comes from taking chances once in a while, too. Sometimes it's wiser to turn your back on the tried and tested. Of course, you had better know who's out there before you try anything. A PD who doesn't study his audience and community is like a race car driver who doesn't familiarize himself with the track. Both can end up off the road and out of the race."

THE PROGRAM DIRECTOR AND THE MUSIC

Not all radio stations have a music director. The larger the station, the more likely it is to have such a person. In any case, it is the PD who is ultimately responsible for the music that goes over the air, even when the position of music director exists. The duties of the music director vary from station to station. Although the title suggests that the individual performing this function would supervise the station's music programming from the selection and acquisition of records to the preparation of playlists, this is not always the case. At some stations, the position is primarily administrative or clerical in nature, leaving the PD to make the major decisions concerning airplay. In this instance, one of the primary duties of the music director might be to improve service from record distributors to keep the station well supplied with the latest releases. A radio station with poor record service may actually be forced to purchase music. This can be prevented to a great extent by maintaining close ties with the various record distributor reps.

Over the years the music industry and the radio medium have formed a

FIGURE 3.19
Radio reflects the lifestyle of its users. Courtesy of Westinghouse Electric.

mutually beneficial alliance. Without the product provided by the recording companies, radio would find itself with little in the way of programming material, since ninety percent of the country's stations feature recorded music. At the same time, radio serves as the principal means by which the recording industry gets word of its new releases to the general public. Succinctly put, radio sells records.

While radio seldom pays for music (cassettes, CDs)—recording companies send demos of their new product to most stations—it must pay annual licensing fees to ASCAP (American Society of Composers, Authors, and Publishers), BMI (Broadcast Music, Inc.), or SESAC (Society of European Stage Authors and Composers) for the privilege of airing recorded music. These fees range from a few hundred dollars at small, noncommercial, educational stations, to tens of thousands of dollars at large, commercial, metro-market stations. The music licensing fees paid by stations are distributed to the artists and composers of the songs broadcast.

When music arrives at the station, the music director (sometimes more appropriately called the music librarian or music assistant) processes them through the system. This may take place after the program director has screened them. Records are categorized, indexed, and eventually added to the library if they suit the station's format. Each station approaches cataloguing in its own fashion. Here is a simple example: an Adult Contemporary outlet receives an album by a popular female

FIGURE 3.20
A host of
computerized
music
programming
services are used
by stations.
Courtesy RCS.

SAMPLER™
The Music Research System.

- POWERFUL
- FLEXIBLE
- RELIABLE
- CONVENIENT
- ECONOMICAL
- SECURE

SAMPLER is powerful...
The competition is tough. If you'd rather not depend entirely on guesswork for your station's success, you need SAMPLER, the computer-aided music research system.

SAMPLER is flexible...
SAMPLER is able to manage your weekly call-outs, auditorium testing, and retail store sales research, even audience listening habits and perceptions. You use the results of its analyses to base your music decisions on information instead of chance.

SAMPLER is reliable...
SAMPLER'S design has been driven by some of the most ambitious and successful PDs in America, for nearly a decade. Its analyses were programmed by a Ph.D. statistician.

SAMPLER is easy to use...
SAMPLER comes to you from the same people who designed SELECTOR, the most widely-used music scheduling system in the world. It looks like SELECTOR, it works like SELECTOR, and you get the famous RCS support.

SAMPLER is confidential and secure...
Because SAMPLER is completely in-house, you don't have to pay outsiders to analyze your sensitive research results; no matter how sophisticated the analysis you need, SAMPLER can handle it. And there's no worry about your data getting into the wrong hands. It's all in your control.

MUSICBASE™
The Music Information System.

MUSICbase is the ultimate song reference for sharp programmers. You'll have the edge because you'll have the information on every song you're playing.

EVERY SONG ?!?!?!?
Probably so, because MUSICbase contains information on about 20,000 songs. That's every song that has charted in the rock era plus hundreds more "Bonus Cuts." These are the album cuts and flipsides most often played by radio, by artists like the Beatles, Bruce Springsteen and Genesis.

WE'VE DONE THE WORK...
On Radio's most requested songs, we've coded the songs to make your job even easier.

COMPLETE CHART HISTORY...
You'll know not just the top 10 or 20, but the entire HOT 100 for any week at the touch of a key.

IT EVEN TALKS !!
To SELECTOR, that is. MUSICbase was compiled by RCS, makers of SELECTOR, the radio industry's standard for music scheduling: SELECTOR and MUSICbase are fully interfaced so you can copy songs right from MUSICbase into your library.

OVER 1000 THEMES...
You can browse through all the love songs, rain songs, car songs, or over a thousand more themes... instantly.

WRITERS/PUBLISHERS/LICENSEE...
No more heavy research for BMI logging periods since MUSICbase contains this information.

IT WORKS FOR EVERY FORMAT...
Because the songs included fit every era and style from 1958 to today.

SIMPLE TO LEARN AND TO USE...
With complete on-line help, MUSICbase has been designed for ease of use. You don't have to be a computer genius to use it, but you'll become a music genius in no time at all.

vocalist whose last name begins with an L. The program director auditions the album and decides to place three cuts into regular on-air rotation. The music director then assigns the cuts the following catalogue numbers: L106/U/F, L106/D/F, and L106/M/F. L106 indicates where the album may be located in the library. In this case, the library is set up alphabetically, then numerically within the given letter that represents the artist's last name. In other words, this would be the 106th album found in the section reserved for female vocalists whose names begin with an L. The next symbol indicates the tempo of the cut: U(p) tempo, D(own) tempo, and M(edium) tempo. The F that follows the tempo symbol indicates the artist's gender: Female.

When a station is computerized (and in this day and age, few are not), this information, including the frequency or rotation of airplay as determined by the PD, will be entered accordingly. Playlists are then assembled and printed by the computer. The music director sees that these lists are placed in the control room for use by the deejays. This last step is eliminated when the on-air studio is equipped with a computer terminal. Deejays then simply punch up the playlists designed for their particular airshifts.

The use of computers in music programming has become widespread, especially in larger markets where the cost of computerization is absorbed more easily. The number of computer companies selling both hardware and software designed for use by programmers has soared. Among those providing computerized music systems are Halper and Associates, Jefferson Pilot Data Systems, Radio Computing Systems, Columbine Systems. Also, some journals, such as *Billboard* and *Radio and Records*, provide up-to-date computerized music research on new singles and albums to assist station programmers.

At those radio stations where the music director's job is less administrative and more directorial, this individual will actually audition and select what songs are to be designated for airplay. However, the music director makes decisions based upon criteria established by the station's programmer. Obviously, a music director must work within the station's prescribed format. If the PD feels that a particular song does not fit the station's sound, he will direct the music director to remove the cut from rotation. Since the PD and music director work closely together, this seldom occurs.

A song's rotation usually is relative to its position on national and local record sales charts. For instance, songs that enjoy top ranking, say those in the top ten, will get the most airplay on hit-oriented stations. When songs descend the charts, their rotation decreases proportionately. Former chart toppers are then assigned another rotation configuration that initially may result in one-tenth of their former airplay and eventually even less. PDs and music directors derive information pertaining to a record's popularity from various trade journals, such as *Billboard, Radio and Records*, and *Monday Morning Replay*, as well as listener surveys, area record store sales, and numerous other sources. Stations that do not program from the current charts compose their

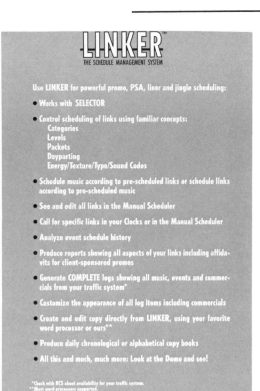

FIGURE 3.20
continued

LINKER
THE SCHEDULE MANAGEMENT SYSTEM

Use LINKER for powerful promo, PSA, liner and jingle scheduling:

• **Works with SELECTOR**

• **Control scheduling of links using familiar concepts:**
  Categories
  Levels
  Packets
  Dayparting
  Energy/Texture/Type/Sound Codes

• **Schedule music according to pre-scheduled links or schedule links according to pre-scheduled music**

• **See and edit all links in the Manual Scheduler**

• **Call for specific links in your Clocks or in the Manual Scheduler**

• **Analyze event schedule history**

• **Produce reports showing all aspects of your links including affidavits for client-sponsored promos**

• **Generate COMPLETE logs showing all music, events and commercials from your traffic system***

• **Customize the appearance of all log items including commercials**

• **Create and edit copy directly from LINKER, using your favorite word processor or ours****

• **Produce daily chronological or alphabetical copy books**

• **All this and much, much more: Look at the Demo and see!**

*Check with RCS about availability for your traffic system.
**Most word processors supported.

playlists of songs that were popular in years gone by. In addition, these stations often remix current hit songs to make them adaptable to their more conservative sound. While Easy Listening stations do not air popular rock songs, they do air softened versions by other artists, usually large orchestras. Critics of this technique accuse the producers of lobotomizing songs to bring them into the fold. In nonhit formats there are no "power-rotation" categories or hit positioning schemata; a song's rotation tends to be more random, although program wheels are used.

Constructing a station playlist is the single most important duty of the music programmer. What to play, when to play it, and how often are some of the key questions confronting this individual. The music director relies on a number of sources, both internal and external, to provide the answers, but also must cultivate an ear for the kind of sound the station is after. Some people are blessed with an almost innate capacity to detect a hit, while most must develop this skill over a period of time.

```
CLASSIC ROCK 95-KXXX
Tuesday, January 21, 1992
THE MOST MUSIC IN THE MORNING
7AM-8AM

CARTED LEGAL I.D.
5B    NO MATTER WHAT  BADFINGER          2:51
A SWEEPER
J16  LIGHTS              JOURNEY         3:09
06:00 "STOP SET"  WEATHER, SPOTS
R-4a RUBY TUESDAY      ROLLING STONES    3:15
Y-11 OWNER OF A LON  YES                 4:27
17:42 STOP SET LINER-PRETEASE
QUICK :10 NEWS CROSS-OVER
20:42 *KLXK NEWSBRIEF*
C-18 WOODSTOCK          C.S.N.           3:44
X-80 SO IN TO YOU    ATLANTA RHYTHM      4:11
B SWEEPER
M-28 TUESDAY AFTERN  MOODY BLUES  (PL    8:15
D-4  SULTANS OF SWI  DIRE STRAITS        5:44
44:36 "STOP SET"  WEATHER, SPOTS
LOST CLASSICS SWEEPER&CALLER
B-7  MAXWELL'S SILV  BEATLES            3:24
C-10 HURTS SO GOOD   JOHN COUGAR         3:30
QUICK :10 NEWS CROSS-OVER
55:30 STOP SET LINER-PRETEASE
56:30 *KLXK NEWSBRIEF*
E-2  LIFE IN THE FA  EAGLES             4:40
18   I HEARD IT THR  C.C.R.             3:40
```

FIGURE 3.21
Computerized music logging has made station programming even more exacting a science. This is just one of the possible logs that can be created. Courtesy RCS.

THE PROGRAM DIRECTOR AND THE FCC

The government is especially interested in the way a station conducts itself on the air. For instance, the program director makes certain that his station is properly identified once an hour, as close to the top of the hour as possible. The ID must include the station's call letters and the town in which it has been authorized to broadcast. Failure to properly identify the station is a violation of FCC rules.

Other on-air rules that the PD must address have to do with program content and certain types of features. For example, profane language, obscenity, sex- and drug-related statements, and even innuendos in announcements, conversations, or music lyrics jeopar-

dize the station's license. Political messages and station editorials are carefully scrutinized by the programmer. On-air contests and promotions must not resemble lotteries in which the audience must invest to win. A station that gets something in return for awarding prizes is subject to punitive actions. Neither the deejays, PD, music director, nor anyone associated with the station may receive payment for plugging a song or album on the air. This constitutes "payola" or "plugola" and was the cause of great industry upheaval in the 1950s. Today, PDs and station managers continue to be particularly careful to guard against any recurrence, although there have been charges that such practices still exist.

The program director must monitor both commercial and noncommercial messages to ensure that no false, misleading, or deceptive statements are aired, something the FCC staunchly opposes. This includes any distortion of the station's ratings survey results. A station that is not number one and claims to be is lying to the public as far as the FCC is concerned, and such behavior is not condoned.

License renewal programming promises must be addressed by the PD. The proportion of nonentertainment programming, such as news and public affairs features, pledged in the station's renewal application must be adhered to, even though such requirements have been all but eliminated. A promise is a promise. If a station claims that it will do something, it must abide by its word.

The PD helps maintain the station's Emergency Broadcast System (EBS), making certain that proper announcements are made on the air and that the EBS checklist containing an authenticator card is placed in the control room area. PDs also instruct personnel in the proper procedures used when conducting on-air telephone conversations to guarantee that the rights of callers are not violated.

The station log (which ultimately is the chief engineer's responsibility) and program log (no longer required

Frank Bell

Advice to Programmers

FIGURE 3.22

Rule #1: *Follow the listeners, not the format.* So many people in radio get caught up in terms like *CHR, New AC, Alternative,* and *AAA* that they lose track of their goal: finding listeners.

Consumers of radio think in terms of "what I like" and "what I don't like." By researching your listeners' tastes and giving them what they want (as opposed to what fits the industry's definition of what they should have), you'll maximize your chances for success.

Rule #2: *Think outside in, not inside out.* The fact that one company may now own several stations in a market and is capable, for example, of skewing one FM toward younger females and the other toward older females does not mean that you will automatically "dominate females."

The only reality that counts is that of the listener. If listeners feel your station serves a meaningful purpose for them, they will happily cast their Arbitron vote in your favor. If they believe you are simply duplicating what is already available elsewhere on the dial, you will be doomed to ratings obscurity.

Rule #3: *Early to bed, early to rise, Advertise, Advertise, Advertise.* In the Arbitron game of unaided recall, the dominant issue is Top-of-Mindness.

The best way to get that is through advertising your name and your station's benefits—on your own air and on any other medium you can afford. Just for fun, here's a diagram I sometimes use to show first-time PD's the various factors which influence their station's ratings:

$$\frac{X - Y}{A} \times B = \text{Your Ratings}$$

X is what your station does.

Y is what your direct competitors do.

A represents "environmental" factors in the market, such as what's on TV during the survey, riots, floods, earthquakes, and (in some cases) the OJ trial.

B is what the rating service does. In the case of Arbitron, this would include the response rate, editing procedures, and distribution of diaries by race, age, and sex.

The most important thing to understand is that as Program Director, the only part of the equation you can control is "X." Do the best you can at keeping your station sounding compelling, entertaining, and focused on its target audience, and don't get an ulcer over those elements you can't control.

---

by the FCC but maintained by most stations anyway) are examined by the PD for accuracy, and he also must see to it that operators have permits and that they are posted in the on-air studio. In addition, the station manager may assign the PD the responsibility for maintaining the station's Public File. If so, the PD must be fully aware of what the file is required to contain. The FCC and many broadcast associations will provide station operators with a Public File checklist upon request. This information is available in the *Code of Federal Regulations* (73.3526) as well.

Additional programming areas of interest to the FCC include procedures governing rebroadcasts, simulcasts, and subcarrier activities. The program director also must be aware that the government is keenly interested in employment practices. The programmer, station manager, and other department heads are under obligation to familiarize themselves with equal employment opportunity (EEO) and affirmative action rules. An annual employment report must be sent to the FCC.

The preceding is only a partial listing of the concerns set forth by the government relative to the program director's position. For a more comprehensive assessment, refer to the CFR Index at the end of Chapter 2.

THE PROGRAM DIRECTOR AND UPPER MANAGEMENT

The pressures of the program director's position should be apparent by now. The station programmer knows well that his or her job entails satisfying the desires of many—the audience, government, air staff, and, of course, management. The relationship between the PD and the station's upper echelon is not always serene or without incident. Although usually a mutually fulfilling and productive alliance, difficulties can and do occur when philosophies or practices clash. "Most inhibiting and detrimental to the PD is the GM who lacks a broad base of experience but imposes his opinions on you anyway.

FIGURE 3.23 All information about a given record can be called up on the screen as shown here. Courtesy TM Century.

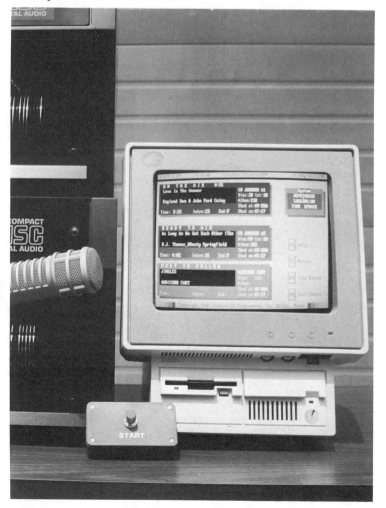

The guy who has come up through sales and has never spent a minute in the studio can be a real thorn in the side. Without a thorough knowledge of programming, management should rely on the expertise of that person hired who does. I don't mean, 'Hey, GM, get out of the way!' What I'm saying is, don't impose programming ideas and policies without at least conferring with that individual who ends up taking the heat if the air product fails to bring in the listeners," says Dick Fatherly.

Station manager Chuck Ducoty contends that managers can enhance as well as inhibit the style of programmers. "I've worked for some managers who give their PDs a great deal of space and others who attempt to control every aspect of programming. From the station manager's perspective, I think the key to a good experience with those who work for you is to find excellent people from the start and then have enough confidence in your judgment to let them do their job with minimal interference. Breathing down the neck of the PD is just going to create tension and resentment."

Programmer Peter Falconi believes that a sincere effort to get to know and understand one another should be exerted by both the PD and manager. "You have to be on the same wavelength, and there has to be an excellent line of communication. When a manager has confidence and trust in his PD, he'll generally let him run with the ball. It's a two-way street. Most problems can be resolved when there is honesty and openness."

Programmer Andy Bloom offers this observation: "Great upper management hires the best players, gives them the tools to do their job, and then leaves them alone. A winning formula."

An adversarial situation between the station's PD and upper management does not have to exist. The station that cultivates an atmosphere of cooperation and mutual respect seldom becomes embroiled in skirmishes that deplete energy—energy better spent raising revenues and ratings.

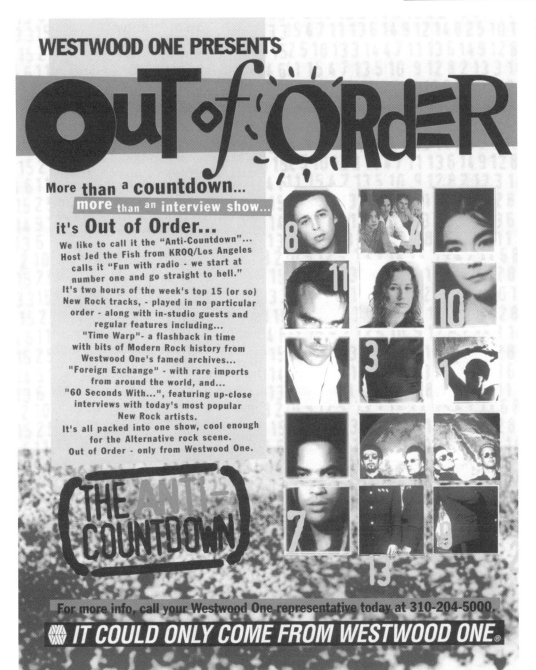

CHAPTER HIGHLIGHTS

1. The Adult Contemporary (A/C) format features current (since the 1970s) pop standards. It appeals to the twenty-five to forty-nine age group, which attracts advertisers. It often utilizes music sweeps and clustered commercials.

2. Contemporary Hit Radio (CHR) features current, fast-selling hits from the Top 40. It targets teens, broadcasts minimal news, and is very promotion/contest oriented. Some CHR's have redirected their playlists to create a Modern Hit or Churban sound.

3. Country is the fastest growing format since the 1970s. More prevalent in the South and Midwest, it attracts a broad age group and offers a variety of subformats.

4. Easy Listening stations evolved from the Beautiful Music stations of the 1960s and 1970s. Featuring mostly instrumentals and minimal talk, many stations have become automated and use prepackaged programming. The primary audience is over age fifty. Its following has dwindled in recent years due to myriad softer AC formats.

5. Album-Oriented Rock (AOR) stations began in the mid-1960s to counter Top 40 stations. Featuring music sweeps with a large airplay library, they play rock album cuts. News is minimal. The format attracts a predominantly male audience aged eighteen to thirty-four. Modern, Alternative, and Classic rock stations have cut into traditional AOR numbers.

6. All-News stations rotate time blocks of local, regional, and national news and features to avoid repetition. The format requires three to four times the staff and budget of most music operations.

7. All-Talk combines discussion and call-in shows. It is primarily a medium and major market format. Like All-News, All-Talk is mostly found on AM but is now finding a home on the FM band.

8. The Nostalgia playlist emphasizes popular tunes from the 1940s and pre-rock 1950s, presenting its music in sweeps with a relatively low deejay profile.

9. The Oldies playlist includes hits from the 1950s and 1960s, relying on fine air personalities. Commercials are randomly placed and songs are spaced to allow deejay patter.

FIGURE 3.25 **Economical, easy-to-use, in-house systems eliminate old-fashioned, time-consuming index card file methods. Courtesy Jefferson Pilot Data Systems.**

```
                      L I B R A R Y   I N F O R M A T I O N
--------------------------------------------------------------------------------
  CART     CATEGORY  LEVEL  PACKET               TITLE
  664         A        1      0       SHE BOP                        LICENSE
                                                                 655    A
--------------------------------------------------------------------------------
    ARTIST ONE                              ROLE           ARTIST TWO
 CYNDI LAUPER                       113      FV
--------------------------------------------------------------------------------
MOOD ........  3 |   APPEAL      |  C H A R T  |          D A Y P A R T
TEMPO  ...... MM |               |             |     R E S T R I C T I O N S
TIMBRE ......    |  AGE   M F    |             |     1         111        11
SOUND CODES FR I |  >18          | CURRENT     |     212345678901212345678901
OPENER ....... Y | 18-24         | LAST ...    |     MAAAAAAAAAAANPPPPPPPPPPP
TYPE .........   | 24-34         | PEAK ...    | M-F           NNNNN
KEY .....    -   | 34-44         |             | SAT           NNNNN
RUNTIME ... 3:45 | 45-54         |  PEAKED IN  | SUN           NNNNN
INTROTIME   8/14 |  55>          |    6/84     ------------------------------
ENDING ....... C |               |             |    DAYPART ROTATION 5252
--------------------------------------------------------------------------------
             DATES            PLAYS IN  |     M O V E   I N F O R M A T I O N
ENTERED    7/20/84            CATEGORY  |     DATE          PLAYS  CAT  LEVEL  PK
LAST PLAY  8/23/84 10 PM        25      |     0/ 0/ 0  OR     0          1     0
--------------------------------------------------------------------------------
```

10. Classic Rock concentrates on tunes primarily featured by AOR stations over the past two decades. Meanwhile, Classic Hit stations fill the gap between Oldies and CHR outlets with playlists that draw from 1970s and 1980s Top 40 charts.

11. Urban Contemporary (UC) is the "melting pot" format, attracting a heterogeneous audience. Its upbeat, danceable sound and hip, friendly deejays attract the eighteen to thirty-four age group. Contests and promotions are important.

12. Classical commercial outlets are few, but they have a loyal audience. Primarily an FM format appealing to a higher-income, college-educated (upscale, twenty-five to forty-nine) audience, Classical features a conservative, straightforward air sound.

13. Religious stations are most prevalent on the AM band. Religious broadcasters usually approach programming in one of two ways. One includes music as a primary part of its presentation, while the other does not.

14. Ethnic stations serve the listening needs of minority groups. Black and Hispanic listeners constitute the largest ethnic audiences.

CAPITAL CITIES/ABC, INC.
STATEMENT
With Respect To Payola

During the term of my employment, neither I nor my spouse, child or other member of my household has accepted, solicited or agreed to accept any money, service, gift, or favor or other thing of value whatsoever from any person which I knew or had any reason to believe was intended to influence any decision by me as to matters to be broadcast. Furthermore, neither I nor my spouse, child or other member of my household has received any social courtesy or gift in a single year having a value exceeding $50.00 from any person, firm or institution involved in any of the following activities:

television or motion picture production, syndication, or distribution;

record manufacturing or distributing;

music publishing;

the creation, production, performance, distribution, manufacture or exploitation of music, films, tapes, recordings, electrical transcriptions, or any live or recorded programming;

the ownership or exploitation of any musical, dramatic, literary or related copyright or performance right;

radio or television broadcasting (including closed circuit, theatre or pay television, cable, SMATV, MMDS and DBS);

book, newspaper or periodical publishing;

advertising and advertising services

concerts and nightclubs;

performers, performing groups, professional sports teams, or any other potential supplier of radio or television program material;

public relations firms, consulting firms, or other firms or individuals that deal in, represent or promote any of the above.

Furthermore, I have not participated in considering, selecting or preparing for broadcast any program material which had as its subject, or which could in any way materially affect, any business concern in which I, or my spouse, child or other member of my household, held a business or financial interest (including any position as officer, director or employee), except those reported to my department head and also listed below.

I have read and understand the provisions of Sections 317 and 507 of the Communications Act of 1934, as amended, copies of which are attached to this statement, and the memorandum on Payola and Conflicts of Interest circulated by Capital Cities/ABC, Inc. and agree to abide by them.

Name Michael Napolitano Position Disc-Jockey

Business Address

Signature [signature] Date 10-19-88

FIGURE 3.26
Payola statement signed by a station employee. Courtesy Michael Napoilitano and Capital Cities/ABC, Inc.

FIGURE 3.27
Computers can
program music and
keep music library
records as well.
Many types of
music directories
can be established,
such as category,
artist, title, and cart
(cartridge number).
Courtesy RCS.

```
                              **********************************
        02/15/86              *          R. C. S.           *    02/P
                              *      ROCKIN' AMERICA         *
                              *       WITH SELECTOR          *
                              **********************************
        CART       T I T L E                    A R T I S T S    TIME/END
        ========================================================================

        ........................................................................
        00:00 00:00    TOP OF THE HOUR ID - KEEP ROCKING!                 0:05
        ........................................................................
        CD-0866  I'M GOIN DOWN            BRUCE SPRINGSTEEN              3:28
                 BORN IN THE USA                                      15/ 0/F
        CT-1056  TALK TO ME               STEVIE NICKS                   4:08
                                                                      5/ 0/F

        ........................................................................
        07:41 08:00    BACKSELL/TEASE/SPOTS/WEATHER/JINGLE INTO MUSIC    3:00
        ........................................................................
        CT-  17  EASY LOVER               PHILLIP BAILEY                 3:32
                                 PHIL COLLINS                        10/ 0/F
```

15. Middle-of-the-Road (MOR) stations rely on the strength of air personalities and features. Mostly an AM format, MOR attempts to be all things to all people, attracting an over-forty audience.

16. Niche formats, like All-Children and Tourist Radio, are popping up all over the dial as the listening audience becomes more diffused.

17. Program directors (PDs) are hired to fit whatever format the station management has selected. They are chosen for their experience, primarily, although education level is important.

18. The PD is responsible for everything that is aired. Second in responsibility to the general manager (when no Operations Manager position exists), the PD establishes programming and format policy; hires and supervises on-air, music, and production personnel; handles the programming budget; develops promotions; monitors the stations and its competitors and assesses research; is accountable for news, public affairs, and sports features; and may even pull an airshift.

19. The PD's effectiveness is measured by ratings in large markets and by sales in smaller markets.

20. The PD determines the content of each sound hour, utilizing program clocks to ensure that each element—commercial, news, promo, weather, music, and so on—is strategically located to enhance flow and optimize impact. Even News/Talk stations need program clocks.

21. PDs must adjust programming to the lifestyle activities of the target audience. They must develop a feel for the area in which the station is located, as well as an understanding of survey information and research data.

22. Finally, the PD must ensure that the station adheres to all FCC regulations pertaining to programming practices, anticipating problems before they occur.

APPENDIX: A STATION OWNER AIRCHECKS HIS PROGRAMMING

To: Steve
Fr: Jay
re: WLKZ Programming

In listening over the past few days, but only to an hour or two total, it appears that much of the "big picture" programming philosophy has been lost. Perhaps Joe hasn't reviewed the old operations manual. But some mechanical changes could dramatically help the sound of the station until the philosophy is recreated or changed.

FIGURE 3.28
Restricted
Radiotelephone
Operator Permit
required by the FCC
for those operating
a broadcast facility.
Permit is good for
an indefinite
period.

FIGURE 3.29
Consultancy firms
offer programming
tips to stations.
Courtesy Shane
Media.

<u>LATEST FROM THE FIELD</u>
By Ed Shane

Looking at research from stations all over the U.S., there
are some patterns emerging.

At Country stations, for example, current songs (hits at the
top of the chart) and recurrents (songs from the past year or
so) are still what people want to hear most. That's followed
by new songs by familiar artists. Next are songs from the
late 80s, tied with new songs by new or up-and-coming
artists.

Even with the popularity of recurrents, Pam has been noticing
less appeal for songs from 1994 in music testing. Songs from
1992 and 1993 score somewhat higher, and in that order.

In addition, she has seen some older songs defy the lower
scores on 80s era testing. Listeners may not want 80s as a
genre, but they're still interested in some of the melodic,
lyric-driven hits from that time.

As Cheryl studies 18-34 year olds for the RADIO X (tm)
alternative rock format, the name Patsy Cline keeps
appearing. No, not a new band, but the Country singer.

Chuck and I have both encountered the attitude among talk
listeners that talk shows bring out information not
available in the newspaper or on TV. There's a growing
cynicism among talk listeners about traditional news sources.

Most talk show listeners never call a talk station. They'd
rather participate mentally. The percentage of people who
call is on the increase, however.

I've noticed in focus groups a tendency for adults to rely
more on TV than on radio for their news coverage. In the
past, I could rely on hearing about radio as a source for
breaking news coverage. Not any more. It's not gone, but
radio's ownership of the image is threatened

Good news: Lots of loyalty out there for radio. Most people
say it would be difficult to get them away from their
favorite stations. Favorites are changing, though, as people
expand their listening. "Scan" is a listening choice, not
just a button on the radio.

1. One part-timer's mike was way too hot compared to the music. Music should always be dominant; personalities should never overwhelm it. The voice should be "in" the music.

2. Song titles were often announced. Oldies listeners know the song titles, and a lot more than our own people do about every record. There is also the danger that our mostly young talent will say something that reveals their age/lack of knowledge when talking about music. This is unnecessary talk and should be eliminated.

3. Conversely, there is almost no local content. This does not mean PSAs about bean suppers; this means the progress on the new building in downtown Wolfeboro, or the time the town Christmas lights will be turned on, for example. It takes more work than announcing a song title, but it's a much better reason to listen.

4. After the weather was given, the personality commented, "at least that's what it says here." Weather is one of the principal reasons people choose a radio station. The forecast (or the Radar Weather franchise) is only as credible as we make it (the information all comes from the same place).

5. Talk between records. The music should never stop unless it has to (commercials). Stopping it down to talk, even for a liner which can be delivered over an intro, kills the momentum.

6. There are literally no prepromotes going into stop sets. The listener is not given any reasons to stick around (and forward momentum again stops).

7. There are recorded PSAs (a corporate no-no). The one I heard tells us to wear seat belts. Every media, every politician, every do-gooder is trying to tell other people what to do. Let's

FIGURE 3.30
Employment
application with
EEO policy
statement.
Courtesy of WHDH-
AM.

RADIO 85

FOUR FORTY-ONE STUART STREET
BOSTON, MASSACHUSETTS 02116
TELEPHONE (617)267-3302

APPLICATION FOR EMPLOYMENT

"Employment discrimination because of race, color, religion, national origin, sex or age is prohibited. Anyone who believes that he or she has been a victim of discrimination in seeking employment at this station is urged to report the matter promptly to the President of WHDH and may notify the Federal Communications Commission or other appropriate agency."

This application will be given every consideration. Its acceptance, however, does not imply that the applicant will be employed. If an applicant is not employed, it will be retained in our active file for one year.

All appointments to positions are on trial and subject to satisfactory work. If it is found that an employee is not adapted to radio work or is likely to prove useful, he or she may be dismissed.

WHDH CORPORATION
EQUAL EMPLOYMENT OPPORTUNITY POLICY

Equal employment opportunity is the policy of WHDH Corporation (WHDH). Prejudice or discrimination, based upon race, color, religion, national origin, or sex, is strictly prohibited by WHDH in recruitment, selection and hiring, placement and promotion, pay, working conditions, demotion, termination, and in all employment practices generally. All personnel are required diligently and affirmatively to observe this policy and are absolutely forbidden to violate, frustrate or diminish that policy. Any applicant for employment, any employee or any other person having reason to believe that there exists any prejudice or discrimination based upon race, color, religion, national origin, or sex, in any area of WHDH's employment practices, is urged to report the matter promptly to the President of WHDH or any other officer, executive or supervisor employed by WHDH Corporation. Any person believing himself to have been a victim of prohibited discrimination in any area of WHDW's employment practices also has the right to notify the Federal Communications Commission or other appropriate federal or state agencies, or all of these, of such incident or incidents. It is requested that all employees and applicants for employment make this policy of equal employment opportunity known to others, whenever appropriate.

just be the medium that informs and entertains. Let's be an escape from all the other pointed fingers. . . .

8. Joe signs off his show and also refers to his listener as "everybody." We should only talk to one listener at a time. And we should never end anything—keep it going, part of forward momentum. Talk about the guy coming up next . . .

Could you send Joe a memo or talk with him on a visit soon? I want to keep that station tight and happening; it's important.

Also, what do you think about adding some Beatles songs (over and above what we have) for the next month or so? And trading/giving away some anthology CD's?

Thanks.

FIGURE 3.30
continued

EDUCATIONAL RECORD	IN THE UNITED STATES (Names of Schools attended)	CITY/TOWN	No. of Years Completed	If Grad. give Mo. and Yr.	If not Grad. give date of leaving
High or Preparatory School					
Commercial School					
College					
Graduate School					

If you did not graduate from school or college, state reason for leaving: _____

Were you ever suspended or expelled from any of the institutions above? _____ If so, state the reason fully: _____

What educational courses are you now taking? _____

PERSONAL RECORD

Date of Birth (OPTIONAL) _____ Are you a citizen of the United States? _____

Have you ever been arrested or convicted for an offense other than a minor traffic violation? _____ If yes, explain _____

Give number of persons in immediate family _____ Does this include (if living) both father and mother? _____

If not living at home, with whom do you live? Relatives _____ Friends _____ Alone _____ Board _____

Number of persons financially dependent upon you? Fully _____ Partially _____

What is your present selective service classification? _____

Have you served in the armed forces of the United States? _____ Branch of service _____

Month and year of induction _____ Date of discharge or transfer to reserve _____

Rank at time of discharge _____

HEALTH RECORD

How much time have you lost from work or school during the last three years on account of illness? _____

What was the nature of the illness? _____

Are you related to any director, officer or employee of WHDH? _____ Relatives name _____

Describe your outside interests or hobbies _____

Please describe any special assets you bring to WHDH _____

List below, five personal references (not relatives, former employers or fellow employees) who have known you well during the past five or more years.

Print Name in Full	Address (Street / City)	No of Years Known	Occupation

THIS SECTION TO BE FILLED OUT ONLY BY APPLICANTS FOR CLERICAL POSITIONS:

What shorthand method do you use? _____ Words per minute _____ Do you type? _____

Words per minute _____ List office machines that you have operated _____

I hereby affirm that my answers to the foregoing questions are true and correct, and that I have not knowingly withheld any fact or circumstance that would, if disclosed, affect my application unfavorably.

Please sign HERE _____

On a separate sheet of paper you may give any further information you desire to explain your qualifications.

SUGGESTED FURTHER READING

Adams, Michael H., and Massey, Kimberly K. *Introduction to Radio: Production and Programming*. Madison, WI.: Brown and Benchmark, 1995.

Armstrong, Ben. *The Electronic Church*. Nashville: J. Nelson, 1979.

Busby, Linda, and Parker, Donald. *The Art and Science of Radio*. Boston: Allyn and Bacon, 1984.

Carroll, Raymond L. and Donald M. Davis. *Electronic Media Programming: Strategies and Decision Making*. New York: McGraw-Hill, 1993.

Chapple, Steve, and Garofalo, R. *Rock 'n' Roll Is Here to Pay*. Chicago: Nelson-Hall, 1977.

Cliff, Charles, and Greer, A. *Broadcasting Programming: The Current Perspective*. Washington, D.C.: University Press of America, 1974 to date, revised annually.

Coddington, R.H. *Modern Radio Programming*. Blue Ridge Summit, Pa.: Tab Books, 1970.

DeLong, Thomas, A. *The Mighty Music Box*. Los Angeles: Amber Crest Books, 1980.

Denisoff, R.S. *Solid Gold: The Popular Record Industry*. New York: Transaction Books, 1976.

Dingle, Jeffrey L. *Essential Radio*. Marblehead, MA.: Peregrine Books, 1995.

Eastman, Susan Tyler. *Broadcast/Cable Programming: Strategies and Practices*, 4th ed. Belmont, Calif.: Wadsworth Publishing, 1992.

Hall, Claude, and Hall, Barbara. *This Business of Radio Programming*. New York: Billboard Publishing, 1977.

Halper, Donna. *Full-Service Radio*. Boston: Focal Press, 1991.

———. *Radio Music Directing*. Boston: Focal Press, 1991.

Keirstead, Phillip A. *All-News Radio*. Blue Ridge Summit, Pa.: Tab Books, 1980.

Keith, Michael C. *Radio Programming: Consultancy and Formatics*. Stoneham, Mass.: Focal Press, 1987.

———. *Signals in the Air: Native Broadcasting in America*. Westport, CT.: Praeger Publishing, 1995.

Lujack, Larry, and Jedlicka, D.A. *Superjock: The Loud, Frantic, Non-Stop World of Rock Radio Deejays*. Chicago: Regnery, 1975.

MacFarland, David T. *The Development of the Top 40 Format*. New York: Arno Press, 1979.

———. *Contemporary Radio Programming Strategies*. Hillsdale, N.J.: Lawrence Erlbaum, 1990.

Maki, Val, and Pederson, Jill. *The Radio Playbook*. St. Louis: Globe Mack, Inc., 1991.

Matelski, Marilyn J. *Broadcast Programming and Promotion Worktext*. Boston: Focal Press, 1989.

Morrow, Bruce. *Cousin Brucie*. New York: Morrow and Company, 1987.

Norberg, Eric. *Radio Programming: Tactics and Strategies*. Boston: Focal Press, 1996.

Passman, Arnold. *The Deejays*. New York: Macmillan, 1971.

Rhoads, B. Eric, et. al., eds. *Programming and Promotion*. West Palm Beach, FL.: 1995.

Routt, Ed., McGrath, James B., and Weiss, Frederic A. *The Radio Format Conundrum*. New York: Hastings House, 1978.

Sklar, Rick. *Rocking America: How the All-Hit Radio Stations Took Over*. New York: St. Martin's Press, 1984.

Vane, Edwin T., and Gross, Lynne S. *Programming for TV, Radio, and Cable*. Boston: Focal Press, 1994.

4 Sales

COMMERCIALIZATION: A RETROSPECTIVE

Selling commercials keeps the radio station on the air. It is that simple, yet not so simple. In the 1920s broadcasters realized the necessity of converting the medium into a sponsor-supported industry. It seemed to be the most viable option and the key to growth and prosperity. Not everyone approved of the method, however. Opponents of commercialization argued that advertising would decrease the medium's ability to effectively serve the public's good, and one United States senator voiced fears that advertisers would turn radio into an on-air pawnshop. These predictions would prove to be somewhat accurate. By the mid-1920s, most radio outlets sold airtime, and few restrictions existed pertaining to the substance and content of messages. Commercials promoting everything from miracle pain relievers to instant hair-growing solutions filled the broadcast day. It was not until the 1930s that the government developed regulations that addressed the issue of false advertising claims. This resulted in the gradual elimination of sponsors peddling dubious products.

Program sponsorships were the most popular form of radio advertising in the 1920s and early 1930s. Stations, networks, and advertising agencies often lured clients onto the air by naming or renaming programs after their products—"Eveready Hour," "Palmolive Hour," "Fleischmann Hour," "Clicquot Club Eskimos," "The Coty Playgirl," and so on. Since formidable opposition to commercialization existed in the beginning, sponsorships, in which the only reference to a product was in a program's title, appeared the best path to take. This approach was known as "indirect" advertising.

As the outcry against advertiser-supported radio subsided, stations became more blatant or "direct" in their presentation of commercial material. Parcels of time, anywhere from one to five, even ten minutes, were sold to advertisers eager to convey the virtues of their labors. Industry advertising revenues soared throughout the 1930s, despite the broken economy. Radio salespeople were among the few who had a salable product. World War II increased the rate of sales revenues twofold. As it became the foremost source of news and information during that bleak period in American history, the value of radio's stock reached new highs, as did the incomes of salespeople. The tide would shift, however, with the advent of television in the late 1940s.

When television unseated radio as the number one source of entertainment in the early 1950s, time sellers for the deposed medium found their fortunes sagging. In the face of adversity, as well as opportunity, many radio salespeople abandoned the old in preference for the new, opting to sell for television. Radio was, indeed, in a dilemma of frightening proportions, but it soon put itself back on course by renovating its programming approach. By 1957 the medium had undergone an almost total transformation and was once again enjoying the rewards of success. Salespeople concentrated on selling airtime to advertisers interested in reaching specific segments of the population. Radio became extremely localized and, out of necessity, the networks diverted their attention to television.

Competition also became keener. Thousands of new outlets began to broadcast between 1950 and 1970. Meanwhile, FM started to spread its wings, preparing to surpass its older rival. AM music stations experienced great difficulties in the face of the mass exodus to FM, which culminated in 1979 when FM exceeded AM's listenership. FM became the medium to sell.

In the first half of the 1980s, radio programming was easily divisible. Talk was found on AM and music on FM. Today, hundreds of AM stations have become stereo in an attempt to improve their marketing potential, while some FM outlets have taken to airing nonmusic formats as a way of surviving the ratings battles.

Selling time in the 1990s is a far cry from what it was like during the medium's heyday. Station account executives no longer search for advertisers willing to sponsor half-hour sitcoms, quiz shows, or mysteries. Today, the advertising dollar generally is spent on spots scheduled for airing during specific dayparts on stations that attract a particular piece of the highly fractionalized radio audience pie. The majority of radio outlets program prerecorded music, and it is that which constitutes the station's product. The salesperson sells the audience, which the station's music attracts, to the advertiser or time-buyer.

SELLING AIRTIME

Airtime is intangible. You cannot see it or hold it in your hand. It is not like any other form of advertising. Newspaper and magazine ads can be cut out by the advertiser and pinned to a bulletin board or taped to a window as tangible evidence of money spent. Television commercials can be seen, but radio commercials are sounds flitting through the ether with no visual component to attest to their existence. They are ephemeral, or fleeting, to use words that are often associated with radio advertising. However, any informed account executive will respond to such terms by stating the simple fact that an effective radio commercial makes a strong and lasting impression on the mind of the listener in much the same way that a popular song tends to permeate the gray matter. "The so-called intangible nature of a radio commercial really only means you can't see it or touch it. There is little doubt, however, that a good spot is concrete in its own unique way. Few of us have gone unaffected or, better

FIGURE 4.1
Courtesy
Communication
Graphics and *Radio
Ink*

still, untouched by radio commercials. If a spot is good, it is felt, and that's a tangible," says radio sales manager Charles W. Friedman.

Initially considered an experimental or novel way to publicize a product, it soon became apparent to advertisers that radio was far more. Early sponsors who earmarked a small portion of their advertising budgets to the new electronic "gadget," while pouring the rest into print, were surprised by the results. Encouraged by radio's performance, advertisers began to spend more heavily. By the 1930s many prominent companies were reallocating substantial portions of their print advertising budgets for radio. To these convinced advertisers, radio was, indeed, a concrete way to market their products.

Yet the feeling that radio is an unconventional mode of advertising continues to persist to some extent even today, especially among small, print-oriented retailers. Usually, the small market radio station's prime competitor for ad dollars is the local newspaper. Many retailers have used papers for years and perceive radio as a secondary or even frivolous means of advertising, contends Friedman. "Retailers who have used print since opening their doors for business are reluctant to change. The toughest factor facing a radio salesperson is the notion that the old way works the best. It is difficult to overcome inertia."

Radio is one of the most effective means of advertising when used correctly. Of course, there is a right way and a wrong way to utilize the medium, and the salesperson who knows and understands the unique character of his or her product is in the best position to succeed. To the extent that a radio commercial cannot be held or taped to a cash register, it is intangible. However, the results produced by a carefully conceived campaign can be seen in the cash register. Consistent radio users, from the giant multinational corporations to the so-called mom and pop shops, know that a radio commercial can capture people's attention as effectively as anything crossing their field of vision. A 1950s promotional slogan says it best: "Radio gives you more than you can see."

BECOMING AN ACCOUNT EXECUTIVE

A notion held by some sales managers is that salespeople are born and not made.

This position holds that a salesperson either "has it" or does not: "it" meaning the innate gift to sell, without which all the schooling and training in the world means little. Although this theory is not embraced by all sales managers, many agree with the view that anyone attempting a career in sales should first and foremost possess an unflagging desire to make money, because without it, failure is almost assured.

According to RAB figures, 70 percent of the radio salespeople hired by stations are gone within three years; another study reveals that 73 percent of new radio salespeople leave the business within a year. While this sounds less than encouraging, it also must be stated that to succeed in broadcast sales invariably means substantial earnings and rapid advancement. True, the battle can be a tough one and the dropout rate is high, but the rewards of success are great.

The majority of newly hired account executives have college training. An understanding of research, marketing, and finance is important. Formal instruction in these areas is particularly advisable for persons considering a career in broadcast sales. Broadcast sales has become a familiar course offering at many schools with programs in electronic media. Research and marketing courses designed for the broadcast major also have become more prevalent since the 1970s. "Young people applying for sales positions here, for the most part, have college backgrounds. A degree indicates a certain amount of tenacity and perseverance, which are important qualities in anyone wanting to sell radio. Not only that, but the candidate with a degree often is more articulate and self-assured. As in most other areas of radio, ten or fifteen years ago fewer people had college diplomas, but the business has become so much more sophisticated and complex because of the greater competition and emphasis on research that managers actually look for salespeople with college training," says KGLD General Manager Richard Bremkamp.

Whether a candidate for a sales position has extensive formal training or not,

FIGURE 4.2
"Radio is the best buy," says RAB. Courtesy RAB.

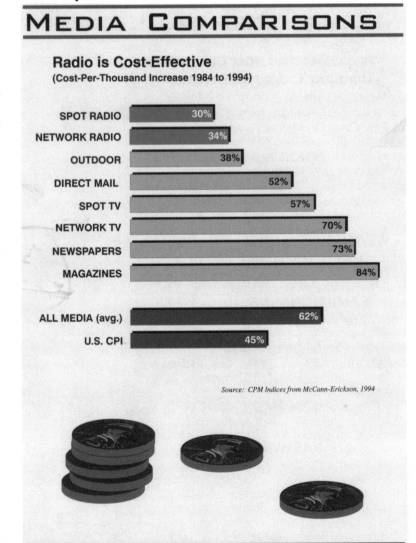

MEDIA COMPARISONS

Radio is Cost-Effective
(Cost-Per-Thousand Increase 1984 to 1994)

SPOT RADIO	30%
NETWORK RADIO	34%
OUTDOOR	38%
DIRECT MAIL	52%
SPOT TV	57%
NETWORK TV	70%
NEWSPAPERS	73%
MAGAZINES	84%
ALL MEDIA (avg.)	62%
U.S. CPI	45%

Source: CPM Indices from McCann-Erickson, 1994

he or she must possess a knowledge of the product in order to be hired. "To begin with, an applicant must show me that she knows something about radio; after all, that is what we're selling. The individual doesn't necessarily have to have a consummate understanding of the medium, although that would be nice, but she must have some product knowledge. Most stations are willing to train to an extent. I suppose you always look for someone with some sales experience, whether in radio or in some other field," says Bob Turley, general sales manager, WQBE AM/FM, Charleston, West Virginia.

Stations do prefer a candidate with sales exposure, be it selling vacuum cleaners door-to-door or shoes in a retail store. "Being out in front of the public, especially in a selling situation, regardless of the product, is excellent training for the prospective radio sales rep. I started in the transportation industry, first in customer service, then in sales. After that I owned and operated a restaurant. Radio sales was a whole new ball game for me when I went to work for WCFR in Springfield, Vermont, in the mid-1970s. Having dealt with the public for two decades served me well. In two years I rose to be the station's top biller, concentrating mainly on direct retail sales. In 1979, WKVT in Brattleboro hired me as sales manager," recalls Charles W. Friedman.

Hiring inexperienced salespeople is a gamble that a small radio station generally must take. The larger outlets almost always require radio sales credentials, a luxury that lesser stations cannot afford. Thus they must hire untested salespeople, and there are no guarantees that a person who has sold lawn mowers can sell airtime, that is, assuming that the newly hired salesperson has ever sold anything at all. In most cases, he or she comes to radio and sales without experience, and the station must provide at least a modicum of training. Unfortunately, many stations fail to provide adequate training, and this too contributes to the high rate of turnover. New salespeople commonly are given two to three months to display their wares and exhibit their potential. If they prove themselves to the sales manager by generating new business, they are asked to stay. On the other hand, if the sales manager is not convinced that the apprentice salesperson has the ability to bring in the accounts, then he or she is shown the door.

During the trial period, the salesperson is given a modest draw against sales, or a "no-strings" salary on which to subsist. In the former case, the salesperson eventually must pay back, through commissions on sales, the amount that he or she has drawn. Thus, after a few months, a new salesperson may well find himself in debt for two or three thousand dollars. As a show of confidence and to encourage the new salesperson who has shown an affinity for radio sales, management may erase the debt. If a station decides to terminate its association with the new salesperson, it must absorb the loss of time, energy, and money invested in the employee.

Characteristics that managers most often look for in prospective salespeople include, among other things, ambition, confidence, energy, determination, honesty, and intelligence. "Ambition is the cornerstone of success. It's one of the first things I look for. Without it, forget it. You have to be hungry. It's a great motivator. You have to be a quick thinker—be able to think on your feet, under pressure, too. Personal appearance and grooming also are important," says sales manager Gene Etheridge.

Joe Martin, general sales manager of WBQW-AM, Scranton, Pennsylvania, places a premium on energy, persistence, creativity, and organization. "Someone who is a self-starter makes life a lot easier. No manager likes having to keep after members of his sales team. A salesperson should take initiative and be adept at planning his day. Too many people make half the calls they should and could. A salesperson without organizational skills is simply not going to bring in the same number of orders as the person who does. There is a correlation between the number of pitches and the number of sales."

Metro market general sales manager Peg Kelly values people skills. "Empathy is right up there. I need someone who

50 Ways to Leave Your Loser

The attributes, habits and characteristics America's top local Radio salespeople say make them winners.

1.	Discipline	28.	Insistence on doing the best possible job — not just "good enough"
2.	Attention to detail	29.	Deliver on promises
3.	Follow-through	30.	Very presentable personal appearance
4.	Honesty	31.	Open minded — never stop growing
5.	Listening	32.	Advance preparation — no winging it
6.	Timeliness, promptness	33.	Rarely forget to ask for the order (as opposed to rarely remembering)
7.	Determination	34.	Regular self-improvement. Read a lot, listen to tapes. Attend seminars
8.	Thoroughness	35.	Strong communication skills: intra-office and with clients mail/phone/in-person
9.	Always prospecting		
10.	Creativeness		
11.	Consistently (re)discovering and fulfilling prospect/client needs	36.	Aggressiveness — stay with prospect/client until the job is done right
12.	Flexibility	37.	Empathy/sincere caring for client's results
13.	Love the business	38.	Results oriented
14.	Sincerely want their clients to succeed	39.	Do-it-now attitude
15.	Knowledge	40.	Keep good records
16.	Strong work ethic	41.	In office/on street early and stay late
17.	50-plus-hour workweek	42.	Under-promise and over-deliver
18.	Organized	43.	Get into the client's shoes, view things from client's perspective
19.	Effective time/territory management	44.	Anticipate and eliminate problems before they develop
20.	Priority management/crisis avoidance techniques	45.	Don't make assumptions
21.	Accessible to clients/peers/management	46.	Don't take anything for granted, especially your station's place in the buy
22.	Faith in God, self, product	47.	Loyalty to company, clients, self
23.	Unwavering enthusiasm	48.	Keep in touch with clients, especially during their schedules
24.	Total personal acceptance of successes and failures	49.	Don't take rejection personally
25.	Focus-focus-focus. Know what you want, what you need to do, do it.	50.	Sell ideas and solutions, not spots, flights, or packages
26.	Understanding and application of the basics of selling		
27.	Persistence		

FIGURE 4.3
Some cogent advice. Courtesy Radio Ink.

possesses the ability to get quick answers, learn the client's business quickly, and relate well to clients. It's a people business, really." Charles W. Friedman also looks for sales personnel who are "people oriented." "My experience has proven that a prospective salesperson had better like people. In other words, be gregarious and friendly. This is something that's hard to fake. Sincerity, or the lack of it, shows. If you're selling airtime, or I guess anything for that matter, you had better enjoy talking with people as well as listening. A good listener is a good salesperson. Another thing that is absolutely essential is the ability to take rejection objectively. It usually takes several 'no's' to get to the 'yes.'"

Friedman believes that it is important for a salesperson to have insight into human nature and behavior. "You really must be adept at psychology. Selling really is a matter of anticipating what the prospect is thinking and knowing how best to address his concerns. It's not so much a matter of outthinking the prospective client, but rather being cognizant of the things that play a significant role in his life. Empathy requires the ability to appreciate the experiences of others. A salesperson who is insensitive to a client's moods or states of mind usually will come away empty handed."

In recent years, sales managers have recruited more heavily from within the radio station itself rather than immediately looking elsewhere for salespeople. For decades, it was felt that programming people were not suited for sales. An inexplicable barrier seemed to separate the two areas. Since the 1970s, however, this attitude has changed to some degree, and sales managers now give serious consideration to on-air people who desire to make the transition into sales. Consolidation and downsizing in the 1990's also inspires multiple role playing at stations. That is to say, the morning deejay becomes the afternoon salesperson. The major advantage of hiring programming people to sell the station is that they have a practical understanding of the product. "A lot of former deejays make good account reps because they had to sell the listener on the product. A deejay really is a sales-person, when you get right down to it," observes WBQW's Joe Martin.

Realizing, too, that sales is the most direct path into station management, programming people often are eager to make the shift. Since the 1980s, there has been a greater trend than ever to recruit managers from the programming ranks. However, a sales background is still preferred.

The salesperson is invariably among the best-paid members of a station. How much a salesperson earns is usually left up to the individual to determine. Contrary to popular opinion, the salesperson's salary generally exceeds the deejay's, especially in the smaller markets. In the larger markets, certain air personalities' salaries are astronomical and even surpass the general manager's income, but major market sales salaries are commonly in the five- and even six-figure range.

Entry-level sales positions are fairly abundant, and stations are always on the lookout for good people. Perhaps no other position in the radio station affords an individual the opportunities that sales does, but most salespeople will never go beyond entry level in sales. Yet for those who are successful, the payoff is worthwhile. "The climb itself can be the most exhilarating part, I think. But you've got to have a lot of reserve in your tanks, because the air can be pretty thin at times," observes Etheridge.

THE SALES MANAGER

The general sales manager directs the marketing of the radio station's airtime. This person is responsible for moving inventory, which in the case of the radio outlet constitutes the selling of spot and feature schedules to advertisers. To achieve this end, the sales manager directs the daily efforts of the station's account executives, establishes sales department policies, develops sales plans and materials, conceives of sales and marketing campaigns and promotions, sets quotas, and also may sell as well.

The organizational structure of a station's sales department customarily includes the positions of national, regional, and local sales managers. The national responsibilities usually are handled by the general sales manager. This includes working with the station's rep company to stimulate business from national advertisers. The regional sales manager is given the responsibility of exploring sales possibilities in a broad geographical area surrounding the station. For example, the regional person for an outlet in New York City may be assigned portions of Connecticut, New Jersey, and Long Island. The local sales manager at the same station would concentrate on advertisers within the city proper. The general sales manager oversees the efforts of each of these individuals.

The size of a station's sales staff varies according to its location and reach. A typical small market radio station employs between two and four account executives, while the medium market station averages about five. Large, top-ranked metropolitan outlets employ as many as eight to ten salespeople, although it is more typical for the major market station to have about a half dozen account executives.

The general sales manager reports directly to the station's general manager and works closely with the program director in developing salable features. Regular daily and weekly sales meetings are scheduled and headed by the sales manager, during which time goals are set and problems addressed. The sales manager also assigns account lists to members of his or her staff and helps coordinate trade and co-op deals.

As mentioned earlier, the head of the sales department usually is responsible for maintaining close contact with the station's rep company as a way of generating income from national advertisers who are handled by advertising agencies. The relationship of the sales manager and rep company is a particularly important one and will be discussed in greater detail later in this chapter. In addition, the sales manager must be adept at working ratings figures to the station's advantage for inclusion in sales

FIGURE 4.4
Stats make the
pitch. Courtesy
RAB.

MAXIMIZE YOUR ADVERTISING WITH RADIO

Why? Because every day your current and potential customers take their favorite Radio station into their cars, their homes, and the workplace. They even include Radio as part of their leisure time activities. To listeners, Radio is more than a medium, it's a special companion with whom they spend nearly three hours each day and more than four hours each weekend.

For the contemporary marketer, there's no other advertising medium that can build personal, one-to-one relationships that can reach and motivate consumers like Radio. Your best potential prospects for each of your most important products and services can be accurately targeted by age, income, gender, race, geography and lifestyle with effective frequency through a system of easily identifiable Radio formats.

YOU'LL REACH ALL YOUR BEST PROSPECTS WITH RADIO

WEEKLY REACH: Virtually every American listens to Radio! Not only do they tune in, they stay tuned — averaging 2 hours and 43 minutes of Radio listening every day.

AT WORK: Radio allows you to influence your best prospects all day long where other media can't: at work. Radio's varied formats appeal to people of all professional backgrounds.

RADIO'S WEEKLY REACH BY AGE AND SEX [28]	
MEN 18+	97%
WOMEN 18+	95%
TEENS 12-17	99%

RADIO'S MIDDAY REACH BY OCCUPATION [29]	
PROFESSIONAL/MANAGERIAL	76.1%
TECH./SALES/CLERICAL	80.3%
OTHER WORKERS	79.4%

YOU'LL MOTIVATE CONSUMERS EVERYWHERE WITH RADIO

AT HOME AND VIRTUALLY EVERYWHERE: Radio reaches your customers no matter where they are at work, shopping or just relaxing at home. Radio's in-car reach is also outstanding, reaching 86% of commuters each week. [30]

Share of Total Radio Listening by Location [31]

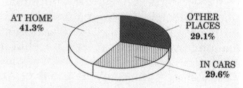

AT HOME 41.3%
OTHER PLACES 29.1%
IN CARS 29.6%

SHARE OF MEDIA TIME: Customers spend more time with Radio than any other media — up to 40% of total time spent with media.

Share of Time Spent With Media
(Total hours per year — Adults 18+)[32]

MAGAZINES 3.0%
NEWSPAPERS 6.0%
TV 38.4%
RADIO 40.0%
CABLE 12.7%

YOU'LL ENJOY THESE BENEFITS WITH RADIO

- *Total Market Reach*
- *Effective Message Frequency*
- *Targetability by Format Options*
- *Positive Advertising Environment*
- *Uncluttered Commercial Platform*
- *Complete Flexibility*
- *Sufficient Competitive Separation*
- *Unparalleled Cost Efficiency* • *Pro-active Promotional Opportunities*

promotional materials that are used on both the national and local level.

All sales come under the scrutiny of the sales manager, who determines if an account is appropriate for the station and whether conditions of the sale meet established standards. In addition, the sales manager may have a pol-

icy that requires credit checks to be made on every new account and that each new client pay for a portion of their spot schedule "up front" as a show of good faith. Again, policies vary from station to station.

It is up to the head of sales to keep abreast of local and national sales and marketing trends that can be used to the station's advantage. This requires that the sales manager constantly survey trade magazines, like *Radio Ink*, *Radio Business Report*, and *ADWEEK*, and attend industry seminars, such as those conducted by the Radio Advertising Bureau. No sales department can operate in a vacuum and hope to succeed in today's dynamic radio marketplace.

Statistics continue to bear out the fact that sales managers are most often recruited to fill the position of general manager. It is also becoming more commonplace for sales managers to have experience in other areas of a station's operations, such as programming and production, a factor that has become increasingly important to the person who hires the chief account executive.

RADIO SALES TOOLS

The fees that a station charges for airtime are published in its rate card. Rates for airtime depend upon the size of a station's listenership, that is, the bigger the audience the higher the rates. At the same time, the unit cost for a spot or a feature is affected by the quantity or amount purchased: the bigger the "buy," the cheaper the unit price. Clients also get discounts for consecutive week purchases over a prescribed period of time, say twenty-six or fifty-two weeks.

The sales manager and station manager work together in designing the rate card, basing their decisions on ratings and what their market can support. A typical rate card will include a brief policy statement concerning terms of payment and commission: "Bills due and payable when rendered. Without prior credit approval, cash in advance required. Commission to recognized advertising agencies on net charges for station time—15%." A statement pertaining to the nature of copy and when it is due at the station also may be included in the rate card: "All programs and announcements are subject to removal without notice for any broadcast which, in our opinion, is not in the public's interest.

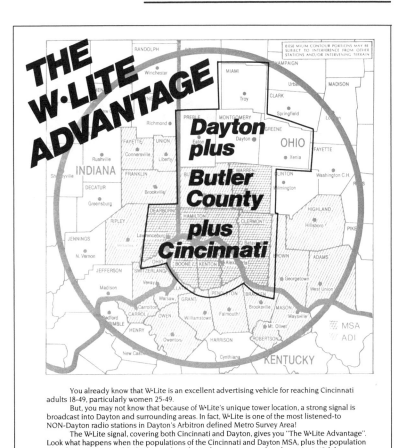

You already know that W·Lite is an excellent advertising vehicle for reaching Cincinnati adults 18-49, particularly women 25-49.

But, you may not know that because of W·Lite's unique tower location, a strong signal is broadcast into Dayton and surrounding areas. In fact, W·Lite is one of the most listened-to NON-Dayton radio stations in Dayton's Arbitron defined Metro Survey Area!

The W·Lite signal, covering both Cincinnati and Dayton, gives you "The W·Lite Advantage". Look what happens when the populations of the Cincinnati and Dayton MSA, plus the population of Butler County, are added! The combined population is greater than the nation's 13th largest radio market!

As a retailer with locations in Cincinnati and Dayton, or in between, you can benefit from "The W·Lite Advantage!"

CINCINNATI MSA 12+	1,155,600	#27 Population Rank
DAYTON MSA 12+	680,100	#48 Population Rank
BUTLER COUNTY 12+	214,878	
TOTAL POPULATION	2,050,578	Greater than #13! (St. Louis)

Copy must be at the station 48 hours prior to broadcast date and before noon on days preceding weekends and holidays." A station's approach to discounting must, of practical necessity, be included in the rate card: "All programs, features, and announcements are provided a 5% discount if on the air for 26 consecutive weeks and a 10% discount if on the air for 52 weeks."

It is important to state as emphatically and clearly as possible the station's position on all possible topics affecting a sale. Most stations provide clients rate protection for a designated period of time should fees for airtime change. This means that if a client purchases a three-month spot schedule in May, and the station raises its rates in

FIGURE 4.5
Coverage maps are used to sell those areas within the station's signal. Courtesy WLLT-FM.

June, the advertiser continues to pay the original rates until the expiration of its current contract.

The rate card also contains its feature and spot rates. Among the most prevalent features that stations offer are traffic, sports, weather, and business reports. Newscasts also are available to advertisers. Features generally include an open (introduction) and a thirty- or sixty-second announcement. They are particularly effective advertising vehicles because listeners tend to pay greater attention. Conditions pertaining to feature buys usually appear in the rate card: "All feature sales are subject to four weeks' notice for renewal and cancellation." A station wants to establish credibility with its features and therefore prefers to maintain continuity among its sponsors. A feature with a regular sponsor conveys stability, and that is what a station seeks.

Since the size of a radio station's audience generally varies depending on the time of day, rates for spots (commercials) or features must reflect that fact. Thus, the broadcast day is divided into time classifications. Six to ten a.m. weekdays is typically a station's prime selling period and therefore may be designated AAA, while afternoon drive-time, usually three to seven p.m., may be called AA because of its secondary drawing power. Under this system, the midday segment, ten to three p.m., would be given a single A designation, and evenings, seven to midnight, a B. Overnights, midnight to six a.m., may be classified as C time. Obviously, the fees charged for spots are established on an ascending scale from C to AAA. A station may charge three hundred dollars for an announcement aired at eight a.m. and forty-five dollars for one aired at two a.m. The difference in the size of the station's audience at those hours warrants the contrast.

As previously mentioned, the more air-time a client purchases, the less expensive the cost for an individual commercial or unit. For instance, if an advertiser buys ten spots a week during AAA time, the cost of each spot would be slightly less than if the sponsor purchased two spots a week. A client must buy a specified number of spots in order to benefit from the frequency discount. A 6X rate, meaning six spots per week, for AAA 60s (sixty-second announcements) may be $75, while the 12X rate may be $71 and the 18X rate $68, and so forth. Thirty-second spots are usually priced at two-thirds the cost of a sixty. Should a client desire that a spot be aired at a fixed time, say 7:10 a.m. daily, the station will tack on an additional charge, possibly 20 percent. Fixed position drivetime spots are among the most expensive in a station's inventory.

Many stations use a grid structure. This gives stations a considerable degree of rate flexibility. For example, if a station has five rate-level grids, it may

FIGURE 4.6
Station rate card with ROS feature. Courtesy WLKZ.

Rate Card

Effective 6/26/95

1) WLKZ-FM Designated Segments - Minute Announcements:

	all rates net to station		
	1-10 Per Week	11-20 Per Week	21+ Per Week
Drive Time Day Part	25	22	20
6a - 8p	20	18	16
R.O.S.	17	15	12

2) Time Designations:

    Drive Time - Commercials that air Monday through Friday 6a-9a and 3p-7p.

    6a - 7p - Commercials that air Monday through Sunday, from 6am - 7pm.

    R.O.S. - Run of Station, commercials that air Monday through Sunday from 6:00am - 12 midnight.

3) Thirty-second Announcements:

    85% of sixty-second rates.

4) Nighttime:

    Commercials that air Monday through Sunday 7pm - 12 midnight are 70% of the R.O.S. rate.

5) Sponsorship:

    WLKZ offers news sponsorships at approximately 15% above the individual Day-part rate.

    WLKZ also offers Music Program Sponsorships. For more information please see individual packages.

21 Production Place • Suite 15 • Gilford, New Hampshire 03246 • (603) 524-0105 • Fax (603) 293-0699

FIGURE 4.7
Rate card using a
grid structure.

KNEW 91 AM/KSAN 95 FM

Vice President &
General Manager
Steve Edwards

General Sales Manager
Joel K. Schwartz

Represented nationally by
Katz Radio

rate card

Weekly 60 Second Spot Rate

GRID	AAA	AA	A	B
I	500	460	450	200
II	450	410	400	150
III	400	360	350	120
IV	350	310	300	100

Time Classifications

AAA	Monday - Saturday	5:00 AM-10:00 AM
	Saturday	10:00 AM- 3:00 PM
AA	Monday-Saturday	3:00 PM- 8:00 PM
A	Monday-Friday	10:00 AM- 3:00 PM
	Sunday	
B	Monday-Saturday	8:00 PM- 1:00 AM
	Sunday	8:00 AM-10:00 AM
		3:00 PM-10:30 PM

The cost per announcement quoted is as if for one station. The schedule is also duplicated on the other station.

Notes:
1. Grid IV subject to immediate preemption without notice.
2. Limitations: Maximum ¼ of all spots in AM drive Monday - Friday.
   Maximum of 2 spots 10:00 AM - 3:00 PM Saturday.
   Any spots over these limits are subject to availability and will only be sold at Grid I.
3. News Sponsorships: Applicable rate plus $20.
4. 30 second rates are 90% of the weekly 60 second rate.
5. 1:00 AM - 5:00 AM Monday - Sunday: No charge with every B time spot purchased.
6. For billing purposes only: AM/FM breakdown is 50/50.
7. KNEW or KSAN only: Deduct $1 from applicable combo rate.

Conditions

Contracts are accepted for a maximum period of one year. Rates quoted herein are guaranteed for a period of 30 days from the effective date of any increase. Rates quoted represent station time only. There is an agency commission of 15% to recognized agencies: no cash discounts. Bills due and payable upon receipt.

All contracts are governed by the conditions detailed on the KNEW/KSAN Broadcast Contract.

All advertising material must be received at least 48 hours before each broadcast. Advertising material must conform to the standards of the station and the station reserves the right to refuse or discontinue any advertising for reasons satisfactory to itself.

Unfulfilled contracts are subject to appropriate short rate.

This rate card is for information purposes and does not constitute an offer on the part of KNEW/KSAN.

● MALRITE COMMUNICATIONS GROUP ● 66 JACK LONDON SQUARE ● OAKLAND, CA 94607 ● (415) 836-0910
YE 1EF 1084

have a range between $20 and $50 for a 60-second spot. Clients would then be given rates at the lower grid if the station had few sponsors on the air, thus creating many availabilities (places to insert commercial messages). As business increased at the station and availabilities became scarcer, the station would ask for rates reflected in the upper grids. Gridding is based on the age-old concept of supply and demand. When availabilities are tight and airtime is at a premium, that time costs more.

Grids are inventory sensitive; they allow a station to remain viable when business is at a low ebb. Certainly when inventory prices reach a bargain level, this encourages business. For instance, during a period when advertiser activity is sluggish, a station that can offer spots at a considerable reduction stands a chance of stimulating buyer interest.

When business is brisk at a station, because of holiday buying, for example, the situation may be exploited in a manner positive to the revenue column. Again the supply-and-demand concept is at work (one idea on which capitalism is based).

Not all stations grid their rate cards, however. Since the late 1970s, this system has gained considerable popularity because of its relevance to the ever fluctuating economy. As apparent in the rate cards exhibited in this chapter, grids are delineated by some sort of scale, usually alphabetical or numerical.

Clients are offered several spot schedule plans suited for their advertising and budgetary needs. For advertisers with limited funds, run-of-station (ROS) or best-time-available (BTA) plans are usually an option. Rates are lower under these plans since no guarantee is given as to what times the spots will be aired. However, most stations make a concerted effort to rotate ROS and BTA spots as equitably as possible, and during periods when commercial loads are light they frequently are scheduled during premium times. Of course, when a station is loaded down with spot schedules, especially around holidays or elections, ROS and BTA spots may find themselves "buried." In the long run, advertisers using these plans receive a more than fair amount of choice times and at rates considerably lower than those clients who buy specific dayparts.

In a recent book (*Radio Advertising's Missing Ingredient: The Optimum Effective Scheduling System*) by Pierre Bouvard and Steve Marx, an innovative system is presented that improves the might of a spot schedule. According to the authors, "Optimum effective scheduling ensures that the effective reach, those hit three or more times, is at least 50 percent of the total reach." The idea behind OES is to strengthen the impact of client buys. The OES formula is designed to heighten the efficiency of a spot buy through a system of schedul-

ing based on ratings performance. This is accomplished by factoring a station's turnover ratio and cume. As of this writing, OES was gaining advocates as well as detractors.

Total audience plan (TAP) is another popular package offered clients by many stations. It is designed to distribute a client's spots among the various dayparts for maximum audience penetration, while costing less than an exclusive prime-time schedule. The rate for a TAP spot is arrived at by averaging out the cost for spots in several time classifications. For example, AAA =$80, AA = $70, A = $58, B = $31; thus, the TAP rate per spot is $59. The advantages are obvious. The advertiser is getting a significant discount on the spots scheduled during morning and afternoon drive periods. At the same time, the advertiser is paying more for airtime during evenings. However, TAP is very attractive because it does expose a client's message to every possible segment of a station's listening audience with a measure of cost-effectiveness.

Bulk or annual discounts are available to advertisers who buy a heavy schedule of commercials over the course of a year. Large companies in particular take advantage of volume discounts because the savings are significant.

Announcements are rotated or "orbited" within time classifications to maximize the number of different listeners reached. If a client buys three drivetime spots per week to be aired on a Monday, Wednesday, and Friday, over a four-week period, the time they are scheduled will be different each day. Here is a possible rotation set-up:

	MON	WED	FRI
Week I	7:15	6:25	9:10
Week II	8:22	7:36	8:05
Week III	6:11	9:12	7:46
Week IV	9:20	8:34	6:52

Rather than purchase a consecutive week schedule, advertisers may choose to purchase time in flights, an alternating pattern of being on one week and off the next. For instance, a client with a seasonal business or one that is geared toward holiday sales may set up a plan in which spots are scheduled at specific times throughout the year. Thus, an annual flight schedule may look something like this:

Feb. 13–19 Washington's Birthday Sale. 10 A 60s

Mar. 14–17 St. Patrick's Day Celebration. 8 ROS 30s

Apr. 16–21 Easter Parade Days. 20 TAP 30s

May 7–12 Mother's Day Sale. 6 AAA and 6 AA 30s

Jun. 1–15 Summer Sale Days. 30 ROS 60s

Aug. 20–30 Back-to-School Sale. 15 A 60s

Sept. 24–Oct. 6 Fall Sale Bonanza. 25 ROS 60s

Nov. 25–Dec. 19 Christmas Sale. 25 AAA 60s and 20 A 30s

Let's take a look at the wisdom behind a few of the preceding flights. The client purchases 10 spots in A time during the week that precedes Washington's Birthday to reach the home female audience. The A time on this station is 10:00 a.m. to 3:00 p.m., so a schedule of spots here does a good job of targeting women who work at home. The client uses a TAP plan to move Easter inventory. This will get the client's spot in all dayparts to help her reach as many different people as possible at the best unit price. During the back-to-school days of late August, the client once again targets the strongest female daypart on the station, and around Christmas time the client purchases the heaviest flight because this is the so-called do-or-die period (the time when business potential reaches its peak). The heavier spot purchase will help assure success during this crucial time. Selling in flights makes a lot of sense to many advertisers because of its calendar relevance.

A rate card is used by salespeople to plan and compute buys. It is generally perceived as a poor idea to simply leave it with a prospective client to figure out, even if such a request is made. First off, few laypersons are really adept at reading rate cards and, secondly, a station does not like to publicize its rates to its

Demographic Rate Card Monday–Sunday 6 A.M.–Midnight			
Demographic	Low	High	Rating
Teens	$230	$500	5.0
12–24 Persons	$235	$400	5.2
12–24 Males	$210	$350	4.8
12–24 Females	$240	$500	5.5
18–24 Persons	$215	$350	3.6
18–24 Males	$200	$330	3.4
18–24 Females	$215	$350	3.6
18–34 Persons	$180	$250	3.0
18–34 Males	$180	$250	2.9
18–34 Females	$180	$250	3.0
18–49 Persons	$170	$225	2.4
18–49 Males	$165	$225	2.2
18–49 Females	$175	$225	2.6
25–34 Persons	$180	$250	2.7
25–34 Males	$180	$250	2.6
25–34 Females	$180	$250	3.0
25–49 Persons	$150	$200	1.9
25–49 Males	$140	$200	1.7
25–49 Females	$150	$200	1.9
25–54 Persons	$130	$190	1.2
25–54 Males	$130	$190	1.1
25–54 Females	$130	$190	1.2

competition, which is what happens when too many station rate cards are in circulation. Granted, it is quite easy for any station to obtain a competitor's sales portfolio, but stations prefer to keep a low profile as a means of retaining a competitive edge.

Radio expert Jay Williams, Jr., contends that the station rate card is on its way to obsolescence. "They're almost gone, these rate cards. Advertisers themselves simplified the increasingly complex grid cards just by asking for the best times at the lowest rate. Sales-people became rate negotiators. Then industry organizations and publishing monopolies, like ASCAP, grabbed the highest rate they could find and used it to calculate dues and payments. Radio stations lost both ways (some even pulled their rates our of SRDS). Taking a cue from the airlines' yield management system, many stations switched to computers loaded with software, like Maximizer and Tapscan, along with sophisticated traffic and inventory projection systems, to generate optimum schedules

FIGURE 4.8
Some stations use a rate card that reflects demographic variations of their audience. This particular card also provides salespeople a negotiating range for selling airtime. A "low" and "high" rate are indicated.

FIGURE 4.9
Optimum Effective
Schedule (OES) is
designed to
intensify the
impact of a client's
spot schedule.
Courtesy
TAPSCAN.

Optimum Effective Scheduling
Radio's "Secret Formula"
For Advertising Success

Here's a typical schedule reach curve for a radio station. The OES spot level for this particular station is 44 commercials, evenly dispersed throughout a seven-day week. Note that the listeners reached _effectively_ (three or more times) comprise the majority of the station's weekly cume, and two-thirds of the schedule reach.

Optimum Effective Scheduling harnesses radio's power as _the_ Effective Reach Medium.

Percent of Cume

100

80

60

40

20

OES

Reach

Effective Reach

S P O T S

0   15   30   45   60   75   90

The Elements of An Optimum Effective Schedule:

▨ Broad-daypart (preferably Monday-Sunday 6AM-Midnight) time period.

▨ Spots evenly dispersed throughout the schedule.

▨ Weekly spot level matched to station's audience characteristics, using the OES formula:
Weekly Schedule = Audience Turnover times 3.29.

The Operative Word Here is "Effective"...
Here's What An Optimum Effective Schedule Will Deliver On Any Station:

▨ Over 50% of the station's weekly audience (cume) will hear the message _at least three times._

▨ _Two-thirds_ of the people exposed to the message (the total schedule reach) will hear it _at least three times._

▨ Almost half of the people reached will hear the message _five or more_ times.

TAPSCAN
INCORPORATED

3000 Riverchase Galleria
Suite 850
Birmingham, Alabama 35244
205-987-7456

© 1991, TAPSCAN, Incorporated

Source material: NAB's "Radio Advertisings's Missing Ingredient: The Optimum Effective Scheduling System."
By Steve Marx and Pierre Bouvard. Copyright 1991. National Association of Broadcasters.

and rates on demand. Now a radio salesperson can meet the advertiser armed with a specific presentation, complete with ratings, schedules, cost and cost per point, reach and frequency, targeted geographic profiles, and other quantitative as well as qualitative data."

POINTS OF THE PITCH

Not all sales are made on the first call; nonetheless, the salesperson does go in with the hopes of "closing" an account. The first call generally is designed to introduce the station to the prospective sponsor and to determine their needs. However, the salesperson should always be prepared to propose a buy that is suitable for the account. This means that some homework must be done relative to the business before an approach is made. "First determine the client's needs, as best as possible. Then address those needs with a schedule built to reach the client's customers. Don't walk into a business cold or without some sense of what the place is about," advises WKVT's Friedman.

Should all go smoothly during the initial call, the salesperson may opt to go for an order there and then. If the account obliges, fine. In the event that the prospective advertiser is not prepared to make an immediate decision, a follow-up appointment must be made. The call-back should be accomplished as close to the initial presentation as possible to prevent the impression made then from fading or growing cold. The primary objective of the return call is to close the deal and get the order. To strengthen the odds, the salesperson must review and assess any objections or reservations that may have arisen during the first call and devise a plan to overcome them. Meanwhile, the initial proposal may be beefed up to appear even more attractive to the client, and a "spec" tape (see the later section in this chapter) for the business can be prepared as further enticement.

Should the salesperson's efforts fail the second time out, a third and even fourth call are made. Perseverance does pay off, and many salespeople admit that just when they figured a situation was hopeless, an account said yes. "Of course, beating your head against the wall accomplishes nothing. You have to know when your time is being wasted. Never give up entirely on an account; just approach it more sensibly. A phone call or a

G r i d O N E

GridONE is the radio industry's original Yield Management system and demand-driven electronic rate card. Quite simply, GridONE provides you with infinite control over the pricing of inventory based on supply and demand, with an eye towards delivering specific financial goals for your station. By combining these two dynamic management principles, GridONE allows stations to incorporate differential pricing for each unit of airtime sold, to yield the greatest possible revenue for the station.

Think of GridONE as a watchdog to help you optimize all revenue possibilities at your station. When demand is high, prices are automatically adjusted according to desired yield levels. When demand is low, prices are maintained at levels which will force demand into lesser-utilized dayparts to even out your overall sellout curve.

An internal spot scheduler will provide your marketing staff with the opportunity to plot an effective schedule for an advertiser using actual commercial availabilities and a current rate card, along with an analysis module to gauge the effectiveness of a campaign. When used in conjunction with Arbitron or Birch ratings, GridONE provides an excellent guidepost for monitoring efficiencies while maintaining optimum spot distribution throughout single or multiple week schedules. Never before has such a powerful management tool been made available in the Radio industry.

Perhaps most importantly, GridONE takes the guesswork out of day to day inventory management. By using GridONE's infinite rate cards fused with actual avails from your business system, you'll more effectively package inventory and increase yield on a daily basis. You'll notice immediate increases in your average unit rates, along with a greater degree of control of spot distribution during peak selling periods.

GridONE uses actual inventory from your traffic system, and interfaces are available for use in conjunction with several of the Radio industry's most popular systems, including TAPSCAN's Director Series Broadcast Management System.

GridONE. The cutting edge in the science of Broadcast Management...from TAPSCAN, worldwide innovators in the science of Broadcast Sales.

TAPSCAN
INCORPORATED

3000 Riverchase Galleria
Suite 850
Birmingham, Alabama 35244
205 987-7456

Boston · Chicago · Los Angeles · Toronto · Vancouver

drop-in every so often keeps you in their thoughts," says general sales manager Ron Piro.

What follows are two checklists. The *Do* list contains some suggestions conducive to a positive sales experience, while the *Don't* list contains things that will have a negative or counterproductive effect.

**FIGURE 4.10
Electronic rate cards employ grid approach for inventory control. Courtesy TAPSCAN.**

Do

- Research advertiser. Be prepared. Have a relevant plan in mind.

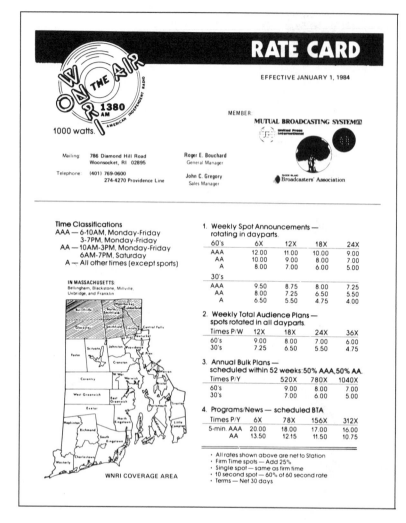

FIGURE 4.11
Rate card featuring TAP, Bulk, and BTA plans. Courtesy WNRI-AM.

- Be enthusiastic. Think positive.
- Display self-confidence. Believe in self and product.
- Smile. Exude friendliness, warmth, and sincerity.
- Listen. Be polite, sympathetic, and interested.
- Tell of station's successes. Provide testimonial material.
- Think creatively.
- Know your competition.
- Maintain integrity and poise.
- Look your best. Check your appearance.
- Be objective and keep proper perspective.
- Pitch the decision maker.
- Ask for the order that will do the job.
- Service account after the sale.

Don't

- Pitch without a plan.
- Criticize or demean client's previous advertising efforts.
- Argue with client. This just creates greater resistance.
- Badmouth competition.
- Talk too much.
- Brag or be overly aggressive.
- Lie, exaggerate, or make unrealistic promises.
- Smoke or chew gum in front of client.
- Procrastinate or put things off.
- Be intimidated or kept waiting an unreasonable amount of time.
- Make a presentation unless you have client's undivided attention.
- Lose your temper.
- Ask for too little. Never undersell a client.
- Fail to follow up.
- Accept a "no" as final.

Checklists like the preceding ones can only serve as basic guidelines. Anyone who has spent time on the street as a station account executive can expand on this or any other such checklist. For the positive-thinking radio salesperson, every call gives something back, whether a sale is made or not.

Overcoming common objections is a necessary step toward achieving the sale. Here are some typical "put-offs" presented to radio sales reps:

1. Nobody listens to radio commercials.
2. Newspaper ads are more effective.
3. Radio costs too much.
4. Nobody listens to your station.
5. We tried radio, and it didn't work.
6. We don't need any more business.
7. We've already allocated our advertising budget.
8. We can get another station for less.
9. Business is off, and we haven't got the money.
10. My partner doesn't like radio.

And so on. There are countless rebuttals for each of these statements, and a knowledgeable and skilled radio salesperson can turn such objections into positives.

LEVELS OF SALES

There are three levels from which the medium draws its sales—retail, local, and national. Retail accounts for the biggest percentage of the industry's income, over 70 percent. Retail, also referred to as *direct*, sales involve the radio station on a one-to-one basis with advertisers within its signal area. In this case, a station's account executive works directly with the client and earns a commission of approximately 15 percent on the airtime he or she sells. An advertiser who spends $1000 would benefit the salesperson to the tune of $150. A newly hired salesperson without previous experience generally will work on a direct retail basis and will not be assigned advertising agencies until he or she has become more seasoned and has displayed some ability. Generally speaking, the smaller the radio station, the more dependent it is on retail sales, although most medium and metro market stations would be in trouble without strong business on this level.

All stations, regardless of size, have some contact with advertising agencies. Here again, however, the larger a market, the more a station will derive its business from ad agencies. This level of station sales generally is classified as "local." The number of advertising agencies in a market will vary depending on its size. A sales manager will divide the market's agencies among his reps as equitably as possible, sometimes using a merit system. In this way, an account executive who has worked hard and produced results will be rewarded for his efforts by being given an agency to work. The top billers, that is, those salespeople who bring in the most business, often possess the greatest number of agencies, or at least the most active. Although the percentage of commission a salesperson is accorded, typically 6 to 8 percent, is less than that derived from retail sales, the size of the agency buys usually is far more substantial.

The third category of station sales comes from the "national" level. In most cases, it is the general sales manager who works with the station's rep company to secure buys from advertising agencies that handle national accounts. Again, national business is greater for the metro station than it is for the rural. Agencies justify a buy on numbers and little else, although it is not uncommon for small market stations, which do not even appear in ratings surveys, to be bought by major accounts interested in maintaining a strong local or community image.

Producer Ty Ford observes that agency involvement has decreased in recent years because of intensified competition among the different media and the unpredictable national economy. "Increased competition from cable, television, radio, and print has forced many ad agencies out of business. Stations now frequently offer 'agency discounts' to direct-retail clients just to close the

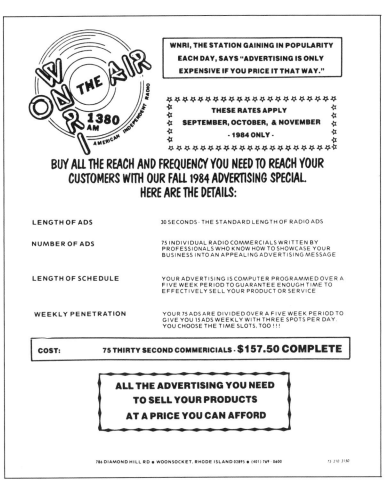

FIGURE 4.12
Flyer offering special rate on a spot schedule. Courtesy *Radio Ink*.

VII. Ten Reasons To Advertise

1. You Must Advertise to Reach New Customers.
Your market changes constantly. New families in the area mean new customers to reach. People earn more money, which means changes in lifestyles and buying habits. The shopper who wouldn't consider your business a few years ago may be a prime customer now.

2. You Must Advertise Continuously.
Shoppers don't have the store loyalty they once did. Shoppers have mobility and freedom of choice. You must keep pace with your competition. The National Retail Merchants Association states: "Mobility and non-loyalty are rampant. Stores must promote to get former customers to return, and to seek new ones."

3. You Must Advertise to Remain with Shoppers Through the Buying Process.
Many people postpone buying decisions. they often go from store to store comparing prices, quality and service. Advertising must reach them steadily through the entire decision-making process. Your name must be fresh in their minds when they decide to buy.

4. You Must Advertise Because Your Competition is Advertising.
There are so many consumers in the market ready to buy at any one time. You have to advertise to keep regular customers, and to counterbalance the advertising of your competition. You must advertise to keep or expand your market share or you will lose to the more aggressive competitors.

5. You Must Advertise Because it Pays Over a Long Period.
Advertising gives you a long-term advantage over competitors who cut back or cancel advertising. A five-year survey of more than 3,000 companies found...
- Advertisers who maintained or expanded advertising over a five-year period saw their sales increase an average of 100%.
- Companies which cut advertising averaged sales increases of 45%.

6. You Must Advertise to Generate Store Traffic.
Continuous store traffic is the first step toward sales increases and expanding your base of shoppers. The more people who come into the store, the more possibilities you have to make sales. For every 100 items that shoppers plan to buy, they make 30 unanticipated "in-the-store" purchases, an NRMA survey shows.

7. You Must Advertise to Make More Sales.
Advertising works. Businesses that succeed are usually strong, steady advertisers. Look around. You'll find the most aggressive and consistent advertisers are almost invariably the most successful merchants.

8. You Must Advertise Because There is Always Business to Generate.
Your doors are open. Salespeople are on the payroll. Even the slowest days produce sales. As long as you're in business, you've got overhead to meet, and new people to reach. Advertising can generate customers now...and in the future.

9. You Must Advertise to Keep a Healthy Positive Image.
In a competitive market, rumors and bad news travel fast. Advertising corrects misleading gossip, punctures "overstated" bad news. Advertising that is vigorous and positive can bring shoppers into the marketplace regardless of the economy.

10. You Must Advertise to Maintain Store Morale.
When advertising and promotion are suddenly cut or cancelled, salespeople may become alarmed and demoralized. They may start false rumors in an honest belief that your business is in trouble. Positive advertising boosts morale. It gives your staff strong additional support.

FIGURE 4.13
Courtesy WOOD
AM/FM and RAB.

sale. Also, more retail companies are forming their own in-house agencies." While this may be true, the ad agency is still an important factor in station revenues.

Each level of sales—retail, local, or national—must be sufficiently cultivated if a station is to enjoy maximum prosperity. To neglect any one of these levels would result in a loss of station revenue.

SPEC TAPES

One of the most effective ways to convince an advertiser to use a station is to provide a fully produced sample commercial, or "spec tape." If prepared properly and imaginatively, a client will find it difficult to deny its potential. Spec tapes often are used in call-backs when a salesperson needs to break down a client's resistance. More than once, a clever spec tape has converted an adamant "no" into an "okay, let's give it a shot." Spec tapes also are used to reactivate the interest of former accounts who may not have spent money on the station for a while and who need some justification to do so.

Specs also are effective tools for motivating clients to "heavy-up" or increase their current spot schedules. A good idea can move a mountain, and salespeople are encouraged by the sales manager to develop spec tape ideas. Many sales managers require that account executives make at least one spec tape presentation each week. The sales manager may even choose to critique spec spots during regularly scheduled meetings.

The information needed to prepare a spec spot is acquired in several ways. If a salesperson already has called on a prospective client, he should have a very good idea of what the business is about as well as the attitude of the retailer toward the enterprise. The station sales rep is then in a very good position to prepare a spot that directly appeals to the needs and perceptions of the would-be advertiser. If a salesperson decides that the first call on a client warrants preparing a spec tape, then he or she may collect information on the business by actually browsing through the store as a customer might. This gives the salesperson an accurate, firsthand impression of the store's environment and merchandise. An idea of how the store perceives itself and specific information, such as address and hours, can be derived by checking its display ad in the Yellow Pages, if it has one, or by examining any ads it may have run in the local newspaper. Flyers that the business may have distributed also provide useful information for the formulation of the copy used in the spec spot. Listening to commercials the advertiser may be running on another station also gives the salesperson an idea of the direction in which to move.

Again, the primary purpose of a spec tape is to motivate a possible advertiser to buy time. A spec that fails to capture the interest and appreciation of the individual for which it has been prepared may be lacking in the necessary ingredients. It is generally a good rule of thumb to avoid humor in a spec, unless the salesperson has had some firsthand experience with the advertiser. Nothing fails as abysmally as a commercial that attempts to be funny and does not come across as such to the client. Thus the saying, "What is funny to one person may be silly or offensive to another."

Although spec spots are, to some extent, a gamble, they should be prepared in such a way that the odds are not too great. Of course, a salesperson who believes in an idea must have the gumption to go with it. Great sales are often inspired by unconventional concepts.

OBJECTIVES OF THE BUY

A single spot on a radio station seldom brings instant riches to an advertiser. However, a thoughtfully devised plan based upon a formula of frequency and consistency will achieve impressive results, contends John Gregory, general sales manager, WNRI-AM, Woonsocket, Rhode Island. "It has to be made clear from the start what a client hopes to accomplish by advertising on your station. Then a schedule that realistically corresponds with the client's goals must be put together. This means selling the advertiser a sufficient number of commercials spread over a specific period of time. An occasional spot here and there doesn't do much in this medium. There's a right way to sell radio, and that isn't it."

Our lists of "dos" and "don'ts" of selling suggested that the salesperson "ask for the order that will do the job." It also said not to undersell an account. Implicit in the first point is the idea that the salesperson has determined what kind of schedule the advertiser should buy to get the results expected. Too often salespeople fail to

ask for what they need for fear the client will balk. Thus they settle for what they can get without much resistance. This, in fact, may be doing the advertiser a disservice, since the buy that the salesperson settles for may not fulfill declared objectives. "It takes a little courage to persist until you get what you think will do the job. There is the temptation just to take what the client hands you and run, but that technique usually backfires when the client doesn't get what he expected. As a radio sales rep, you should know how best to sell the medium. Don't be apologetic or easily compromised. Sell the medium the way it should be sold. Write enough of an order to get the job done," says Sales Manager Ron Piro.

Inflated claims and unrealistic promises should never be a part of a sales presentation. Avoid "If you buy spots on my station, you'll have to hire additional salespeople to handle the huge crowds." Salespeople must be honest in their projections and in what a client may expect from the spot schedule he purchases. "You will notice a gradual increase in store traffic over the next few weeks as the audience is exposed to your commercial over WXXX" is the better approach. Unfulfilled promises ruin any chances of future buys. Too often salespeople, caught up in the enthusiasm of the pitch, make claims that cannot be achieved. Radio is a phenomenally effective advertising medium. This is a proven fact. Those who have successfully used the medium can attest to the importance of placing an adequate order. "An advertiser has to buy a decent schedule to get strong results. Frequency is essential in radio," notes Piro. A radio sales axiom says it best: "The more spots aired, the more impressions made, and the more impressions made, the more impressed the client."

PROSPECTING AND LIST BUILDING

When a salesperson is hired by a radio station, he or she is customarily pro-

Ralph Guild

On the Future Market for Radio

The radio industry has undergone radical but positive change in recent years and will continue to experience significant evolutionary changes between now and the end of the decade. Ironically, these changes are, in many ways, the result of a constant that we have witnessed throughout the history of the medium. Radio

FIGURE 4.14

consistently has demonstrated a tremendous ability to adapt to a variety of market conditions, whether they be economic or competitive in nature. It is this flexibility, a byproduct of intelligent industry leadership, that has fueled radio's growth in the past and will continue to elevate the industry to new heights in the years to come.

During the past decade or so, we have witnessed a number of significant developments in radio. It is important to note these changes so that we may better understand and appreciate what the future holds. Some of the significant developments of recent years include the following:

The consolidation of the radio industry. The atmosphere of business deregulation that was prevalent in the United States in the late 70s and early 80s caused The Interep Radio Store to hold a week-long think tank session to predict and chart the future of our business. We identified an opportunity to start new rep companies and acquire some existing firms. In so doing, we ultimately demonstrated to the radio industry and the government that one organization can sell advertising on multiple radio stations in the same city, and that such an arrangement could be a positive force.

Years later and continuing today, we witnessed the rise of duopolies, where one radio group owns two stations in the same market. Soon there will be an unlimited ownership in most cities and the Interep model of team selling will be the norm in station management. The successful advertising sales consolidation in the 1980s and early 90s—which proved that *big* could be synonymous with *good*—has paved the way for the station consolidation that we are seeing today.

The consolidation of advertising agencies. This has had a strong impact on the radio industry. Agencies tend to have smaller media departments than they did before, and there are fewer major agencies because of mergers. As a result, fewer salespeople are required and radio is able to strategically redeploy its personnel to more effectively serve the advertising community.

The stunning rise of new technology. This is related to the trend cited above, because a significant result of re-deployment of personnel has been the ability to provide advertisers and agencies with greater support and more timely services. And it is the availability of new and powerful technology that serves as the tools by which we can harvest much of this support, particularly in the areas of research and electronic transfer of contracts and invoices between buyer and seller.

The introduction of *Radio 2000* in 1990. The major radio new-business initiative, launched by Interep, caused the entire industry to focus on the importance of bringing new advertisers into the medium. To date, more than $100 million has been generated by the *Radio 2000* effort, and it has helped radio to think "out of the box" in terms of identifying possible clients and ways to attract them. The industry is more centrally focused and radio broad-casters see the wisdom of selling the medium.

What does the future hold for radio? What innovations and changes can we expect to see by the next millennium? A few predictions:

The FCC will approve satellite-to-receiver transmission. This will create a new level of competition for listeners and national advertising dollars. The national network radio programmers will be most effected, and it is likely they will respond by getting into the satellite transmission business in much the same way that broadcast television networks have gotten into cable (satellite) as a means of protecting their franchise.

Advertisers will take greater control of media decisions, wresting much of the process away from their agencies. The new focus will be on cost-per-customer, not cost-per-point. Media, including radio, will need to become highly accountable and result oriented.

Most of the stations in the top fifty markets will be owned by a relative handful of major companies. Many of today's small, mid-sized, and, in some cases, even large station groups won't be around by decade's end. Many of the industry leaders of 2000 will be names not even on the scene today. They will be unique hybrids of broadcasters/marketers/financiers. Pure broadcasters will not be able to compete. The survivors will be companies that are market-driven organizations with broadcasting knowledge and strong financial expertise.

Continued technological advancements will lead to even greater interactivity between radio and its listeners. Radio was the original interactive medium, but new technology will make radio more interactive than ever before. People will be able to access radio stations through home computers and, for that matter, will be able to access particular programming selections from their favorite stations. Let's call it "Radio on Demand"—if listeners want to hear a particular music or talk show that aired recently on their local station, they will be able to do so through their computer terminal at any time of day or night.

New forms of interactivity will be seen in other ways. For example, listeners will be able to "save" special retail discount information on credit card–like "smart-cards" that can be swiped at retail tactics for instant savings. The stored information will be generated by special radios in conjunction with particular commercials, creating a new level of measurability for the effectiveness of radio advertising. This is no pie-in-the-sky dream. Interep has been participating in a test of a technology called CouponRadio, which does exactly what is described above.

Changes are in the wind for the radio industry. The one certainty is that the changes will be in response to, and anticipation of, the needs and interests of listeners and advertisers. Radio's time-tested ability to move forward intelligently and aggressively will be reinforced once again as the medium accelerates onto the fast lane of the information superhighway of the future.

---

vided with a list of accounts to which airtime may be sold. For an inexperienced salesperson, this list may consist of essentially inactive or dormant accounts, that is, businesses that either have been on the air in the past or those that have never purchased airtime on the station. The new sales rep is expected to breathe life into the list by selling spot schedules to those accounts listed, as well as by adding to the list by bringing in new business. This is called *list building*, and it is the primary challenge facing the new account executive.

A more active list, one that generates commissions, will be given to the more experienced radio salesperson. A salesperson may be persuaded to leave one station in favor of another based upon the contents of a list, which may include large accounts and prominent advertising agencies. Lists held by a station's top billers invariably contain the most enthusiastic radio users. Salespeople cultivate their lists as a farmer does his fields. The more the account list yields, the more commissions in the salesperson's pocket.

New accounts are added to a sales rep's list in several ways. Once the status of the list's existing accounts is determined, which is accomplished through a series of in-person calls and presentations, a salesperson must begin prospecting for additional business. Area newspapers are a common source. When a salesperson finds an account that he wishes to add to his list, the account must be "declared." This involves consulting the sales manager for approval to add the account to the salesperson's existing list. In some cases the account declared may already belong to another salesperson. If it is an "open" account, the individual who comes forward first is usually allowed to add it to his list.

PRESENTING SPONSOR (ONE AVAILABLE)
MIX 98.5 FALL FEST '95

Exclusive Package Includes:

☐ Inclusion in a minimum of 150 promotional
   announcements to air on MIX 98.5 3-1/2 weeks prior
   to the festival.
       VALUE: $75,000

☐ Inclusion in MIX 98.5 live broadcast during Fall Fest.
   Sixteen (16) :10 commercials (one each hour) over
   two days.   **VALUE: $3,200**

☐ Inclusion in a minimum of 75 promotional announcements
   to air on Eagle 93.7/WRKO/WEEI two weeks prior to Fall Fest.
       VALUE: $30,000

☐ Bank of forty spots to air on MIX 98.5 from October 1-
   October 31, 1995.
       VALUE: $16,000

☐ Corporate logo on Fall Fest '95 signage at main stage,
   entry points and VIP tent.
       VALUE: $5,000

☐ 30 reserved seats for viewing of main stage entertainment
   on Saturday and Sunday (can be used exclusively by sponsor
   for trade).   **VALUE: $750**

116 Huntington Avenue, Boston, MA 02116 (617) 236-6800 FAX (617) 236-6832

**FIGURE 4.15
Stations sell
sponsored
programs, often at
a premium rate.
Courtesy WHKL.**

Other sources for new accounts include the *Yellow Pages*, television stations, and competing radio outlets, and today, station salespeople are also tapping the Internet and using e-mail to enhance their search for clients. In the first case, the *Yellow Pages*, every business in the area is listed in this directory, and many have display ads that provide useful information. Local television stations are viewed with an eye toward its advertisers. Television can be an expensive proposition, even in smaller markets, and businesses that spend money on it may find radio's rates more palatable. On the other hand, if a business can afford to buy television, it often can afford to embellish its advertising campaign with radio

spots. Many advertisers place money in several media—newspaper, radio, television—simultaneously. This is called a *mixed media* buy and is a proven advertising formula for the obvious reason that the client is reaching all possible audiences. Finally, accounts currently on other stations constitute good prospects since they obviously already have been sold on the medium.

In the course of an average workday, a salesperson will pass hundreds of businesses, some of which may have just opened their doors, or are about to do so. Sales reps must keep their eyes open and be prepared to make an impromptu call. The old saying "The early bird gets the worm" is particularly relevant in radio sales. The first account executive into a newly launched business often is the one who gets the buy.

A list containing dozens of accounts does not necessarily assure a good income. If those businesses listed are small spenders or inactive, little in the way of commissions will be generated and billing will be low. The objective of list building is not merely to increase the number of accounts, but rather to raise the level of commissions it produces. In other words, a list that contains thirty accounts, of which twenty-two are active, is preferable to one with fifty accounts containing only twelve that are doing business with the station. A salesperson does not get points for having a lot of names on his list.

It is the sales manager's prerogative to shift an account from one salesperson's list to another's if he believes the account is being neglected or handled incorrectly. At the same time, certain in-house accounts, those handled by the sales manager, may be added to a sales rep's list as a reward for performing well. A salesperson's account list also may be pared down if the sales manager concludes that it is disproportional with the others at the station. The attempt to more equitably distribute the wealth may cause a brouhaha with the account person whose list is being trimmed. The sales manager attempting this feat may find himself losing a top biller; thus, he must consider the ramifications of such a move and proceed accordingly. This may even mean letting things remain as they are. The top biller often is responsible for as much as 30 to 40 percent of the station's earnings.

PLANNING THE SALES DAY

A radio salesperson makes between seventy-five and one hundred in-person calls a week, or on the average of fifteen to twenty each day. This requires careful planning and organization. Ron Piro advises preparing a day's itinerary the night before. "There's nothing worse than facing the day without an idea of where to go. A salesperson can spare himself that dreaded sensation and a lot of lost time by preparing a complete schedule of calls the night before."

When preparing a daily call sheet a salesperson, especially one whose station covers a vast area, attempts to centralize, as much as possible, the businesses to be contacted. Time, energy, and gas are needlessly expended through poor planning. A sales rep who is traveling ten miles between each presentation can only get to half as many clients as the person with a consolidated call sheet. Of course, there are days when a salesperson must spend more time traveling. Not every day can be ideally plotted. It may be necessary to make a call in one part of the city at 9 a.m. and be in another part at 10 a.m. A salesperson must be where he or she feels the buys are going to be made. "Go first to those businesses likeliest to buy. The tone of the day will be sweetened by an early sale," contends Piro.

Sales managers advise their reps to list more prospects than they expect to contact. In so doing, they are not likely to run out of places to go should those prospects they had planned to see be unavailable. "You have to make the calls to make the sales. The more calls you make, the more the odds favor a sale," points out Gene Etheridge.

Itineraries should be adhered to

Connecticut's Best Oldies

KOOL 96.7 SPONSORSHIP OPPORTUNITIES

KOOL Information Update Reports

Every weekday morning at, 6:00am, 6:30am, 7:00am, 7:30am, 8:00am, and 8:30am, during the KOOL Morning Show KOOL 96.7 listeners can depend on the news and information they need to start the day.

Traffic Updates

18 times every weekday morning between 6:00am and 9:00am, KOOL 96.7 provides accurate, traffic updates. KOOL 96.7 also provides 9 traffic updates each weekday afternoon between 4:15pm and 6:30pm. Including Fairfield County's ONLY On-Air FM traffic plane.

Morning Programs and Contests

Each weekday morning at 6:10am we play the "TV Time Trivia" game where callers are asked to identify a "snippet of a popular TV show theme song. At 6:40 am it's "Beatle Juice" two from the Fab Four to start your day. At 7:10am we ask "Wha'da ya know?" a brain buster trivia question, and at 8:10am it's the Blockbuster Movie clue game where callers identify a movie from a sound clip of the video.

Mid-Day Programs and Contests

At 10:35 Kool Joe Ray plays "My Three Songs" a block of songs requested by a listener.

The KOOL KAFE

Each weekday from Noon to 1:00pm, Kool Joe Ray hosts the KOOL KAFE, a listener request program. Once each month the KOOL Kafe "hits the road" and broadcasts from Noon to 2:00pm from an area restaurant. One lucky office wins lunch for ten and joins us.

regardless of whether a sale is made early in the day, says WNRI's John Gregory. "You can't pack it in at ten in the morning because you've closed an account. A salesperson who is easily satisfied is one who will never make much money. You must stay true to your day's game plan and follow through. No all-day coffee klatch at the local Ho-Jo's or movie matinee because you nailed an order after two calls."

The telephone is one of the salesperson's best tools. While it is true that a client cannot sign a contract over the phone, much time and energy can be saved through its effective use. Appointments can be made and a client can be qualified via the telephone. That is to say, a salesperson can ascertain

FIGURE 4.15
continued

<u>Afternoon   Programs   and   Contests</u>

Each weekday afternoon at 4:35pm J.C. Haze presents the KOOL Top Ten Trivia hits game, and at 6:00pm J.C. plays the "Top Six at Six" a "themed"block of six songs.

<u>Dick   Bartley's   American   Gold</u>

Broadcast Sundays from 10:00am - 2:00pm. Syndicated radio host Dick Bartley hosts America's #1 oldies countdown program.

<u>Dick Clark's   "Rock, Roll and Remember"</u>

Broadcast Sunday mornings from 6:00am - 10:00am. Dick Clark takes listeners on a nostalgic trip through the music, times and events that shaped the early days of Rock-N-Roll.

SEASONAL OPPORTUNITIES

<u>KOOL   Storm   Center</u>

Activated when the weather is bad. Sponsors receive 2 mentions per hour during the Storm Center.

<u>KOOL   Ski   Reports</u>

Broadcast from mid-November through March, Wednesday thru Saturday.

<u>KOOL   Beach   and   Boating   Report</u>

Broadcast from Memorial Day through Labor Day.

<u>KOOL   Oldies   Band</u>

Makes appearances at area beaches, office complexes and restaurants

Ask your KOOL 96.7 Account Executive for more information about any of these Sponsorship Opportunities!

FIGURE 4.15
continued

when the decision maker will be available. "Rather than travel twenty miles without knowing if the person who has the authority to make a buy will be around, take a couple of minutes and make a phone call. As they say, 'Time is money.' In the time spent finding out that the store manager or owner is not on the premises when you get there, other, more productive calls can be made," says Charles Friedman.

If a client is not available when the salesperson appears, a call-back should be arranged for either later the same day or soon thereafter. The prospective advertiser should never be forgotten or relegated to a call three months hence. If the sales rep is able to rearrange his schedule to accommo-

date a return visit the same day, given that the person to see is available, then he should do so. However, it is futile to make a presentation to someone who cannot give full attention. The sales rep who arrives at a business only to find the decision maker overwhelmed by distractions is wise to ask for another appointment. In fact, the client will perceive this as an act of kindness and consideration. Timing is important.

A record of each call should be kept for follow-up purposes. When calling on a myriad of accounts, it is easy to lose track of what transpired during a particular call. Maintaining a record of a call requires little more than a brief notation after it is made. Notes may then be periodically reviewed to help determine what action should be taken on the account. Follow-ups are crucial. There is nothing more embarrassing and disheartening than to discover a client, who was pitched and then forgotten, advertising on another station. Sales managers usually require that salespeople turn in copies of their call sheets on a daily or weekly basis for review purposes.

SELLING WITH AND WITHOUT NUMBERS

Not all stations can claim to be number one or two in the ratings. In fact, not all stations appear in any formal ratings survey. Very small markets are not visited by Arbitron or Birch for the simple reason that there may only be one station broadcasting in the area. An outlet in a nonsurvey area relies on its good reputation in the community to attract advertisers. In small markets, salespeople do not work out of a ratings book, and clients are not concerned with cumes and shares. In the truest sense of the word, an account person must sell the station. Local businesses often account for more than 95 percent of a small market station's revenue. Thus, the stronger the ties with the community, the better. Broadcasters in rural markets must foster an image of good citizenship in order to make a living.

Civic-mindedness is not as marketable a commodity in the larger markets as are ratings points. In the sophisticated multistation urban market, the ratings book is the bible. A station without numbers in the highly competitive environment finds the task of earning an income a difficult one, although there are numerous examples of low-rated stations that do very well. However, "no-numbers" pretty much puts a metro area station out of the running for agency business. Agencies almost invariably "buy by the book." A station without numbers "works the street," to use the popular phrase, focusing its sales efforts on direct business.

An obvious difference in approaches exists between selling the station with ratings and the one without. In the first case, a station centers its entire presentation around its high ratings. "According to the latest Arbitron, WXXX-FM is number one with adults 24 to 39." Never out of the conversation for very long are the station's numbers, and at advertising agencies the station's standing speaks for itself. "We'll buy WXXX because the book shows that they have the largest audience in the demos we're after."

The station without ratings numbers sells itself on a more personal level, perhaps focusing on its unique features and special blend of music and personalities, and so forth. In an effort to attract advertisers, nonrated outlets often develop programs with a targeted retail market in mind; for example, a home "how-to" show designed to interest hardware and interior decor stores, or a cooking feature aimed at food and appliance stores.

The salesperson working for the station with the cherished "good book" must be especially adept at talking numbers, since they are the key subject of the presentation in most situations. "Selling a top-rated metro station requires more than a pedestrian knowledge of numbers, especially when dealing with agencies. In big cities, retailers have plenty of book savvy, too," contends Piro.

Selling without numbers demands its own unique set of skills, notes WNRI's Gregory. "There are really two different types of radio selling—with numbers and without. In the former instance, you'd better know your math, whereas in the latter, you've got to be really effective at molding your station to suit the desires of the individual advertiser. Without the numbers to speak for you, you have to do all the selling yourself. Flexibility and ingenuity are the keys to the sale."

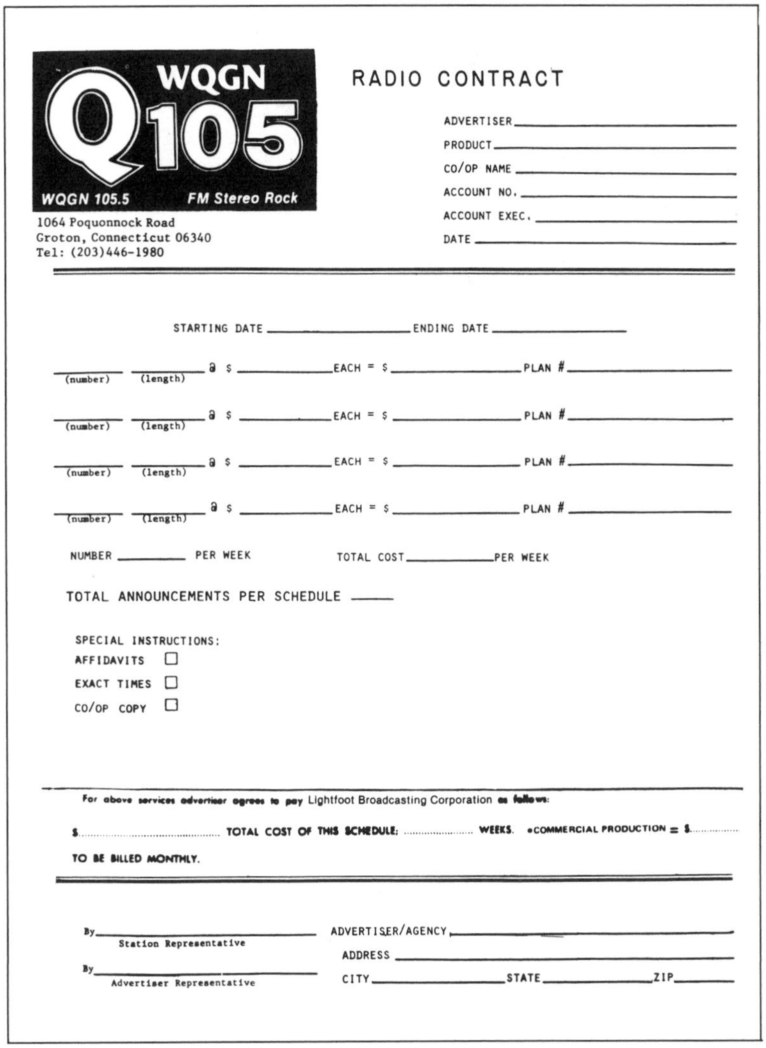

WQGN
Q105
WQGN 105.5 FM Stereo Rock
1064 Poquonnock Road
Groton, Connecticut 06340
Tel: (203)446-1980

RADIO CONTRACT

ADVERTISER ____
PRODUCT ____
CO/OP NAME ____
ACCOUNT NO. ____
ACCOUNT EXEC. ____
DATE ____

STARTING DATE ____ ENDING DATE ____

____ (number) ____ (length) @ $ ____ EACH = $ ____ PLAN # ____

____ (number) ____ (length) @ $ ____ EACH = $ ____ PLAN # ____

____ (number) ____ (length) @ $ ____ EACH = $ ____ PLAN # ____

____ (number) ____ (length) @ $ ____ EACH = $ ____ PLAN # ____

NUMBER ____ PER WEEK TOTAL COST ____ PER WEEK

TOTAL ANNOUNCEMENTS PER SCHEDULE ____

SPECIAL INSTRUCTIONS:
AFFIDAVITS ☐
EXACT TIMES ☐
CO/OP COPY ☐

For above services advertiser agrees to pay Lightfoot Broadcasting Corporation as follows:

$.......... TOTAL COST OF THIS SCHEDULE; .... WEEKS. •COMMERCIAL PRODUCTION = $..........

TO BE BILLED MONTHLY.

By ____ Station Representative
By ____ Advertiser Representative

ADVERTISER/AGENCY ____
ADDRESS ____
CITY ____ STATE ____ ZIP ____

FIGURE 4.16 Station contract for a spot schedule. Courtesy WQGN.

ADVERTISING AGENCIES

Advertising agencies came into existence more than a century ago and have played an integral role in broadcasting since its inception. During

FIGURE 4.17
Sponsors are
convinced to target
their particular
clientele through
the use of radio
advertising.
Courtesy RAB.

YOU have to maximize your advertising investment by reaching your best prospects...

RADIO reaches the big spenders!

Upscale individuals with above-average incomes and affluent lifestyles spend more of their disposable income on "big-ticket" items—and spend more of their time with Radio.

Household Income $50,000+

	Weekly Reach	Average Daily Time Spent Listening
Adults 18+	96.9%	2:53
Men 18+	97.6%	2:57
Women 18+	96.1%	2:47

Want professionals and managers? Radio delivers! These decision-makers have money to spend ... and Radio is the best marketing medium to reach these influential men and women.

Professional/Managerial, $50,000+Household Income

	Weekly Reach	Average Daily Time Spent Listening
Adults 18+	97.5%	2:37
Men 18+	98.1%	2:44
Women 18+	98.0%	2:30

Source: RADAR ® 50, Fall 1994, © Copyright Statistical Research, Inc.

radio's famed heyday, advertising agencies were omnipotent. Not only did they handle the advertising budgets of some of the nation's largest businesses, but they also provided the networks with fully produced programs. The programs were designed by the agencies for the specific satisfaction of their clients. If the networks and certain independent stations wanted a company's business, they had little choice but to air the agency's program. This practice in the 1920s and 1930s gave ad agencies unprecedented power. At one point, advertising agencies were the biggest supplier of network radio programming. By the 1940s, agencies were forced to abandon their direct programming involvement, and the industry was left to its own devices, or almost. Agencies continued to influence the content of what was aired. Their presence continues to be felt today, but not to the extent that it was prior to the advent of television.

Agencies annually account for hun-

dreds of millions in radio ad dollars. The long, and at times turbulent, marriage of radio and advertising agencies was, and continues to be, based on the need of national companies to convey their messages on the local level and the need of the local broadcaster for national business. It is a two-way street.

Today nearly two thousand advertising agencies use the radio medium. They range in size from mammoth to minute. Agencies such as Young and Rubicam, J. Walter Thompson, Dancer Fitzgerald Sample, and Leo Burnett bill in the hundreds of millions annually and employ hundreds. More typical, however, are the agencies scattered throughout the country that bill between one-half and two-and-a-half million dollars each year and employ anywhere from a half dozen to twenty people. Agencies come in all shapes and sizes and provide various services, depending on their scope and dimensions.

The process of getting national business onto a local station is an involved one. The major agencies must compete against dozens of others to win the right to handle the advertising of large companies. This usually involves elaborate presentations and substantial investments by agencies. When and if the account is secured, the agency must then prepare the materials—audio, video, print—for the campaign and see to it that the advertiser's money is spent in the most effective way possible. Little is done without extensive marketing research and planning. The agency's media buyer oversees the placement of dollars in the various media. Media buyers at national agencies deal with station and network reps rather than directly with the stations themselves. It would be impossible for an agency placing a buy on four hundred stations to personally transact with each.

There basically are three types of agencies: *full-service agencies*, which provide clients with a complete range of services, including research, marketing, and production; *modular agencies*, which provide specific services to advertisers; and *in-house agencies*, which handle the advertising needs of their own business.

```
===============                   ------------------------
== BreakOut ==                    MULTI-STATION CPM GRID
===============                   ------------------------

          < PROJECTED RATES BASED ON: ADULTS 25-54, MON-FRI   6A-7P >

                 [ DALLAS/FT WORTH ARB METRO: SPRNG      ]


            AQH    --------------------------- CPMs ----------------------
  # STATION  PER    3.50    4.00    4.50    5.00    5.50    6.00    6.50
  ---------  -----  ------  ------  ------  ------  ------  ------  ------


                   ------------------ RESULTING RATES ------------------

  1 KVIL-FM  39,400 137.90  157.60  177.30  197.00  216.70  236.40  256.10
  2 KSCS-FM  26,900  94.15  107.60  121.05  134.50  147.95  161.40  174.85
  3 KPLX-FM  25,700  89.95  102.80  115.65  128.50  141.35  154.20  167.05
  4 KRLD-AM  21,300  74.55   85.20   95.85  106.50  117.15  127.80  138.45
  5 WBAP-AM  19,600  68.60   78.40   88.20   98.00  107.80  117.60  127.40
  6 KMEZ-FM  16,900  59.15   67.60   76.05   84.50   92.95  101.40  109.85
  7 KKDA-FM  13,000  45.50   52.00   58.50   65.00   71.50   78.00   84.50
  8 KMGC-FM  12,700  44.45   50.80   57.15   63.50   69.85   76.20   82.55
  9 KOAX-FM  12,000  42.00   48.00   54.00   60.00   66.00   72.00   78.00
 10 KZEW-FM  11,800  41.30   47.20   53.10   59.00   64.90   70.80   76.70
 11 KNOK-FM  10,300  36.05   41.20   46.35   51.50   56.65   61.80   66.95
 12 KEGL-FM   9,800  34.30   39.20   44.10   49.00   53.90   58.80   63.70
 13 KLVU-FM   8,300  29.05   33.20   37.35   41.50   45.65   49.80   53.95
 14 KTXQ-FM   7,500  26.25   30.00   33.75   37.50   41.25   45.00   48.75
 15 KAAM-AM   6,900  24.15   27.60   31.05   34.50   37.95   41.40   44.85
 16 KAFM-FM   6,800  23.80   27.20   30.60   34.00   37.40   40.80   44.20
 17 KKDA-AM   6,400  22.40   25.60   28.80   32.00   35.20   38.40   41.60
 18 WFAA-AM   6,400  22.40   25.60   28.80   32.00   35.20   38.40   41.60
 19 KESS-FM   5,700  19.95   22.80   25.65   28.50   31.35   34.20   37.05
 20 KIXK-FM   5,200  18.20   20.80   23.40   26.00   28.60   31.20   33.80


NOTE: Population for "ADULTS 25-54" is  1,337,900

[ BreakOut Report  copyright 1983  Jefferson-Pilot Data Systems  Charlotte NC ]
```

> IF YOU KNOW YOUR CPM AND WANT TO SEE EQUIVALENT RATES FOR THE REST OF THE
> MARKET, THIS REPORT WILL SHOW THEM TO YOU. SOME PEOPLE CALL THIS A "REVERSE
> CPM". YOU PICK THE CPM LEVELS AND INTERVALS AND ANY OR ALL STATIONS AND THIS
> REPORT WILL DO THE REST. THIS WILL HELP YOU JUSTIFY YOUR RATES WHENEVER YOU
> NEED TO. YOU MAY ALSO DISPLAY RATES BASED ON COST-PER-POINT VALUES.

FIGURE 4.18 Computerized breakouts show a client how well a station performs. Computers have become an integral part of radio sales. Courtesy Jefferson Pilot Data Systems.

The standard commission that an agency receives for its service is 15 percent on billing. For example, if an agency places $100,000 on radio, it earns $15,000 for its efforts. Agencies often charge clients additional fees to cover production costs, and some agencies receive a retainer from clients.

The business generated by agencies constitutes an important percentage of radio's revenues, especially for medium and large market stations. However, compared to other media, such as television for example, radio's allocation is diminutive. The nation's top three agencies invest over 80 percent of their broadcast budgets in television. Nonetheless, hundreds of millions of dollars are channeled into radio by agencies that recognize the effectiveness of the medium.

REP COMPANIES

Rep companies are the industry's middlemen. Rep companies are given the task of convincing national agency media buyers to place money on the

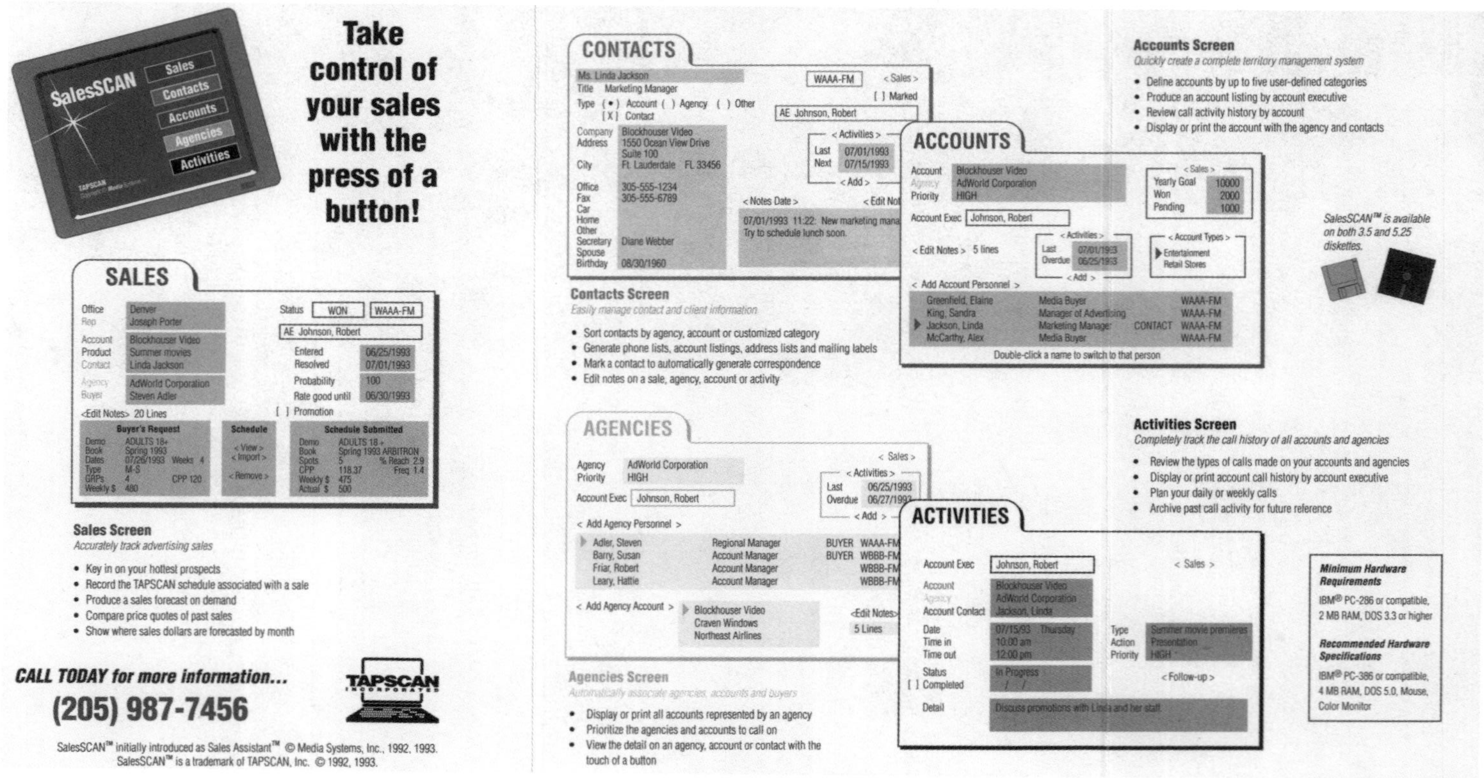

FIGURE 4.19
The computer is an
integral part of most
sales departments.
Courtesy TAPSCAN.

stations they represent. Without their existence, radio stations would have to find a way to reach the myriad of agencies on their own—an impossible feat.

With few exceptions, radio outlets contract the services of a station rep company. Even the smallest station wants to be included in buys on the national level. The rep company basically is an extension of a station's sales department. The rep and the station's sales manager work together closely. Information about a station and its market are crucial to the rep. The burden of keeping the rep fully aware of what is happening back at the station rests on the sales manager's shoulders. Since a rep company based in New York or Chicago would have no way of knowing that its client-station in Arkansas has decided to carry the local college's basketball games, it is the station's responsibility to make the information available. A rep cannot sell what it does not know exists. Of course, a good rep will keep in contact with a station on a regular basis simply to keep up on station changes.

There are far fewer radio station reps than there are ad agencies. Approximately 125 reps handle the nine thousand plus commercial stations around the country. Major rep firms, such as Katz, Eastman, Blair, and Torbet, pitch agencies on behalf of hundreds of client stations. The large and very successful reps often refuse to act as the envoy for small market stations because of their lack of earning potential. A rep company typically receives a commission of between 5 and 12 percent on the spot buys made by agencies, and since the national advertising money usually is directed first to the medium and large markets, the bigger commissions are not to be made from handling small market outlets. Many rep companies specialize in small market stations, however.

While a rep company may work the agencies on behalf of numerous stations, it will seldom handle two radio outlets in the same market. Doing so could result in a rep company being placed in the untenable position of competing with itself for a buy, thus creating an obvious conflict of interest.

In the past couple of years, some rep firms have taken on multiple clients in the same market when they do not target identical audience demographics.

The majority of station reps provide additional services. In recent years many have expanded into the areas of programming and management consultancy, while almost all offer clients audience research data, as well as aid in developing station promotions and designing sales materials such as rate cards.

CO-OP SALES

It is estimated that over $600 million in radio revenue comes from co-op advertising—no small piece of change, indeed. As a consequence of a negative economy, however, the co-op market went a bit flat in the first half of the 1990s, says WTIC's Peter Drew. "Actually there is less co-op nowadays.

Media Plan

PLAN-A

1400 WSTC-AM

The   schedule:

- 6 sixty-second commercials each day @ $350

Delivers 12,834 gross impressions

5:30am - 9:00am	2 commercials
9:00am - 12noon	2 commercials
12noon - 5:00pm	2 commercial

KOOL 96.7-FM

The   schedule:

- 6 sixty-second commercials each day @ $425

Delivers 10,400 gross impressions

5:30am - 10:00am	2 commercials
10:00am - 12noon	2 commercials
3:00pm - 7:00pm	2 commercials

Combo Plan $700.00

Combo plan delivers 23,234 gross impressions

FIGURE 4.20 Stations offer clients a variety of different plans. Courtesy WSTC.

Media Plan

PLAN-B

1400 WSTC-AM

• 10 sixty-second commercials each day @ $550

Delivers 20,011 gross impressions

The schedule:

5:30am - 9:00am	3 commercials
9:00am - 12noon	3 commercials
12noon - 5:00pm	4 commercials

KOOL 96.7-FM

• 10 sixty-second commercials each day @ $700

Delivers 16,800 gross impressions

The schedule:

5:30am - 10:00am	4 commercials
10:00am - 12noon	2 commercials
12noon - 7:00pm	4 commercials

Combo Plan $1,200.00

Combo plan delivers 36,811 gross impressions

Clients may cap the amount of expenditure

FIGURE 4.20
continued

Things are tight on every level, of course. It is more difficult today to qualify and collect. This is the result of higher accruals and inventory requirements. Still, co-op money is worth pursuing. A little more effort is required to acquire it." Co-op advertising involves the cooperation of three parties: the retailer whose business is being promoted, the manufacturer whose product is being promoted, and the medium used for the promotion. In other words, a retailer and manufacturer get together to share advertising expenses. For example, Smith's Sporting Goods is informed by the Converse Running Shoes representative that the company will match, dollar for dollar up to $5000, the money that the retailer invests in radio advertising. The only stipulation of the deal is that Converse be promoted in the commercials on which the money is spent. This means that no competitive product can be mentioned. Converse demands exclusivity for its contribution.

Manufacturers of practically every conceivable type of product, from lawn mowers to mobile homes, establish co-op advertising budgets. A radio salesperson can use co-op to great advantage. First the station account executive must determine the extent of co-op subsidy a client is entitled to receive. Most of the time the retailer knows the answer to this. Frequently, however, retailers do not take full advantage of the co-op funds that manufacturers make available. In some instances, retailers are not aware that a particular manufacturer will share radio advertising expenses. Many potential advertisers have been motivated to go on the air after discovering the existence of co-op dollars. Midsized retailers account for the biggest chunk of the industry's co-op revenues. However, even the smallest retailer likely is eligible for some subsidy, and a salesperson can make this fact known for everyone's mutual advantage.

The sales manager generally directs a station's co-op efforts. Large stations often employ a full-time co-op specialist. The individual responsible for stimulating co-op revenue will survey retail trade journals for pertinent information about available dollars. Retail associations also are a good source of information, since they generally possess manufacturer co-op advertising lists. The importance of taking advantage of co-op opportunities cannot be overstressed. Some stations, especially metro market outlets, earn hundreds of thousands of dollars in additional ad revenue through their co-op efforts.

From the retailer's perspective, co-op advertising is not always a great bargain. This usually stems from copy constraints imposed by certain manufacturers, which give the retailer a ten-second tagout in a thirty- or sixty-second commercial. Obviously, this does not please the retailer who has split the cost of advertising fifty-fifty. In recent years, this type of copy domination by the manufacturer has decreased somewhat, and a more equitable approach, whereby both parties share evenly the exposure and the ex-pense, is more commonplace.

Co-op also is appealing to radio stations since they do not have to modify their billing practices to accommodate the third party. Stations simply bill the retailer and provide an affidavit attesting to the time commercials aired. The retailer, in turn, bills the manufacturer for its share of the airtime. For its part, the manufacturer requires receipt of an affidavit before making payment. In certain cases, the station is asked to mail affidavits directly to the manufacturer. Some manufacturers stipulate that bills be sent to audit houses, which inspect the materials before authorizing payment.

TRADE-OUTS

Stations commonly exchange airtime for goods, although top-rated outlets, whose time is sold at a premium, are less likely to swap spots for anything other than cash. Rather than pay for needed items, such as office supplies and furnishings, studio equipment, meals for clients and listeners, new cars, and so forth, a station may choose to strike a deal with merchants in which airtime is traded for merchandise. There are advertisers who only use radio on a trade basis. A station may start out in an exclusively trade relationship with a client in the hope of eventually converting him to cash. Split contracts also are written when a client agrees to provide both money and merchandise. For example, WXXX-FM needs two new office desks. The total cost of the desks is $800. An agreement is made whereby the client receives a $1400 ROS spot schedule and $600 cash in exchange for the desks. Trade-outs are not always this equitable. Stations often provide trade clients with airtime worth two or three times the merchandise value in order to get what is needed. Thus, the saying "need inspires deals."

```
          KXXX-FM CAll Sheet

AE:  Lindy Bergen                    Date:  March 2, 1995


Bela's Jewelry
Ms. Bela not interested now.  Wants to do something in the
Summer.  Will return in late April with a plan.

Libid Hardwares
Sold 20 BTA 60's for week of March 17.  Will do more in April.

S and M Travel
Still not ready to budge.  Sold on newspapers.  Will return with
a spec tape and special plan next week.  Contact: Susanne Riette.

Sum Yung Gi Gardens
Interested in drive-time.  On KZZZ three years now.  Not happy
with results though.  Likes our format.  Is warming.  Will
follow-up end of week for the close.  Contact: Mr. Yung Gi.

Bedroomarama
Business slow.  Won't move on proposal.  Maybe in May.  Checking
for co-op bucks.  Call back appointment April 23.  Contact: Mr.
Hugest.
```

FIGURE 4.21
Excerpts from a salesperson's call sheet.

Many sales managers also feel that it makes good business sense to write radio trade contracts to fill available and unsold airtime, rather than let it pass unused. Once airtime is gone, it cannot be retrieved, and yesterday's unfilled availability is a lost opportunity.

CHAPTER HIGHLIGHTS

1. Selling commercials keeps radio stations on the air. Between 1920 and today, advertising revenues and forms reflected the ebb and flow of radio's popularity. Today, advertising dollars are selectively spent on spots aired during times of the day and on stations that attract the type of audience the advertiser wants to reach.

2. An effective radio commercial makes a strong and lasting impression on the mind of the listener.

3. A successful account executive needs an understanding of research

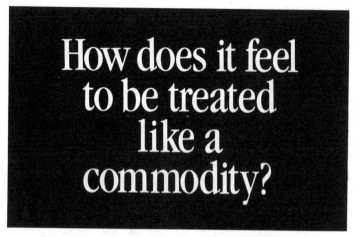

How does it feel to be treated like a commodity?

Not very good, we're sure.

There's no way to avoid this feeling when you're just one of 10, 12 or 20 stations that are being represented by one of the megareps.

So if you're beginning to get that sinking feeling of becoming just a supplier of inventory, we have a way for you to escape the "numbers game."

Move to a place where you don't have to compete for attention.

At Eastman, you're an individual station with a unique personality, not a commodity. We don't blur the competitive differences between stations.

In fact, our people work overtime to understand your individual market, station and audience. They look for ideas and values that contribute to your getting a higher price for your product. In the top 20 markets, where Eastman reps a leading station, the market cost per point has increased 11 percent in two years.

Does your rep really sell your station the way you want it sold. . .or is it just price and ratings? If you're ready to be sold on quality, not just quantity, give Eastman a call at (212) 581-0800.

Number of stations represented.

	Megarep A.	Megarep B.
New York	20	12
Los Angeles	8	17
Chicago	12	10
San Francisco	9	12
Boston	7	11
Washington	10	12

Source: SRDS, July 1, 1988.

Number of stations represented.

	Eastman Radio.
New York	2
Los Angeles	2
Chicago	1
San Francisco	1
Boston	1
Washington	1

EASTMAN RADIO
The alternative to the megarep.

FIGURE 4.22
A rep company's advertisement espousing the virtues of their approach to station representation. Courtesy of Eastman Radio.

methods, marketing, finance; some form of sales experience; and such personal traits as ambition, confidence, honesty, energy, determination, intelligence, and good grooming.

4. Since the 1970s, programming people have made successful job transitions to sales because they have a practical understanding of the product they are selling.

5. Although an increasing number of station managers are being drawn from programming people, a sales background is still preferred.

6. The sales manager, who reports directly to the station's general manager, oversees the account executives, establishes departmental policies, develops sales plans and materials, conceives campaigns and promotions, sets quotas, works closely with the pro-

gram director to develop salable features, and sometimes sells.

7. Rates for selling airtime vary according to listenership and are published on the station's rate card. The card lists terms of payment and commission, nature of copy and due dates, station's approach to discounting, rate protection policy, as well as feature and spot rates.

8. Station listenership varies according to time of day, so rate card daypart classifications range from the highest-costing AAA (typically 6 to 10 a.m. weekdays) to C (usually midnight to 6 a.m.). Fixed-position drivetime spots are usually among the most expensive to purchase.

9. Many stations use a grid structure because it allows for considerable rate flexibility. Grids are inventory-sensitive, and they let a station remain viable when business is slow.

10. For advertisers with limited funds, run-of-station (ROS), best time available (BTA), or total audience plan (TAP) are cost-effective alternatives.

11. Since few accounts are "closed" on the first call, it is used to introduce the station to the client and to determine its needs. Follow-up calls are made to offset reservations and, if necessary, to improve the proposal. Perseverance is essential.

12. Radio sales are drawn from three levels: retail, local, and national. Retail sales are direct sales to advertisers within the station's signal area. Local sales are obtained from advertising agencies representing businesses in the market area. National sales are obtained by the station's rep company from agencies representing national accounts.

13. A fully produced sample commercial (spec tape) is an effective selling tool. It is used to break down client resistance on call-backs, to interest former clients who have not bought time recently, and to encourage clients to increase their schedules.

14. The salesperson should commit the advertiser to sufficient commercials, placed properly, to ensure achieving the advertiser's objectives. Underselling is as self-defeating as overselling.

15. New accounts are added to a salesperson's list by "prospecting": searching newspapers, *Yellow Pages*, television ads, competing radio station ads, and new store openings. Only "open" accounts may be added (those not already "declared" by another salesperson at the same station).

16. Because a salesperson must average fifteen to twenty in-person calls each day, when preparing a daily call sheet it is important to logically sequence and centralize the businesses to be contacted. Also, advance telephone contacts can eliminate much wasted time.

17. E-mail is becoming a useful sales and prospecting tool for account executives in the last half of the 1990s.

18. A salesperson at a station with a high rating has a decided advantage when contacting advertisers. Stations with low or no "numbers" must focus on retail sales (work the street), developing programs and programming to attract targeted clients. Stations in nonsurvey areas must rely on their image of good citizenship and strong community ties.

19. Ad agencies annually supply hundreds of millions of dollars in advertising revenue to stations with good ratings. Media buyers at the agencies deal directly with station and network reps.

20. A station's rep company must convince national agency media buyers to select their station as their advertising outlet for the area. Therefore, the station's sales manager and the rep must work together closely.

21. Co-op advertising involves the sharing of advertising expenses by the retailer of the business being promoted and the manufacturer of the product being promoted.

22. Rather than pay for needed items or to obtain something of value for unsold time, a station may trade (trade-out) advertising airtime with a merchant in exchange for specific merchandise.

APPENDIX: A STATION OWNER CONVEYS HIS SALES PHILOSOPHY TO HIS MANAGER

To: Rich Krezwick
Fr: Jay Williams, Jr.
Re: Sales

The Advertiser's Guide To Using Radio Co-op

Discover how co-operative advertising on Radio builds traffic and sales for your business, without breaking your budget.

Radio Advertising Bureau, Inc. • 304 Park Avenue South, New York, NY 10010
1 (800) 252-RADIO • **FAX** (212) 254-8713
New York, Atlanta, Chicago, Detroit, Houston, Los Angeles

FIGURE 4.23
RAB's co-op material provides pertinent facts. Courtesy RAB.

Although I am not familiar with what happens every day, I do think I have a general idea of the sales philosophy and structure that has gotten us where we are, both positively and negatively. As I see the declining sales as a percentage of goal in a marketplace that is hotter now than at any time in the past four years, I think it's time we continue our discussion on a more specific level. To that end, I thought I would outline what I see from here (to see where you agree/disagree or can supply additional information) and suggest some solutions (as I have certainly seen similar trends many times before). First my suggestions.

1. Let's get Casey out of the station for a few days (your suggestion, actually). I think he needs some perspective, and there's nothing like seeing another problem to see your own better.
2. Let's give Casey a sales call quota. Sales managers learn both real problems (not self-manufactured) on the street, and they learn solu-

tions. More importantly, they're on the front lines with their sales people and can better relate to them and teach them. He needs to be more of a player-coach. Perhaps a minimum of twenty calls a week.

3. Let's do real meetings that teach the "consultant sell." This, in my opinion, is the philosophy that has always been missing (in the past year or so) that (a) makes us look as if we don't care about the client's problems and (b) puts downward pressure on rates. The consultant sell, which I do every day for DMR and frequently for WLKZ, talks about clients needs and revolves around solutions. I never, as a result, and I do mean *never*, get into a rate discussion. Conversely, WXLO is now selling spots and inventory. We're like a beer distributor, competing against other beers. It's totally the wrong way to sell.

4. Let's figure out a way to hire someone to really handle vendor, interactive phones, and other nonspot revenues that you and I both know can be sold at a premium instead of the despicable "value added." Perhaps you should be the sales manager of this special sales branch, starting with one or two people. The two-check philosophy in action. Get us in the direction you and I have talked about—as a marketing group instead of spot salespeople (or really, spot whores as I see it).

5. Let's put a stop to any sales criticism of programming, even if it's warranted, At Fairbanks broadcasting, I learned how little criticisms can become a culture if unchecked. In reality, they are more than critiquing the product they must turn around and sell, they (sales people) are giving themselves an excuse for poor sales, low rates, and giveaways to clients. I was brought up with the philosophy that only the best could sell for station XXXX; if you can't, get the hell out of the way and we'll hire someone who can. It changed the paradigm—and the success of the stations I worked with.

6. Let's hire at least one more, maybe two more, salespeople and specifically target metrowest and southern New Hampshire. I know people like Peter's Auto Sales in Nashua, and other major advertisers, that need business from south and west of Nashua. We can help, but we don't have the horses to develop that business, With not enough sales people competing for the available spots, it also lessens demand and that, in WXLO's "beer distributor" mentality, lowers rates.

As we've talked about much of this, I'm sure you will agree with much of it, But I think we must now act quickly. Fourth-quarter buys and rates are being set; too much later and the die will be cast and we won't be hitting the cash flow numbers you want and the station needs. I offer these as my first steps to changing the structure. (I'm referring to a $4,000/person seminar taught by a consultant to the Motorola management last week. In shorthand, most manage events. Some manage the patterns caused by the events, yet these both yield poor results as they are reactions to past events. Smart managers manage structure.) Our structure, and the philosophy you generated and that I generated in programming, is being changed in sales. This change in structure (because of the coterminous change in philosophy) is changing our patterns (sales curves) and events (individual sales, rates, complaints, bonuses to clients, low-rate packages, etc). That's why I have proposed these solutions versus mere package or rate changes, etc. Let me know what you think.

SUGGESTED FURTHER READING

Aaker, David A., and Myers, John G. *Advertising Management*, 2nd ed. Englewood Cliffs, N.J.: Prentice-Hall, 1982.

Barnouw, Erik. *The Sponsor: Notes on a Modern Potentate*. New York: Oxford University Press, 1978.

Bergendorf, Fred. *Broadcast Advertising*. New York: Hastings House, 1983.

Bouvard, Pierre, and Marx, Steve. *Radio Advertising's Missing Ingredient: The Optimum Effective Scheduling System*. Washington, D.C.: NAB, 1991.

Bovee, Courtland, and Arena, William F. *Contemporary Advertising*. Homewood, Ill.: Irwin, 1982.

Broadcast Marketing Company. *Building Store Traffic with Broadcast Advertising*. San Francisco: Broadcast Marketing Company, 1978.

Burton, Philip Ward, and Sandhusen, Richard. *Cases in Advertising*. Columbus, Ohio: Grid Publishing, 1981.

Culligan, Matthew J. *Getting Back to the Basics of Selling*. New York: Crown, 1981.

Cundiff, Edward W., Still, Richard R., and Govoni, Norman A.P. *Fundamentals of Modern Marketing*, 3rd ed. Englewood Cliffs, N.J.: Prentice-Hall, 1980.

Delmar, Ken. *Winning Moves: The Body Language of Selling*. New York: Warner, 1984.

Dunn, W. Watson, and Barban, Arnold M. *Advertising: Its Role in Modern Marketing*, 4th ed. Hinsdale, Ill.: Dryden Press, 1978.

Gardner, Herbert S., Jr. *The Advertising Agency Business*. Chicago: Crain Books, 1976.

Gilson, Christopher, and Berkman, Harold W. *Advertising Concepts and*

Strategies. New York: Random House, 1980.

Heighton, Elizabeth J., and Cunningham, Don R. *Advertising in the Broadcast and Cable Media*, 2nd ed. Belmont, Calif.: Wadsworth Publishing, 1984.

Herweg, Godfrey W., and Herweg, Ashley Page. *Making More Money Selling Radio Advertising Without Numbers*. Washington: NAB, 1995.

Hoffer, Jay, and McRae, John. *The Complete Broadcast Sales Guide for Stations, Reps, and Ad Agencies*. Blue Ridge Summit, Pa.: Tab Books, 1981.

Johnson, J. Douglas. *Advertising Today*. Chicago: SRA, 1978.

Jugenheimer, Donald W., and Turk, Peter B. *Advertising Media*. Columbus, Ohio: Grid Publishing, 1980.

Keith, Michael C. *Selling Radio Direct*. Boston: Focal Press, 1992.

Kleppner, Otto. *Advertising Procedure*, 7th ed. Englewood Cliffs, N.J.: Prentice-Hall, 1979.

McGee, William L. *Broadcast Co-op, the Untapped Goldmine*. San Francisco: Broadcast Marketing Company, 1975.

Montgomery, Robert Leo. *How to Sell in the 1980's*. Englewood Cliffs, N.J.: Prentice-Hall, 1980.

Murphy, Jonne. *Handbook of Radio Advertising*. Radnor, Pa.: Chilton Books, 1980.

National Association of Broadcasters. *Think Big: Event Marketing for Radio*. Washington: NAB, 1994.

Rhoads, B, Eric., et. al., eds. *Sales and Marketing*. West Palm Beach, FL.: Streamline Press, 1995.

Schultz, Don E. *Essentials of Advertising Strategy*. Chicago: Crain Books, 1981.

Sissors, Jack Z., and Surmanek, Jim. *Advertising Media Planning*, 2nd ed. Chicago: Crain Books, 1982.

Standard Rate and Data Service: Spot Radio. Skokie, Ill.: SRDS, annual.

Warner, Charles, and Buchman, Joseph. *Broadcasting and Cable Selling*, 2nd ed. Belmont, CA.: Wadsworth Publishing, 1993.

Willing, Si. *How to Sell Radio Advertising*. Blue Ridge Summit, Pa.: Tab Books, 1977.

5 NEWS

NEWS FROM THE START

The medium of radio was used to convey news before news of the medium had reached the majority of the general public. Ironically, it was the sinking of the *Titanic* in 1912 and the subsequent rebroadcast of the ship's coded distress message that helped launch a wider awareness and appreciation of the new-fangled gadget called the *wireless telegraph*. It was not until the early 1920s, when the "wireless" had become known as the "radio," that broadcast journalism actually began to evolve.

A historical benchmark in radio news is the broadcast of the Harding-Cox election results in 1920 by stations WWJ in Detroit and KDKA in Pittsburgh, although the first actual newscast is reported to have occurred in California a decade earlier. Despite these early ventures, news programming progressed slowly until the late 1920s. By then, two networks, NBC and CBS, were providing national audiences with certain news and information features.

Until 1932 radio depended on newspapers for its stories. That year newspapers officially perceived the electronic medium as a competitive threat. Fearing a decline in readership, they imposed a blackout. Radio was left to its own resources. The networks put forth substantial efforts to gather news and did very well without the wire services that they had come to rely on. Late in 1934, United Press (UP), International News Services (INS), and Associated Press (AP) agreed to sell their news services to radio, thus ending the boycott. However, by then the medium had demonstrated its ability to fend for itself.

Radio has served as a vital source of news and information throughout the most significant historical events. When the nation was gripped by economic turmoil in the 1930s, the incumbent head-of-state, Franklin D. Roosevelt, demonstrated the tremendous reach of the medium by using it to address the people. The majority of Americans heard and responded to the president's talks.

Radio's status as a news source reached its apex during World War II. On-the-spot reports and interviews, as well as commentaries, brought the war into the nation's living rooms. In contrast to World War I, when the fledgling wireless was exclusively used for military purposes, during World War II radio served as the primary link between those at home and the foreign battlefronts around the globe. News programming during this troubled period matured, while the public adjusted its perception of the medium, casting it in a more austere light. Radio journalism became a more credible profession.

The effects of television on radio news were wide ranging. While the medium in general reeled from the blow dealt it by the *enfant terrible*, in the late 1940s and early 1950s news programming underwent a sort of metamorphosis. Faced with drastically reduced network schedules, radio stations began to localize their news efforts. Attention was focused on area news events rather than national and international. Stations that had relied almost exclusively on network news began to hire newspeople and broadcast a regular schedule of local newscasts. By the mid-1950s the transformation was nearly complete, and radio news had become a local programming matter. Radio news had undergone a 180-degree turn, even before the medium gave up trying to directly compete on a program-for-program basis with television. By the time radio set a new and revivifying course for itself by programming for

Radio Is The First Morning News Source				
Morning(6 AM-10 AM)	Radio	TV	News-papers	Other/None
Persons 12 +	49%	29%	15%	7%
Teens 12-17	60	21	12	7
Adults 18 +	48	30	16	6
Adults 18-34	53	28	13	6
Adults 25-54	50	29	16	5
College Grads.	46	27	22	5
Prof./Mgr. Males	55	19	20	6
F/T Working Women	56	27	11	6
$50K+ Income	49	28	18	5

Radio Is The First News Source At Midday				
Midday(10 AM-3 PM)	Radio	TV	News-papers	Other/None
Persons 12 +	36%	24%	14%	26%
Teens 12-17	30	24	17	29
Adults 18 +	37	24	14	25
Adults 18-34	43	19	18	20
Adults 25-54	41	18	15	26
College Grads.	36	17	15	32
Prof./Mgr. Males	36	9	23	32
F/T Working Women	48	9	12	31
$50K+ Income	44	10	17	29

Radio Is The Major Source of News				
Morning (6AM-10 AM)	Radio	TV	News-papers	Other/None
Persons 12 +	42%	31%	18%	9%
Teens 12-17	54	29	9	8
Adults 18 +	41	31	19	9
Adults 18-34	45	28	20	7
Adults 25-54	43	29	20	8
College Grads.	38	24	30	8
Prof./Mgr. Males	48	16	27	9
Working Women	49	28	15	8
$50K+ Income	45	21	27	7

Radio Is The First Source For Local Emergency News			
	Radio	TV	Other/None
Persons 12 +	50%	48%	2%
Teens 12-17	50	49	1
Adults 18 +	50	48	2
Adults 18-34	50	49	1
Adults 25-54	54	44	2
College Grads.	56	42	2
Prof./Mgr. Males	57	40	3
F/T Working Women	51	47	2

FIGURE 5.1 News is a prominent feature on most radio stations. Courtesy RAB.

specific segments of the listening audience, local newscasts were the norm.

Since its period of reconstruction in the 1950s, radio has proven time and time again to be the nation's first source of information about major news events. The majority of Americans first heard of the assassination of President Kennedy and the subsequent shootings of Martin Luther King, Jr., and Senator Robert Kennedy over radio. In 1965 when most of the Northeast was crippled by a power blackout, battery-powered radios literally became a lifeline for millions of people by providing continuous news and information until power was restored.

During the 1970s and 1980s, news on both the world and local levels reached radio listeners first. Today the public knows that radio is the first place to turn for up-to-the-minute news about occurrences halfway around the world, as well as in its own backyard.

NEWS AND TODAY'S RADIO

More people claim to listen to radio for music than for any other reason. Somewhat surprising, however, is that most of these same people admit to relying on the medium for the news they receive. Recent studies have found that while most of those surveyed tuned in to radio for entertainment, three-quarters considered news programming important. These surveys also ascertained that radio is the first morning news source for two-thirds of all full-time working women.

According to the Radio Advertising Bureau (RAB), over 50 percent of young adults get their first news of the day from radio. In comparison, only 16 percent of adults rely on newspapers as the first source of daily news. Practically all of the nation's 9,000

FIGURE 5.2
A metro market newsroom typically contains several desks, phones, tape recorders, typewriters or computer terminals, monitoring equipment, and teletype machines. Courtesy WMJX-FM.

commercial stations program news to some extent. Radio's tremendous mobility and pervasiveness has made it an instant and reliable news source for over 170 million Americans.

THE NEWSROOM

The number of individuals working in a radio station newsroom will vary depending on the size of the station and its format. On the average, a station in a small market employs one or two full-time newspeople. Of course, some outlets find it financially unfeasible to hire newspeople. These stations do not necessarily ignore news, rather they delegate responsibilities to their deejays to deliver brief newscasts at specified times, often at the top of the hour. Stations approaching news in this manner make it necessary for the on-air person to collect news from the wire service during record cuts and broadcast it nearly verbatim. Little, if any, rewrite is done, because the deejay simply does not have the time to do it. About the only thing that persons at "rip 'n' read" outlets can and must do is examine wire copy before going on the air. This eliminates the likelihood of mistakes. Again, all this is accomplished while the records are spinning.

NPR reporter Corey Flintoff warns against neglecting examining wire copy before air time. "We've all been caught with stuff that appears to scan at first sight but turns out to be incomprehensible when you read it."

Music-oriented stations in larger markets rarely allow their deejays to do news. Occasionally the person jocking the overnight shift will be expected to give a brief newscast every hour or two, but in metro markets this is fairly uncommon. There is generally a newsperson on duty around the clock. A top-rated station in a medium market typically employs four full-time newspeople; again, this varies depending on the status of the outlet and the type of programming it airs. For example, Easy Listening stations that stress music and deemphasize talk may employ only one or two newspeople. Meanwhile, an MOR station in the same market may have five people on its news staff in an attempt to promote itself as a heavy news and information outlet, even though its primary product is music. Certain music stations in major markets hire as many as a dozen news employees. This figure may include not only on-air newscasters, but writers, street reporters, and technical people as well. Stringers and interns also swell the figure.

During the prime listening periods when a station's audience is at its maximum, newscasts are programmed with greater frequency, sometimes twice as often as during other dayparts. The newsroom is a hub of activity as newspeople prepare for newscasts scheduled every twenty to thirty minutes. Half a dozen people may be involved in assembling news, but only two may actually enter the broadcast booth. A prime-time newscast schedule may look something like this:

A.M. Drive Coverage		P.M. Drive Coverage	
Smith	6:25 A.M.	Lopez	3:25 P.M.
Bernard	7:00 A.M.	Gardner	4:00 P.M.
Smith	7:25 A.M.	Lopez	4:25 P.M.
Bernard	8:00 A.M.	Gardner	5:00 P.M.
Smith	8:25 A.M.	Lopez	5:25 P.M.
Bernard	9:00 A.M.	Gardner	6:00 P.M.
Smith	9:25 A.M.	Lopez	6:25 P.M.

Midday and evening are far less frenetic in the newsroom, and one person per shift may be considered sufficient.

A standard-size newsroom in a medium market will contain several pieces of audio equipment, not to mention office furniture such as desks, computers, typewriters, file cabinets, and so on. Reel-to-reel recorders and cassette and cartridge machines are important tools for the newsperson. The newsroom also will be equipped with various monitors to keep newspeople on top of what is happening at the local police and fire departments and weather bureau. Various wireservice machines provide the latest news, sports, stock, and weather information, as well as a host of other data. Depending on the station's budget, two or more news services may be used. Stations with a genuine commitment to news create work areas that are designed for maximum efficiency and productivity.

In situations where newsrooms have been combined and consolidated due to LMA's and the relaxation of duopoly rules, more personnel, equipment, and space may be in evidence, since the plant itself may be serving myriad signals.

THE ALL-NEWS STATION

Stations devoted entirely to news programming arrived on the scene in the mid-1960s. Program innovator Gordon McLendon, who had been a key figure in the development of two music formats, Beautiful Music and Top 40, implemented All-News at WNUS-AM ("NEWS") in Chicago. In 1965, Group W, Westinghouse Broadcasting, changed WINS-AM in New York to All-News, and soon did the same at more of its metro outlets—KYW-AM, Philadelphia, and KFWB-AM, Los Angeles. While Group W was converting several of its outlets to nonmusic programming, CBS decided that All-News was the way to go at WCBS-AM, New York; KCBS-AM, San Francisco; and KNX-AM, Los Angeles.

Not long after KCBS in San Francisco began its All-News programming, another Bay City station, KGO-AM, introduced the hybrid News/Talk format in which news shares the microphone with conversation and interview features. Over the years, the hybrid approach has caught on and leads the pure All-News format in popularity.

Because of the exorbitant cost of running a news-only operation, it has remained primarily a metro market endeavor. It often costs several times as much to run an effective All-News station as it does one broadcasting music. This usually keeps small market outlets out of the business. Staff size in All-News far exceeds that of formats that primarily serve up music. Whereas a lone deejay is needed at an Adult Contemporary or Top 40 station, All-News requires the involvement of several people to keep the air sound credible.

While the cost of running a News station is high, the payback can more than justify expenditures. However, this is one format that requires a sizable initial investment, as well as the financial wherewithal and patience to last until it becomes an established and viable entity. Considerable planning takes place before a station decides to convert to News, since it is not simply a matter of hiring new jocks and updating the music library. Switching from a music format to News is dramatic and anything but cosmetic.

AM has always been the home of the All-News station. There are only a handful of FM News outlets. The format's prevalence on AM has grown considerably since the late 1970s when FM took the lead in listeners. The percentage of News and News/Talk formats on AM continued to increase in the 1980s as the band lost more and more of its music listeners to FM. However, News stations in many metro markets keep AM at the top of the ratings charts. In the early 1990s, it was common to find one AM outlet among the leaders, and almost invariably it programmed nonmusic. Some media observers predict that All-News will make significant inroads into FM.

FIGURE 5.3
Prominent talk
personalities often
constitute an
important
programming
ingredient of
information-
oriented stations.
Courtesy Premiere
Radio Networks.

THE ELECTRONIC NEWSROOM

The use of computers in the radio newsroom has increased significantly since their industry debut in December 1980 at KCBS in San Francisco. Computers linked to the various wire and information services are used to access primary and background data on fast-breaking stories and features. Many stations have installed video display terminals (VDTs) in on-air studios. Instead of hand-held copy, newscasters simply broadcast off the screens. Gradually desktop computers are replacing typewriters in the newsrooms at larger stations. The speed and agility with which copy can be produced and edited makes a computer the perfect tool for broadcast journalists.

In 1986 Boston All-News station WEEI-AM installed Media Touch's Touchstone system, thus converting their entire operation to computer. News people access and store data and even activate equipment simply by touching a computer screen.

Computers are slowly appearing at nonmetro market outlets. "There's not much resistance remaining against computers in the newsroom, especially since the advent of cheaper PCs and the improvement of programs like Newspro, which is the system we use here at NPR. It allows us to access and manipulate about a half-dozen wire services. We can split screen and write stories while searching the wires, and so on. There are many more sophisticated systems too. We're considering buying the D-CART system that is used by the Australian Broadcasting Corporation. This system incorporates digital audio right onto the screen," comments Cory Flintoff.

Producer Ty Ford observes, "Computers are an integral part of radio newsrooms now, and on the increase is auto-download of wire copy to word processing terminals, as well as search-by-word or topic search, auto-word count and digital archiving of sound bites with a computer data base for retrieval. To put it in the contemporary lexicon, radio news is 'on line.'"

More and more newsrooms utilize the Internet and e-mail. It makes sense that the information highway be accessed by a medium determined to keep its listening public informed and up-to-date.

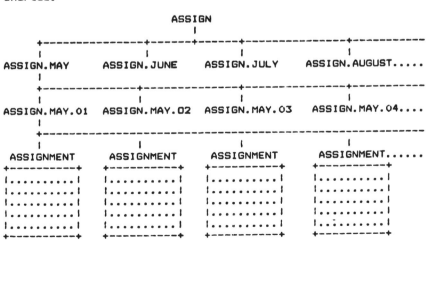

FIGURE 5.4
Computerized news assignments file. Computers have transformed the radio newsroom. Courtesy Jefferson Pilot Data Systems.

FIGURE 5.5
Even small market newsrooms employ computers.

While some news directors resist the idea of a computerized newsroom for various reasons, most industry experts predict that shortly computers will be standard equipment in even the tiniest radio station newsroom. Larry Jewett, news director of WTOD-AM, Toledo, Ohio, suggests that today's newspeople, and individuals anticipating careers in radio news, become computer knowledgeable. "Computers are a fact of life now. They cannot be ignored. Newspeople, and would-be newspeople, should learn all they can about the new technology because it is fast becoming an integral part of the profession."

THE NEWS DIRECTOR

News directors, like other department heads, are responsible for developing and implementing policies pertaining to their area, supervising staff members, and handling budgetary concerns. These are basic to any managerial position. However, the news department poses its own unique challenges to the individual who oversees its operation. These challenges must be met with a considerable degree of skill and know-how. Education and training are important. Surveys have concluded that station managers look for college degrees when hiring news directors. In addition, most news directors have, on the average, five years of experience in radio news before advancement to the managerial level.

The news director and program director work closely. At most stations, the PD has authority over the news department, since everything going over the air or affecting the air product is his direct concern and responsibility. Any changes in the format of the news or in the scheduling of newscasts or newscasters may, in fact, have to be approved by the station's programmer. For example, if the PD is opposed to the news director's plans to include two or more taped reports (actualities) per newscast, he may withhold approval. While the news director may feel that the reports enhance the newscasts, the PD may argue that they create congestion and clutter. In terms of establishing the on-air news schedule, the PD works with the news director to ensure that the sound of a given newsperson is suitably matched with the time slot he or she is assigned.

Getting the news out rapidly and accurately is a top priority of the news director. "People tune radio news to find out what is happening right now. That's what makes the medium such a key source for most people. While it is important to get news on the air as fast as possible, it is more important that the stories broadcast be factual and correct. You can't sacrifice accuracy for the sake of speed. As news director at KAPE my first responsibility is to inform our audience about breaking events on the local level. That's what our listeners want to hear," says Judy Smith, who functions as a one-person newsroom at the San Antonio station. "Because I'm the only newsperson on duty, I have to spend a lot of time verifying facts on the phone and recording actualities. I don't have the luxury of assigning that work to someone else, but it has to be done."

Larry Jewett perceives his responsibilities similarly. "First and foremost, the news director's job is to keep the listener informed of what is happening in the world around him. A newsperson

```
          Tue Oct 09 12:36  page  1

SLUG                  ANCHOR    WRITER    CART NO.   COPY  TAPE   TOTAL    CUME
===============================================================================
OPENER                                             0:08  00:00  0:08   - 8:00

NEWS TEASER                                        0:18  00:00  0:18   - 7:52

COMMERCIAL ONE                                     0:00  01:00  1:00   - 7:34

NEWS SEGMENT                                       0:00  00:00  0:00   - 6:34

DEBATE REAX                     cundiff   12       0:32  00:45  1:17   - 6:34

PRESIDENTIAL VISIT              cundiff   21A/B    0:34  00:30  1:04   - 5:17

I77 FATAL                       cundiff   NA       0:28  00:00  0:28   - 4:13

TRANSIT MALL UPDATE             cundiff   17       0:18  00:42  1:00   - 3:45

SPORTS                                             0:06  01:00  1:06   - 2:45

COMMERCIAL TWO                                     0:00  01:00  1:00   - 1:39

WEATHER                                            0:30  00:00  0:30   - 0:39

CLOSE                                              0:09  00:00  0:09   - 0:09
```

is a gatherer and conveyor of information. News is a serious business. A jock can be wacky and outrageous on the air and be a great success. On the other hand, a newsperson must communicate credibility or find another occupation."

Gathering local news is the most time-consuming task facing a radio news director, according to news director Cecilia Mason. "To do the job well you have to keep moving. All kinds of meetings—governmental, civic, business—have to be covered if you intend being a primary source of local news. A station with a news commitment must have the resources to be where the stories are, too. A news director has to be a logistical engineer at times. You have to be good at prioritizing and making the most out of what you have at hand. All too often there are just too many events unfolding for a news department to effectively cover, so you call the shots the best way that you can. If you know your business, your best shot is usually more than adequate."

In addition to the gathering and reporting of news, public affairs programming often is the responsibility of the news director. This generally includes the planning and preparation of local information features, such as interviews, debates, and even documentaries.

WHAT MAKES A NEWSPERSON

College training has become an increasingly important criterion to the radio news director planning to hire personnel. It is not impossible to land a news job without a degree, but formal education is a definite asset. An individual planning to enter the radio news profession should consider pursuing a broadcasting, journalism, or liberal arts degree. Courses in political science, history, economics, and literature give the aspiring newsperson the kind of well-rounded background that is most useful. "Coming into this field, especially in the 1990s, a college degree is an attractive, if not essential, credential. There's so much that a newsperson has to know. I think an education makes the kind of difference you can hear, and that's what our business is about. It's a fact that most people are more cognizant of the world and write better after attending college.

FIGURE 5.7
Women initially found greater employment opportunities in news than in other areas of radio programming. Today women hold approximately 30 percent of top radio news jobs.
Courtesy WMJX-FM.

Credibility is crucial in this business, and college training provides some of that. A degree is something that I would look for in prospective newspeople," says Cecilia Mason.

While education ranks high, most news directors still look for experience first. "As far as I'm concerned, experience counts the most. I'm not suggesting that education isn't important. It is. Most news directors want the person that they are hiring to have a college background, but experience impresses them more. I believe a person should have a good understanding of the basics before attempting to make a living at something. Whereas a college education is useful, a person should not lean back and point to a degree. Mine hasn't gotten me a job yet, though I wouldn't trade it for the world," notes WTOD's Jewett. KAPE's Smith agrees, "The first thing I think most news directors really look for is experience. Although I have a bachelor of arts degree myself, I wouldn't hold out for a person with a college diploma. I think if it came down to hiring a person with a degree versus someone with solid experience, I'd go for the latter."

Gaining news experience can be somewhat difficult, at least more so than acquiring deejay experience. Small stations, where the beginner is most likely to break into the business, have slots for several deejays but seldom more than one for a newsperson. It becomes even more problematical when employers at small stations want the one person that they hire for news to bring some experience to the job. Larger stations place even greater emphasis on experience. Thus the aspiring newsperson is faced with a sort of Catch 22 situation, in which a job cannot be acquired without experience and experience cannot be acquired without a job.

News director Frank Titus says that there are ways of gaining experience that will lead to a news job. "Working in news at high school and college stations is very valid experience. That's how Dan Rather and a hundred other newsmen got started. Also working as an intern at a commercial radio station fattens out the resume. If someone comes to me with this kind of background and a strong desire to do news, I'm interested."

Among the personal qualities that most appeal to news directors are enthusiasm, aggressiveness, energy, and inquisitiveness. "I want someone with a strong news sense and unflagging desire to get a story and get it right. A person either wants to do news or doesn't. Someone with a pedestrian interest in radio journalism is more of a hindrance to an operation than a help," contends Mason. WARA's Titus wants someone who is totally devoted to the profession. "When you get right down to it, I want someone on my staff who eats, drinks, and sleeps news."

On the practical side of the ledger, WBCN's Sherman Whitman says typing or keyboard skills are essential. "If you can't type, you can't work in a newsroom. It's an essential ability, and the more accuracy and speed the better. It's one of those skills basic to the job. A candidate for a news job can come in here with two degrees, but if that person can't type, that person won't be hired. Broadcast students should learn to type." Meanwhile, Jewett stresses the value of possessing a firm command of the English language. "Proper punctuation, spelling, and syntax make a news story intelligible. A newsperson doesn't have to be a grammarian, but he or she had better know where to put a comma and a period and how to compose a good clean sentence. A copy of Strunk and White's *Elements of Style* is good to have around."

An individual who is knowledgeable about the area in which a station is located has a major advantage over those who are not, says Whitman. "A newsperson here at WBCN has to know Boston inside out. I'd advise anybody about to be interviewed for a news position to find out as much as possible about the station's coverage area. Read back issues of newspapers, get socioeconomic stats from the library or chamber of commerce, and study street directories and maps of the town or city in which the station is located. Go into the job interview well informed, and you'll make a strong impression."

Unlike a print journalist, a radio newsperson also must be a performer. In addition to good writing and news-gathering skills, the newsperson in radio must have announcing abilities. Again, training is usually essential. "Not only must a radio newsperson be able to write a story, but he or she has to be able to present it on the air. You have to be an announcer, too. It takes both training and experience to become a really effective newscaster. Voice performance courses can provide a foundation," says KAPE's Smith. Most colleges with broadcasting programs offer announcing and newscasting instruction.

Entry-level news positions pay modestly, while newspeople at metro market stations earn impressive incomes. With experience come the better paying jobs. Finding that first full-time news position often takes patience and determination. Several industry trade journals, such as *Broadcasting*, the Radio-Television News Directors Association's (RTNDA) *Communicator*, *Radio and Records*, and others, list news openings.

PREPARING THE NEWS STORY

Clean copy is imperative. News stories must be legible and intelligible and designed for effortless reading by the newscaster, or several different newscasters. Typos, mispunctuation, awkward phrasing, and incorrect spelling are anathema to the person at the microphone. Try reading the following news story aloud and imagine yourself in a studio broadcasting to thousands of perplexed listeners.

PRESIDENTCLINTON STATD TODXAY THAT

HE WILL SEED RELECTION TO A SECOND

TURMIN OFFICE, DEPICE ROOMERS THAT

HE WILT LEAVE WASHINDON TO PURDUE

A QUEER IN HOLLYWOOD.

Going on the air with copy riddled with errors is inviting disaster. About the only things right about the preceding news copy are that it is typed in upper case and double-spaced. Here are a few suggestions to keep in mind when preparing a radio news story:

1. Type neatly. Avoid typos and x-outs. Eliminate a typing error completely. If it is left on the page, it could trip you up during a broadcast.
2. Use UPPERCASE throughout the story. It is easier to read. Don't forget, the story you are writing is going to be read on the air.
3. Double-space between lines for the same reason upper case is used—copy is easier to read. Space between lines of copy keeps them from merging together when read aloud.
4. Use one-inch margins. Don't run the copy off the page. Uniformity eliminates errors. At the same time, try not to break up words.
5. Avoid abbreviations, except for those meant to be read as such: YMCA, U.S.A., NAACP, AFL/CIO.
6. Write out numbers under ten, and use numerals for figures between 10

and 999. Spell out thousand, million, and so forth. For example, 21 million people, instead of 21,000,000 people. Numbers can be tricky, but a consistent approach prevents problems.

7. Use the phonetic spelling for words that may cause pronunciation difficulties, and underline the stressed syllable: Monsignor (Mon-seen-yor).

8. Punctuate properly. A comma out of place can change the meaning of a sentence.

9. When in doubt, consult a standard style guide. In addition, both AP and UPI publish handbooks on newswriting.

Notice how much easier it is to read a news story that is correctly prepared:

PRESIDENT CLINTON STATED TODAY THAT HE WILL SEEK REELECTION TO A SECOND TERM IN OFFICE, DESPITE RUMORS THAT HE WILL LEAVE WASHINGTON TO PURSUE A CAREER IN HOLLYWOOD.

Since radio news copy is written for the ear and not for the eye, its style must reflect that fact. In contrast to writing done for the printed page, radio writing is more conversational and informal. Necessity dictates this. Elaborately constructed sentences containing highly sophisticated language may effectively communicate to the reader but create serious problems for the listener, who must digest the text while it is being spoken. Whereas the reader has the luxury to move along at his or her own pace, the radio listener must keep pace with the newscaster or miss out on information. Radio writing must be accessible and immediately comprehensible. The most widely accepted and used words must be chosen so as to prevent confusion on the part of the listener, who usually does not have the time or opportunity to consult the dictionary. "Keep it simple and direct. No com-pound-complex sentences with dozens of esoteric phrases and terms. Try to picture the listener in your mind. He is probably driving a car or doing any number of things. Because of the nature of the medium, writing must be concise and conversational," contends Judy Smith.

Corey Flintoff agrees. "Copy should be adapted to a conversational style. Titles should be simplified and numbers rounded off."

News stories must be well structured and organized. This adds to their level of understanding. The journalist's five W's—who, what, when, where, and why—should be incorporated into each story. If a story fails to provide adequate details, the listener may tune in elsewhere to get what radio commentator Paul Harvey calls "the rest of the story."

When quoting a source in a news story, proper attribution must be made. This increases credibility while placing the burden of responsibility for a statement on the shoulders of the person who actually made it:

THE DRIVER OF THE CAR THAT STRUCK THE BUILDING APPEARED INTOXICATED, ACCORDING TO LISA BARNES, WHO VIEWED THE INCIDENT.

Observes Flintoff on the matter of attribution, "I prefer to identify a source at the beginning of the sentence on the theory that its more conversational. Thus 'Lisa Barnes, a witness, said the driver . . .' or 'Lisa Barnes saw the car crash into the building. She said the driver . . .' Incidently, I think there's a danger of legal problems using a witness's speculation that someone is drunk. I've had similar problems in the past, and I never use anything about potential intoxication unless there's a police test for drugs or alcohol."

Uncorroborated statements can make a station vulnerable to legal actions. The reliability of news sources must be established. When there are doubts concerning the facts, the newsperson has a responsibility to seek verification.

ORGANIZING THE NEWSCAST

News on music-oriented radio stations commonly is presented in five-minute blocks and aired at the top or bottom of the hour. During drivetime periods stations often increase the length and/or frequency of newscasts. The five minutes allotted news generally is divided into segments to accommodate the presentation of specific information. A station may establish a format that allows for two minutes of local and regional stories, one minute for key national and international stories, one minute for sports, and 15 seconds for weather information. A 30- or 60-second commercial break will be counted as part of the five-minute newscast.

The number of stories in a newscast may be preordained by program management or may vary depending on the significance and scope of the stories being reported. News policy may require that no stories, except in particular cases, exceed 15 seconds. Here the idea is to deliver as many stories as possible in the limited time available, the underlying sentiment being that more is better. In five minutes, 15 to 20 items may be covered. In contrast, other stations prefer that key stories be addressed in greater detail. As few as five to ten news items may be broadcast at stations taking this approach.

Stories are arranged according to their rank of importance, the most significant story of the hour topping the news. An informed newsperson will know what stories deserve the most attention. Wire services weigh each story and position them accordingly in news roundups. The local radio newsperson decides what wire stories will be aired and in what order.

Assembling a five-minute newscast takes skill, speed, and accuracy. Stories must be updated and rewritten to keep news broadcasts from sounding stale.

This often requires that telephone calls be made for late-breaking information. Meanwhile, on-the-scene voicers (actualities) originating from audio news services (UPI, AP) or fed by local reporters must be taped and slotted in the newscast. "Preparing a fresh newscast each hour can put you in mind of what it must have been like to be a contestant on the old game show 'Beat the Clock.' A conscientious newsperson is a vision of perpetual motion," observes Cecilia Mason.

Finally, most newspeople read their news copy before going on the air. "Reading stories cold is foolhardy and invites trouble. Even the most seasoned newscasters at metro market stations take the time to read over their copy before going on," comments WBCN's Whitman. Many newspeople read copy aloud in the news studio before airtime. This gives them a chance to get a feel for their copy. Proper preparation prevents unpleasant surprises from occurring while on the air.

THE WIRE SERVICES

Without the aid of the major broadcast news wire services, radio stations would find it almost impossible to cover news on national and international lev-

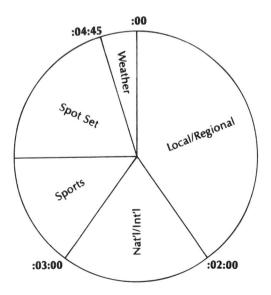

FIGURE 5.8 Five-minute newscast format clock.

FIGURE 5.9
Wire-service teletype machines provide the bulk of a radio newsroom's broadcast copy. At many stations the teletypes have been replaced by computers and printers.

els. The wire services are a vital source of news information to nearly all of the nation's commercial radio stations. Both large and small stations rely on the news copy fed them by either Associated Press or United Press International, the two most prominent news wire services.

Broadcast wire services came into existence in the mid-1930s, when United Press (which became United Press International in 1958 after merging with International News Service) began providing broadcasters with news copy. Today UPI and AP serve over 7,500 broadcast outlets.

Both news sources supply subscriber stations with around-the-clock coverage of national and world events. Over 100,000 stringers furnish stories from across the globe. AP and UPI also maintain regional bureaus for the dissemination of local news. Each wire service transmits over twenty complete news summaries daily. In addition, they provide weather, stock-market, and sports information, as well as a formidable list of features and data useful to the station's news and programming efforts. Rates for wire service vary depending on the size of the radio market, and audio service is available for an additional fee. Some 1,800 stations use UPI and AP audio news feeds.

Broadcasters are evenly divided over the question as to which is the best wire service. Each news service has about the same number of radio stations under contract.

Both major wire services have kept pace with the new technologies. In the mid-1980s, UPI alone purchased 6,000

Z-15 desktop computers from Zenith Data Systems. Satellites also are utilized by the two news organizations for the transmission of teletype, teletext, and audio. The wire services have become as wireless as the wireless itself.

RADIO NETWORK NEWS

During the medium's first three decades, the terms *networks* and *news* were virtually synonymous. Most of the news broadcast over America's radio stations emanated from the networks. The public's dependence on network radio news reached its height during World War II. As television succeeded radio as the mainstay for entertainment programming in the 1950s and 1960s, the networks concentrated their efforts on supplying affiliates with news and information feeds. This approach helped the networks regain their footing in radio after a period of substantial decline. By the mid-1960s, the majority of the nation's stations utilized one of the four major networks for news programming.

In 1968, ABC decided to make available four distinct news formats designed for compatibility with the dominant sounds of the day. American Contemporary Radio Network, American FM Radio Network, American Entertainment Radio Network, and American Information Network each offered a unique style and method of news presentation. ABC's venture proved enormously successful. In the 1970s, over 1,500 stations subscribed to one of ABC's four news networks.

In response to a growing racial and ethnic awareness, the Mutual Broadcasting System launched two minority news networks in 1971. While the network's Black news service proved to be a fruitful venture, its Spanish news service ceased operation within two years of its inception. Mutual discovered that the ethnic group simply was too refracted and diverse to be effectively serviced by one network and that the Latin listeners they did attract did not constitute the numbers necessary to justify operation.

In 1973 the network also went head-to-head with ABC by offering a network news service (Mutual Progressive Network) that catered to rock-oriented stations. Mutual's various efforts paid off by making it second only to ABC in number of affiliates.

The News and Information Service (NIS) was introduced by NBC in 1975 but ended in 1977. NIS offered client stations an All-News format. Fifty minutes of news was fed to stations each hour. The venture was abandoned after only moderate acceptance. CBS, which has offered its member stations World News Roundup since 1938, and NBC have under 300 affiliates apiece.

Several state and regional news networks do well, but the big four, ABC, MBS, NBC, and CBS, continue to dominate. Meanwhile, independent satellite news and information networks have joined the field and more are planned.

The usual length of a network newscast is five minutes, during which affiliates are afforded an opportunity to insert local sponsor messages at designated times. The networks make their money by selling national advertisers spot availabilities in their widely broadcast news. Stations also pay the networks a fee for the programming they receive.

According to Metro Network president David Saperstein, today "more and more stations are realizing the benefits that exist in outside news services, which provide the information that listeners would otherwise seek elsewhere. This allows the station to focus its marketing dollars, thus directing resources towards optimizing and maintaining what draws and keeps listeners."

Of course, this latter trend has raised additional concerns about the decline in local news coverage as cited in recent RTNDA surveys.

RADIO SPORTSCASTS

Sports is most commonly presented as an element within newscasts (see Fig. 5.8 on p. 157). While many stations air sports as programming features unto themselves, most stations insert infor-

FIGURE 5.10 Network features are a popular ingredient of station news and public affairs programming. Courtesy of Westwood One.

mation, such as scores and schedules of upcoming games, at a designated point in a newscast and call it sports. Whether a station emphasizes sports largely depends on its audience. Stations gearing their format for youngsters or women often all but ignore sports. Adult-oriented stations, such as Middle-of-the-Road, will frequently offer a greater abundance of sports information, especially when the station is located in an area that has a major-league team.

Stations that hire individuals to do sports, and invariably these are larger outlets since few small stations can afford a full-time sportsperson, look for someone who is well versed in athletics. "To be good at radio sports, you have to have been involved as a participant somewhere along the line. That's for

FIGURE 5.11
Associated Press
promotional piece.
AP claims more
subscribers than
any other wire
service. Courtesy
AP.

AP has always given you more service for your money. Here's how to make more money from our service.

If your station has the AP Radio Wire, you've got access to the world's most credible, up-to-date, and salable news.

But along with your news, you also get hundreds of authoritative, timely and equally salable feature programs each week.

And if you don't use those features, if you don't at least weigh them against what you're programming now, you may be letting thousands of profitable commercial avails slip through your fingers.

AP professionalism pays off in audience loyalty, and salability.

The standards of quality that apply to our news scripts also apply to our other programming. Whether it's sports, business, agriculture or lifestyle features.

The result? Unmatched audience loyalty. People tune in to hear our features, most of which provide time for commercial breaks.

That loyalty is what makes AP feature programming such a logical media buy for clothiers and car dealers. For stock brokers and sporting good stores.

And that's important to you. Because the more your feature programming appeals to prospective advertisers, the easier it is for you to sell time. And boost profits.

Just as important, all of our feature programming is designed to be compatible with your particular format.

Have a profitable business lunch with your AP Radio Wire Machine this Monday.

Along with all of the other attractive features of the AP Radio Wire, we feature a list of coming attractions. That's an AP exclusive.

If you've never seen it before, check your Radio Wire around noon, on Monday. And brace yourself for a shock.

Because it will do more than just tell you what's in store for the week ahead.

It will convince you that you've been sitting on a gold mine...of information.

To find out more about how our Radio Wire features can help you improve your profit picture, call Glenn Serafin collect at the Broadcast Services Division of The Associated Press (202)955-7214.

AP Associated Press Broadcast Services. Without a doubt.

starters, in my opinion. This doesn't mean that you have to be a former major leaguer before doing radio sports, but to have a feel for what you're talking about, it certainly helps to have been on

FIGURE 5.12
Network-affiliated
All-News stations,
such as WCBS,
invest heavily in
promotion in
efforts to establish
a strong image in
their markets.
Courtesy of WCBS.

In August 1967, WCBS Radio became WCBS NEWSRADIO. During those 15 years, WCBS has become the station millions of people rely on for radio news. And today, WCBS is listened to by more adults than any other radio station in the country.

For radio news, the biggest...and.... the best in the business.

WCBS NEWSRADIO
New York

New York Arbitron, Spring '82   TSA TOTAL WEEK CUME ESTIMATES, ADULTS 18 +

As seen in the September 6, 1982 issue of TELEVISION/RADIO AGE

the field or court yourself. A good sportscaster must have the ability to accurately analyze a sport through the eyes and body of the athlete," contends John Colletto, sports director, WPRO-AM, Providence, Rhode Island.

Unlike news that requires an impartial and somewhat austere presentation, sportscasts frequently are delivered in a casual and even opinionated manner. "Let's face it, there's a big difference between nuclear arms talks between the U.S. and the Soviets and last night's Red Sox/Yankees score. I don't think sports reports should be treated in a style that's too solemn. It's entertainment, and sportscasters should exercise their license to comment and analyze," says Colletto.

Although sports is presented in a less heavy-handed way than news, credibility is an important factor, contends Colletto. "There is a need for radio sportscasters to establish credibility just as there is for newspeople to do so. If you're not believable, you're not listened to. The best way to win the respect of your audience is by demonstrating a thorough knowledge of the game and by sounding like an insider, not just a guy reading the wire copy. Remember, sports fans can be as loyal to a sportscaster as they are to their favorite team. They want to hear the stories and scores from a person they feel comfortable with."

The style of a news story and a sports story may differ considerably. While news is written in a no-frills, straightforward way, sports stories often contain colorful colloquialisms and even popular slang. Here is an example by radio sportswriter Roger Crosley:

THE DEAN COLLEGE RED DEMON FOOTBALL TEAM RODE THE STRONG RUNNING OF FULLBACK BILL PALAZOLLO YESTERDAY TO AN 18–16 COME FROM BEHIND VICTORY OVER THE AMERICAN INTERNATIONAL COLLEGE JUNIOR VARSITY YELLOW JACKETS. PALAZOLLO CHURNED OUT A TEAM HIGH 93 YARDS ON TWENTY-FIVE CARRIES AND SCORED ALL THREE TOUCHDOWNS ON BLASTS OF 7, 2, AND 6 YARDS. THE DEMONS TRAILED THE HARD-HITTING CONTEST 16–6 ENTERING

THE FINAL QUARTER. PALAZOLLO CAPPED A TWELVE PLAY 81 YARD DRIVE WITH HIS SECOND SIX-POINTER EARLY IN THE STANZA AND SCORED THE CLINCHER WITH 4:34 REMAINING. THE DEMONS WILL PUT THEIR 1 AND 0 RECORD ON THE LINE NEXT SUNDAY AT 1:30 AGAINST THE ALWAYS TOUGH HOLY CROSS JAYVEES IN WORCESTER.

Sportscasters are personalities, says WPRO's Colletto, and as such must be able to communicate on a different level than newscasters. "You're expected to have a sense of humor. Most successful sportscasters can make an audience smile or laugh. You have to be able to ad-lib, also."

The wire services and networks are the primary source for sports news at local stations. On the other hand, information about the outcome of local games, such as high school football and so forth, must be acquired firsthand. This usually entails a call to the team's coach or a direct report from a stringer or reporter.

RADIO NEWS AND THE FCC

The government takes a greater role in regulating broadcast journalism than it does print. Whereas it usually maintains a hands-off position when it comes to newspapers, the government keeps a watchful eye on radio to ensure that it meets certain operating criteria. Since the FCC perceives the airways as public domain, it expects broadcasters to operate in the public's interest.

The FCC requires that radio reporters present news factually and in good faith. Stories that defame citizens through reckless or false statements may not only bring a libel suit from the injured party but action from the FCC, which views such behavior on the part of broadcasters as contrary to the public's interest. While broadcasters are protected under the First Amendment and therefore have certain rights, as public trustees they are charged with the additional responsibility of acting in a manner that benefits rather than harms members of society.

Broadcasters are free to express opinions and sentiments on issues through editorials. However, to avoid controversy, many radio stations choose not to editorialize even though the FCC encourages them to do so.

NEWS ETHICS

The highly competitive nature of radio places unusual pressure on newspeople. In a business where being first with the story is often equated with being the best, certain dangers exist. Being first at all costs can be costly, indeed, if information and facts are not adequately verified. As previously mentioned, it is the radio journalist's obligation to get the story straight and accurate before putting it on the air. Anything short of this is unprofessional.

The pressures of the clock can, if allowed, result in haphazard reporting. If a story cannot be sufficiently prepared in time for the upcoming news broadcast, it should be withheld. Getting it on is not as important as getting it on right. Accuracy is the newsperson's first criterion. News accounts should never be

FIGURE 5.13 Sports on radio in 1923. Today sports programming generates a significant percentage of the medium's annual income. Courtesy Westinghouse Electric.

Broadcasting from Station WJZ

fudged. It is tantamount to deceiving and misleading the public.

News reporters must exhibit discretion not only in the newsroom but also when on the scene of a story. It is commendable to assiduously pursue the facts and details of a story, but it is inconsiderate and insensitive to ignore the suffering and pain of those involved. For example, to press for comments from a grief-stricken parent whose child has just been seriously injured in an accident is callous and cruel and a disservice to all concerned, including the station the newsperson represents. Of course, a newsperson wants as much information as possible about an incident, but the public's right to privacy must be respected.

Objectivity is the cornerstone of good reporting. A newsperson who has lost his or her capacity to see the whole picture is handicapped. At the same time, the newsperson's job is to report the news and not create it. The mere presence of a member of the media can inspire a disturbance or agitate a volatile situation. Staging an event for the sake of increasing the newsiness of a story is not only unprofessional, but illegal. Groups have been known to await the arrival of reporters before initiating a disturbance for the sake of gaining publicity. It is the duty of reporters to remain as innocuous and uninvolved as possible when on an assignment.

Several industry associations, such as RTNDA and Society of Professional Journalists, have established codes pertaining to the ethics and conduct of broadcast reporters. One such set of criteria dealing with the responsibilities of radio newspeople is reproduced in full in Figure 5.15 (p. 164).

TRAFFIC REPORTS

Traffic reports are an integral part of drivetime news programming at many metropolitan radio stations. Although providing listeners with traffic condition updates can be costly, especially air-to-ground reports which require the use of a helicopter or small plane, they can help strengthen a station's community service image and also generate substantial revenue. To avoid the cost involved in airborne observation, stations sometimes employ the services of local auto clubs or put their own mobile units out on the roads. A station in Providence, Rhode Island, broadcasts traffic conditions from atop a twenty-story hotel that overlooks the city's key arteries.

Says David Saperstein, "Companies like Metro Network provide stations with outside traffic reporting services in a manner that is more cost- and quality-effective than a station handling it themselves."

Traffic reports are scheduled several times an hour throughout the prime commuter periods on stations primarily catering to adults, and they range in length from 30 to 90 seconds. The actual reports may be done by a station employee who works in other areas of programming when not surveying the roads, or a member of the local police department or auto club may be hired for the job. Obviously, the prime criterion for such a position is a thorough knowledge of the streets and highways of the area being reported.

NEWS IN MUSIC RADIO

In the 1980s, the FCC saw fit to eliminate the requirement that all radio stations devote a percentage of their broadcast day to news and public affairs programming. Opponents of the decision argued that such a move would mark the decline of news on radio. In contrast, proponents of the deregulation commended the FCC's actions that allow for the marketplace to determine the extent to which nonentertainment features are broadcast. In the late 1980s, RTNDA expressed the concern that local news coverage had declined. This, they said, had resulted in a decrease in the number of news positions around the country. Supporting their contention they pointed out that several major stations, such as KDKA, WOWO, and WIND, had cut back their news budgets.

At that time, RTNDA's Bob Priddy noted, "There has been a perceived decline in the amount of news broadcast. I don't see this as a cold-hearted act on the part of station managers, but rather one frequently inspired by economics. The decline in news programming is particularly alarming when you realize that it is at a time when a number of new stations are entering the airwaves."

In 1992, RTNDA's president, Dave Bartlett, declared, "Deregulation really hasn't taken news off radio. News is far from dead on the medium. The vast majority do news. All deregulation did was allow the marketplace to adjust at will. A lot of shifting has occurred, but the aggregate is the same."

A couple of years later, a survey published in the association's newsletter, *Communicator*, told a different story. The report revealed that hundreds of radio newsrooms had, in fact, closed down, and it suggested many more would likely occur. In 1994, *Radio World* reported that over 1,100 radio news operations had closed since the deregulation of the medium.

WBCN's Sherman Whitman believes that the radio audience wants news even when a station's primary product is music. "The public has come to depend on the medium to keep it informed. It's a volatile world and certain events affect us all. Stations that aim to be full-service cannot do so without a solid news schedule."

Cecilia Mason says that economics alone will help keep news a viable entity at many radio stations. "While a lot of stations consider news departments expense centers, news is a money maker. This is especially true during drivetime periods when practically everyone tuned wants information, be it weather, sports, or news headlines. I don't see a growing movement to eliminate news. However, I do see a movement to soften things up, that is, to hire voices instead of radio journalists. In the long run, this means fewer news jobs, I suppose. Economics again.

FIGURE 5.14 Information features are popular elements of news and public affairs programming. Courtesy CRN.

While stations, for the most part, are not appreciably reducing the amount of time devoted to airing news, I suspect that some may be thinning out their news departments. Hopefully, this is not a prelude to a measurable cutback. News is still big business, though."

Responsible broadcasters know that it is the inherent duty of the medium to keep the public apprised of what is going on," claims Larry Jewett. "While radio is primarily an entertainment medium, it is still one of the country's foremost sources of information. Responsible broadcasters—and most of us are—realize that we have a special obligation to fulfill. The tremendous reach and immediacy that is unique to radio forces the medium to be something more than just a jukebox."

While WARA's Frank Titus believes that stations will continue to broadcast news in the future. "There might be a tendency to invest less in news operations, especially at more music-oriented outlets, as the result of the regulation change, but news is as much a part of what radio is as are the deejays and songs. What it comes right

FIGURE 5.15
Traffic reports are often a central element of news broadcasts.
Courtesy Metro Traffic.

down to is people want news broadcasts, so they're going to get them. That's the whole idea behind the commission's FCC's actions. There's no doubt in my mind that the marketplace will continue to dictate the programming of radio news."

Roger Nadel, director of news and programming for WWJ-AM, concurs. "As the age of the average listener increases, even people tuning in 'music' stations find themselves in need of at least minimal doses of news. So long as those stations are doing well financially, owners can be content to maintain some kind of a news operation. News is not likely to disappear; not even at music stations."

CHAPTER HIGHLIGHTS

1. Although the first newscast occurred in 1910, broadcast journalism did not evolve until the early 1920s. The broadcast of the Harding-Cox election results in 1920 was a historical benchmark.

2. Because newspapers perceived radio as a competitive threat, United Press (UP), International News Service (INS), and Associated Press (AP) refused to sell to radio outlets from 1932 until 1934. Radio, however, proved it could provide its own news sources.

3. The advent of television led radio outlets to localize their news content, which meant less reliance on news networks and the creation of a station news department.

4. Surveys by the National Association of Broadcasters and the Radio Advertising Bureau found that more people tune in to radio news for their first daily source of information than turn to television or newspapers.

5. The size of a station's news staff depends upon the degree to which the station's format emphasizes news, the station's market size, and the emphasis of its competition. Small stations often have no newspeople and require deejays to "rip 'n' read" wireservice copy.

6. Large news staffs may consist of newscasters, writers, street reporters, and tech people, as well as stringers and interns.

7. Computers in radio newsrooms can be used as links to the various wire, news, and data base services, as display terminals for reading news copy on the air, and as word processors for writing and storing news.

8. The news director, who works with and for the program director, supervises news staff, develops and implements policy, handles the budget, assures the gathering of local news, is responsible for getting out breaking news stories rapidly and accurately, and plans public affairs programming.

9. News directors seek personnel with both college education and experience. However, finding a news slot at a small station is difficult, since their news staffs are small, so internships and experience at high school and college stations are important. Additionally, such personal qualities as enthusiasm, aggressiveness, energy, inquisitiveness, keyboarding skills, a knowledge of the area where the station is located, announcing abilities, and a command of the English language are assets.

10. News stories must be legible, intelligible, and designed for effortless reading. They should sound conversational, informal, simple, direct, concise, and organized.

11. Actualities (on-the-scene voicers) are taped from news service feeds or recorded by station personnel at the scene.

12. Wire services are the primary, and often only, sources of national and international news for most stations. Associated Press and United Press International are the largest.

13. The FCC expects broadcasters to report the news in a balanced and impartial manner. Although protected under the First Amendment, broadcasters making reckless or false statements are subject to both civil and FCC charges.

14. Ethically, newspersons must maintain objectivity, discretion, and sensitivity.

15. Despite the FCC's deregulation of news and public affairs programming in the early 1980s, industry professionals believe that the demands of the listeners will ensure the continued importance of news on most commercial stations.

SUGGESTED FURTHER READING

Bartlett, Jonathan, ed. *The First Amendment in a Free Society.* New York: H.W. Wilson, 1979.

Bittner, John R., and Bittner, Denise A. *Radio Journalism.* Englewood Cliffs, N.J.: Prentice-Hall, 1977.

Bliss, Edward J. *Now the News.* New York: Oxford Press, 1991.

————, and Patterson, John M. *Writing News for Broadcast,* 2nd ed. New York: Columbia University Press, 1978.

Block, Mervin. *Broadcast Newswriting.* Washington: NAB Publications, 1994.

Boyer, Peter J. *Who Killed CBS?* New York: Random House, 1988.

Charnley, Mitchell. *News by Radio.* New York: Macmillan, 1948.

Culbert, David Holbrook. *News for Everyman: Radio and Foreign Affairs in Thirties America.* Westport, Conn.: Greenwood Press, 1976.

Day, Louis A. *Ethics in Media Communications.* Belmont, Calif.: Wadsworth Publishing, 1991.

Fang, Irving. *Radio News/Television News,* 2nd ed. St. Paul, Minn.: Rada Press, 1985.

————. *Those Radio Commentators.* Ames: Iowa State University Press, 1977.

Friendly, Fred W. *The Good Guys, The Bad Guys, and the First Amendment: Free Speech vs. Fairness in Broadcasting.* New York: Random House, 1976.

Garvey, Daniel E. *Newswriting for the Electronic Media.* Belmont, Calif.: Wadsworth Publishing, 1982.

Gibson, Roy. *Radio and Television Reporting.* Boston: Allyn and Bacon, 1991.

Gilbert, Bob. *Perry's Broadcast News Handbook.* Knoxville, Tenn.: Perry Publishing, 1982.

Hall, Mark W. *Broadcast Journalism: An Introduction to News Writing.* New York: Hastings House, 1978.

Hitchcock, John R. *Sportscasting.* Boston: Focal Press, 1991.

Hood, James R., and Kalbfeld, Brad, eds. *The Associated Press Handbook.* New York: Associated Press, 1982.

Hunter, Julius K. *Broadcast News.* St. Louis: C.V. Mosby Company, 1980.

Johnston, Carla. *Election Coverage: Blueprint for Broadcasters.* Boston: Focal Press, 1991.

Keirstead, Phillip A. *All-News Radio.* Blue Ridge Summit, Pa.: Tab Books, 1980.

Nelson, Harold L. *Laws of Mass Communication.* Mineola, N.Y.: The Foundation Press, 1982.

Simmons, Steven J. *The Fairness Doctrine and the Media.* Berkeley: University of California Press, 1978.

Stephens, Mitchell. *Broadcast News: Radio Journalism and an Introduction to Television.* New York: Holt, Rinehart and Winston, 1980.

UPI Stylebook: A Handbook for Writing and Preparing Broadcast News. New York: United Press International, 1979.

White, Ted. *Broadcast Newswriting.* New York: Macmillan, 1984.

Wulfemeyer, K. Tim. *Broadcast Newswriting.* Ames: Iowa State University Press, 1983.

6 Research

WHO IS LISTENING

As early as 1929, the question of listenership was of interest to broadcasters and advertisers alike. That year Cooperative Analysis of Broadcasting (CAB), headed by Archibald M. Crossley, undertook a study to determine how many people were tuned to certain network radio programs. Information was gathered by phoning a preselected sample of homes. One of the things the survey found was that the majority of listening occurred evenings between seven and eleven. This became known as radio's "prime time" until the 1950s.

On the local station level, various methods were employed to collect audience data, including telephone interviews and mail-out questionnaires. However, only a nominal amount of actual audience research was attempted during the late 1920s and early 1930s. For the most part, just who was listening remained somewhat of a mystery until the late 1930s.

In 1938 C. E. Hooper, Inc., began the most formidable attempt up to that time to provide radio broadcasters with audience information. Like Crossley's service, Hooper also utilized the telephone to accumulate listener data. While CAB relied on listener recall, Hooper required that interviewers make calls until they reached someone who was actually listening to the radio. This approach became known as the "coincidental" telephone method. Both survey services found their efforts limited by the fact that 40 percent of the radio listening homes in the 1930s were without a telephone.

As World War II approached, another major ratings service, known as The Pulse, began to measure radio audience size. Unlike its competitors, Pulse collected information by conducting face-to-face interviews.

Interest in audience research grew steadily throughout the 1930s and cul-

minated in the establishment of the Office of Radio Research (ORR) in 1937. Funded by a Rockefeller Foundation grant, the ORR was headed by Paul F. Lazarsfeld, who was assisted by Hadley Cantril and Frank Stanton. The latter would go on to assume the presidency of CBS in 1946 and would serve in that capacity into the 1970s. Over a ten-year period, the ORR published several texts dealing with audience research findings and methodology. Among them were Lazarsfeld and Stanton's multivolume *Radio Research*, which covered the periods of 1941-1943 and 1948-1949. During the same decade, Lazarsfeld also published booklength reports on the public's attitude toward radio: *The People Look at Radio* (1946) and *Radio Listening in America* (1948). Both works cast radio in a favorable light by concluding that most listeners felt the medium did an exemplary job.

The Pulse and Hooper were the prevailing radio station rating services in the 1950s as the medium worked at regaining its footing following the meteoric rise of television. In 1965 Arbitron Ratings began measuring radio audience size through the use of a diary, which required respondents to document their listening habits over a seven-day period. By the 1970s Arbitron reigned as the leading radio measurement company, while Hooper and Pulse faded from the scene. To provide the radio networks and their affiliates and advertisers with much-needed ratings information, Statistical Research, Inc., of New Jersey, introduced Radio's All Dimension Audience Research (RADAR) in 1968. The company gathers its information through telephone interviews to over six thousand households. In the 1990s, Arbitron retains its hold on first place among services measuring radio audiences, especially since the demise of Birch/Scarborough, which gained considerable acceptance following its debut in the late 1970s. In 1991, this audience measurement company became yet another victim of the economic malaise.

Ratings companies must be reliable. Credibility is crucial to success. Therefore, measurement techniques must be tried and true. Information must be accurate, since millions of dollars are at stake. In 1963 the Broadcast Rating Council was established to monitor, audit, and accredit the various ratings companies. The council created performance standards to which rating services are expected to adhere. Those that fail to meet the council's operating criteria are not accredited. A nonaccredited ratings service will seldom succeed. In 1982 the Broadcast Rating Council was renamed the Electronic Media Planning Council to reflect a connection with the ratings services dealing with the cable television industry.

THE RATINGS AND SURVEY SERVICES

The extreme fragmentation of today's listening audience, created by the almost inestimable number of stations and formats, makes the job of research a complex but necessary one. All stations, regardless of size, must put forth an effort to acquaint themselves with the characteristics of the audience, says Edward J. Noonan, co-director, Survey Research Associates. "A station cannot operate in a vacuum. It has to know who is listening and why before making any serious programming changes." Today this information is made available through several ratings services and research companies. More stations depend on Arbitron audience surveys than any other.

Since the collapse of Birch/Scarborough, broadcasters have had little choice but to subscribe to Arbitron—that or go without the listening estimates on which so many agencies and advertisers rely.

However, the radio audience survey industry has begun to expand, if only slowly. For example, in recent years, additional listener/ratings services

have begun to emerge. AccuRatings is one example. It began to measure audiences in major metropolitan areas in the middle 1990s. Similar companies were beginning to surface to the relief of many broadcasters concerned with Arbitron's hold. Arbitron's dominance has indeed caused anxiety in the radio community. Many managers and programmers were more than disturbed by the failure of the alternative measurement service. Arbitron covers over 250 markets ranging in size from large to small. Arbitron claims over 2,700 radio clients and a staff of 3,000 interviewers who collect listening information from 2 million households across the country. All markets are measured at least once a year during the spring; however, larger markets are measured on an ongoing basis year round. Until the

FIGURE 6.1
Arbitron listening location analysis. Courtesy Arbitron.

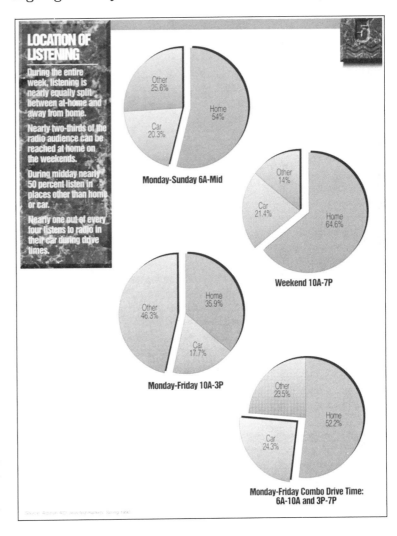

FIGURE 6.2
First page of an Arbitron Radio Market Report containing an explanation of what constitutes the survey area. Courtesy Arbitron.

REPORT CONTENTS AND MAP PAGE

This is the first page of an Arbitron Radio Market Report. This page, in addition to showing a map of the Radio Market survey area and table of Report contents, contains a basic introduction to the Report, restrictions on the use of the Report, a copyright infringement warning, the survey time period, frequency of measurement and Arbitron's schedule of radio surveys for the year.

1▶ The map shows the survey area(s) included in a market definition and for which audience estimates are reported. Areas in grey are not included in a market definition (unless it is a part of an ADI to be reported for that particular market) and will have no audience estimates included in the Report.

2▶ The Metro Survey Area (MSA) of the reported market is shown by horizontal hatching.

3▶ The Total Survey Area (TSA) of the reported market is shown in white.

4▶ The Areas of Dominant Influence (ADI) of the reported market, if applicable, is shown by diagonal hatching.

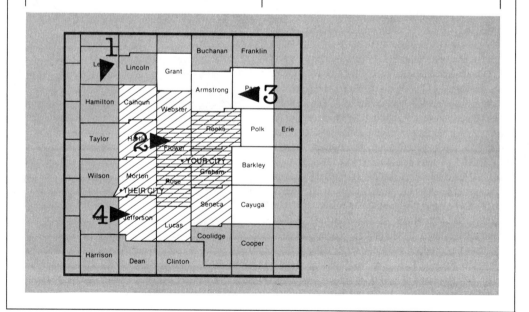

FIGURE 6.3
Arbitron sample
explanation.
Courtesy Arbitron.

POPULATION ESTIMATES AND SAMPLE DISTRIBUTION BY AGE/SEX GROUP

1▶ The estimated populations of reported demographic categories within survey area(s) are reported for the Metro Survey Area, Area of Dominant Influence (where applicable), and the Total Survey Area.

2▶ The ratio (expressed as a percentage) of each reported demographic's estimated population to the estimated population of all persons age 12 or older within the relevant survey area.

3▶ The number of unweighted diaries in-tab, from persons in reported demographics, expressed as a percentage of the total number of unweighted in-tab diaries from all persons age 12 or older.

4▶ The distribution of in-tab diaries after weighting for reported demographics expressed as a percentage of the total number of tabulated diaries from all persons age 12 or older.

5▶ The total persons age 12 or older in the Metro and the percent of those persons living in military housing, college dormitories and other group quarters.

6▶ The number of estimated residences in the predesignated sample, number of homes contacted, number of homes in which diaries were placed, number of persons with whom diaries were placed and the number of persons who returned usable diaries (see The Sample, Section II, for further definition of these terms) are reported for Metro, ADI (if applicable), and TSA. Standard and ESF sample statistics are reported separately along with a total sample.

Total Survey Area

	1 Estimated Population	2 Estimated Population as Percent of Tot. Persons 12 +	3 Percent of Unweighted In-Tab Sample	4 Percent of Weighted In-Tab Sample
Men 18-24	115,900	7.2	6.4	7.2
Men 25-34	139,900	8.7	8.5	8.7
Men 35-44	107,300	6.6	6.7	6.6
Men 45-49	49,400	3.1	2.5	3.1
Men 50-54	54,600	3.4	2.9	3.4
Men 55-64	96,300	6.0	7.0	6.0
Men 65 +	94,700	5.9	4.2	5.9

Area of Dominant Influence

Men 18-24	104,200	7.2	6.3	7.2
Men 25-34	124,900	8.6	8.1	8.6
Men 35-44	96,500	6.6	7.1	6.6
Men 45-49	44,800	3.1	2.3	3.1
Men 50-54	49,300	3.4	2.7	3.4
Men 55-64	86,300	5.9	7.0	5.9
Men 65 +	84,000	5.8	4.1	5.8

Metro Survey Area

Men 18-24	74,800	7.2	6.8	7.1
Men 25-34	90,900	8.6	8.5	8.6
Men 35-44	71,500	6.8	7.3	6.8
Men 45-49	33,500	3.2	2.4	3.2
Men 50-54	36,800	3.5	2.5	3.5
Men 55-64	62,300	5.9	6.7	5.9
Men 65 +	57,600	5.4	3.8	5.4

Metro Persons Living in Group Quarters

5▲

	% Military	% College	% Other Group Quarters
Total Persons 12 + 13,476,300	.1	.4	1.7

Diary Placement and Return Information

6▶

	Metro	ADI	TSA
Standard Residential Listings in Designated Sample	751	1,033	1,185
ESF Residential Listings in Designated Sample	359	380	383
Total Residential Listings in Designated Sample	1,110	1,413	1,568
Standard Contacts (homes in which telephone was answered)	740	1,023	1,568
ESF Contacts (homes in which telephone was answered)	353	359	374
Total Contacts (homes in which telephone was answered)	1,093	1,382	11,514
Standard Homes in Which Diaries Were Placed	657	920	1,037
ESF Homes in Which Diaries Were Placed	253	257	283
Total Homes in Which Diaries Were Placed	910	1,177	1,320
Standard Individuals Who Were Sent a Diary	1,562	2,210	2,487
ESF Individuals Who Were Sent a Diary	588	600	625
Total Individuals Who Were Sent a Diary	2,150	2,810	3,112
Standard Individuals Who Returned a Usable Diary (In-Tab)	1,031	1,421	1,610
ESF Individuals Who Returned a Usable Diary (In-Tab)	360	374	398
Total Individuals Who Returned a Usable Diary (In-Tab)	1,391	1,795	2,008

FIGURE 6.3
continued

POPULATION ESTIMATES AND TABULATED DIARIES BY SAMPLING UNIT

1 ▶ The specific designation of the Arbitron Sampling Unit and a coded indicator of its relationship to the market definition. The codes "M," "T" and, if applicable, "A" are used to identify Metro, TSA and ADI sampling units, respectively. Where a sampling unit qualifies for inclusion in more than one area, a code will appear for each area.

2 ▶ Estimated Population: The estimated population of persons 12 years or older in a sampling unit as determined by Market Statistics, Inc. (MSI), based on the 1980 U.S. Census estimates. These population estimates are updated annually as of the first of January, by MSI.

3 ▶ In-Tab: The number of usable diaries returned from diarykeepers in a sampling unit.

4 ▶ The county/sampling unit name and state included in the survey area.

5 ▶ Occasionally, no county estimated population and in-tab information will appear next to a county name if the county's in-tab has been combined (clustered). For Market Report production purposes, population and in-tab for the county have been included with the preceding population estimates and in-tab numbers, respectively. A footnote at the bottom of page 2 of the Radio Market Report pertaining to this "clustering" procedure has been added effective with the Winter 1983 radio survey. Specific county data breakouts are obtainable through Arbitron's Radio Policies and Procedures Department.

Population Estimates and Tabulated Diaries by Sampling Unit

1 2 3 4

Area	Estimated Population	In-Tab	Counties	State
MTA	76,100	67	GRAHAM	US
MTA	501,600	514	FLOWER	US
MTA	698,900	750	ROOKS	US
MTA	202,100	186	ROSE	US
TA	103,800	118	WEBSTER	US
			CALHOUN	US
TA	78,700	51	HARLAN	US
	◀5▶		MORTON	US
			JEFFERSON	US
TA	142,500	132	LUCAS	US
TA	202,100	153	SENECA	US
T	41,900	20	GRANT	US
T	248,200	208	ARMSTRONG	US
T	61,300	30	PAGE	US
T	48,000	82	POLK	US
			BARKLEY	US
			CAYUGA	US

Area	Estimated Population	In-Tab	Counties	State

M — METRO SAMPLING UNIT T — TSA SAMPLING UNIT A — ADI SAMPLING UNIT early 1980s, metro markets traditionally were rated in the spring and fall. However, six months between surveys was considered too long in light of the volatile nature of the radio marketplace.

To determine a station's ranking, Arbitron follows an elaborate procedure. First the parameters of the area to be surveyed are established. Arbitron sees fit to measure listening both in the city or urban center, which it refers to as the Metro Survey Area (MSA), and in the surrounding communities or suburbs, which it classifies Total Survey Area (TSA). Arbitron classifies a station's primary listening locations as its Areas of Dominant Influence (ADI).

Once the areas to be measured have been ascertained, the next thing Arbitron does is select a sample base composed of individuals to be queried regarding their listening habits. Metromail provides Arbitron with computer tapes that contain telephone and mailing lists from which the rating company derives its randomly selected sample. Arbitron conducts its surveys over a three-to-four-week period, during which time new samples are selected weekly.

When the sample has been established, a letter is sent to each targeted household. The replacement letter informs members of the sample that they have been selected to participate in a radio listening survey and asks their cooperation. Within a couple of days after the letter has been received, an Arbitron interviewer calls to describe the purpose of the survey as well as to determine how many individuals 12 or older reside in the household. Upon receiving the go-ahead, Arbitron mails its seven-day survey diary, which requires respondees to log their listening habits. An incentive stipend of a dollar or two accompanies the document. The diary is simple to deal with, and the information it requests is quite basic: time (day/part) tuned to a station, station call letters or program name, whether AM or FM, where listening occurred—car, home, elsewhere. Although the diary asks for information pertaining to age, sex, and residence, the actual identity or name of those participating is not requested.

Prior to the start of the survey, a representative of Arbitron makes a presurvey follow-up call to those who have agreed to participate. This is done to make certain that the diary has been received and that everyone involved understands how to maintain it. Another follow-up call is made during the middle of the survey week to ascertain if the diary is being kept and to remind each participant to promptly return it upon completion of the survey. Outside the metro area, follow-ups take the form of a letter. The diaries are mailed to Beltsville, Maryland, for processing and computation.

Arbitron claims that 65 out of every 100 diaries it receives are usable. Diaries that are inadequately or inaccurately filled out are not used. Upon arriving at Arbitron headquarters, diaries are examined by editors and rejected if they fail to meet criteria. Any diary received before the conclusion of the survey period is immediately voided, as are those that arrive more than twelve days after the end of the survey period. Diaries with blank or ambiguous entries also are rejected. Those diaries that survive the editors' scrutiny are then processed through the computer, and their information is tabulated. Computer printouts showing audience estimates are sent to subscribers. Stations receive the "book" within a few weeks after the last day of the survey.

The most recent research tool made available by Arbitron to its subscribers is a telephone-delivered radio market service called Arbitrends, which utilizes the IBM-XT microcomputer to feed data to stations similarly equipped. Information regarding a station's past and current performances and those of competitors is available at the touch of a finger. Breakouts and tailor-made reports are provided on an ongoing basis by Arbitrends to assist stations in the planning of sales and marketing strategies. The survey company has over 13 billion characters reserved on computer disk packs. Arbitron also makes available its Arbitrends Rolling Average Printed Reports to those stations without computers. To date, Arbitron has prepared over 380,000 radio market reports.

FIGURE 6.4
Instructions for filling out a diary. Accuracy is important. Courtesy Arbitron.

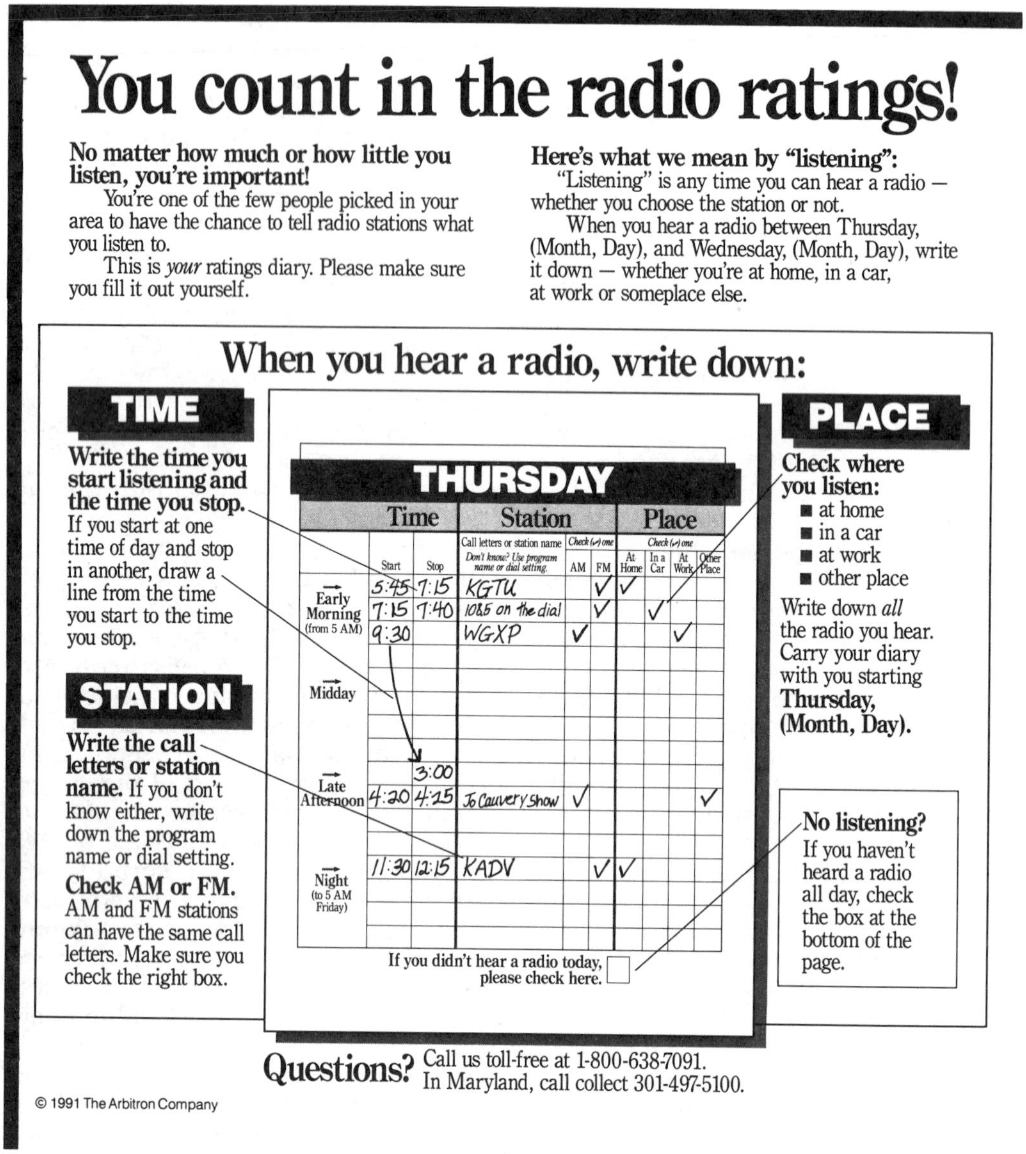

You count in the radio ratings!

No matter how much or how little you listen, you're important!

You're one of the few people picked in your area to have the chance to tell radio stations what you listen to.

This is *your* ratings diary. Please make sure you fill it out yourself.

Here's what we mean by "listening":

"Listening" is any time you can hear a radio — whether you choose the station or not.

When you hear a radio between Thursday, (Month, Day), and Wednesday, (Month, Day), write it down — whether you're at home, in a car, at work or someplace else.

When you hear a radio, write down:

TIME

Write the time you start listening and the time you stop.
If you start at one time of day and stop in another, draw a line from the time you start to the time you stop.

STATION

Write the call letters or station name. If you don't know either, write down the program name or dial setting.

Check AM or FM. AM and FM stations can have the same call letters. Make sure you check the right box.

THURSDAY

Time		Station	Place				
		Call letters or station name *Don't know? Use program name or dial setting.*	*Check (✓) one*		*Check (✓) one*		
Start	Stop		AM	FM	At Home	In a Car	At Work	Other Place
Early Morning (from 5 AM)	5:45	7:15	KGTU		✓	✓		
	7:15	7:40	108.5 on the dial	✓			✓	
	9:30		WGXP	✓				✓
Midday								
Late Afternoon		3:00						
	4:20	4:25	Jo Cauvery show	✓				✓
Night (to 5 AM Friday)	11:30	12:15	KADV		✓	✓		

If you didn't hear a radio today, please check here. ☐

PLACE

Check where you listen:
- at home
- in a car
- at work
- other place

Write down *all* the radio you hear. Carry your diary with you starting **Thursday, (Month, Day).**

No listening?
If you haven't heard a radio all day, check the box at the bottom of the page.

Questions? Call us toll-free at 1-800-638-7091. In Maryland, call collect 301-497-5100.

Arbitron's most formidable rival in recent years was Birch/Scarborough, headquartered in New Jersey. As a radio audience measurement service, Birch provided clients with both quantitative and qualitative data on local listening patterns, audience size, and demographics. Birch interviewers telephoned a prebalanced sample of households during the evening hours, seven days a week, to acquire the information they needed. "Respondents aged twelve or older were randomly selected from both listed and non-listed telephone households. These calls were made from highly supervised company WATTS facilities," notes Phil Beswick, vice president of the defunct Birch/ Scarborough broadcast services. The sample sizes varied depending on the size of the market being surveyed. For example, Birch/Scarborough contacted approximately 1,100 households in a medium market and between 2,000 and 8,000 in major metro markets.

A wide range of reports were available to clients, including the *Quarterly Summary Report*: estimates of listening by location, county by county, and other detailed audience information; *Standard Market Report*: audience analysis especially designed for small market broadcasters; *Capsule Market Report*: listening estimates in the nation's smallest radio markets; *Condensed Market Report*: designed specifically for radio outlets in markets not provided with regular syndicated measurements and where cost was a key consideration; *Monthly Trend Report*: an ongoing picture of the listening audience so that clients might benefit from current shifts in the marketplace; *Prizm*: lifestyle-oriented radio ratings book that defined radio audiences by lifestyle characteristics in more than eighty-five markets; and *BirchPlus*: microcomputerized system (IBM-PC) ratings retrieval and analysis. Fees for Birch/Scarborough services were based upon market size. Birch provided subscribers with general

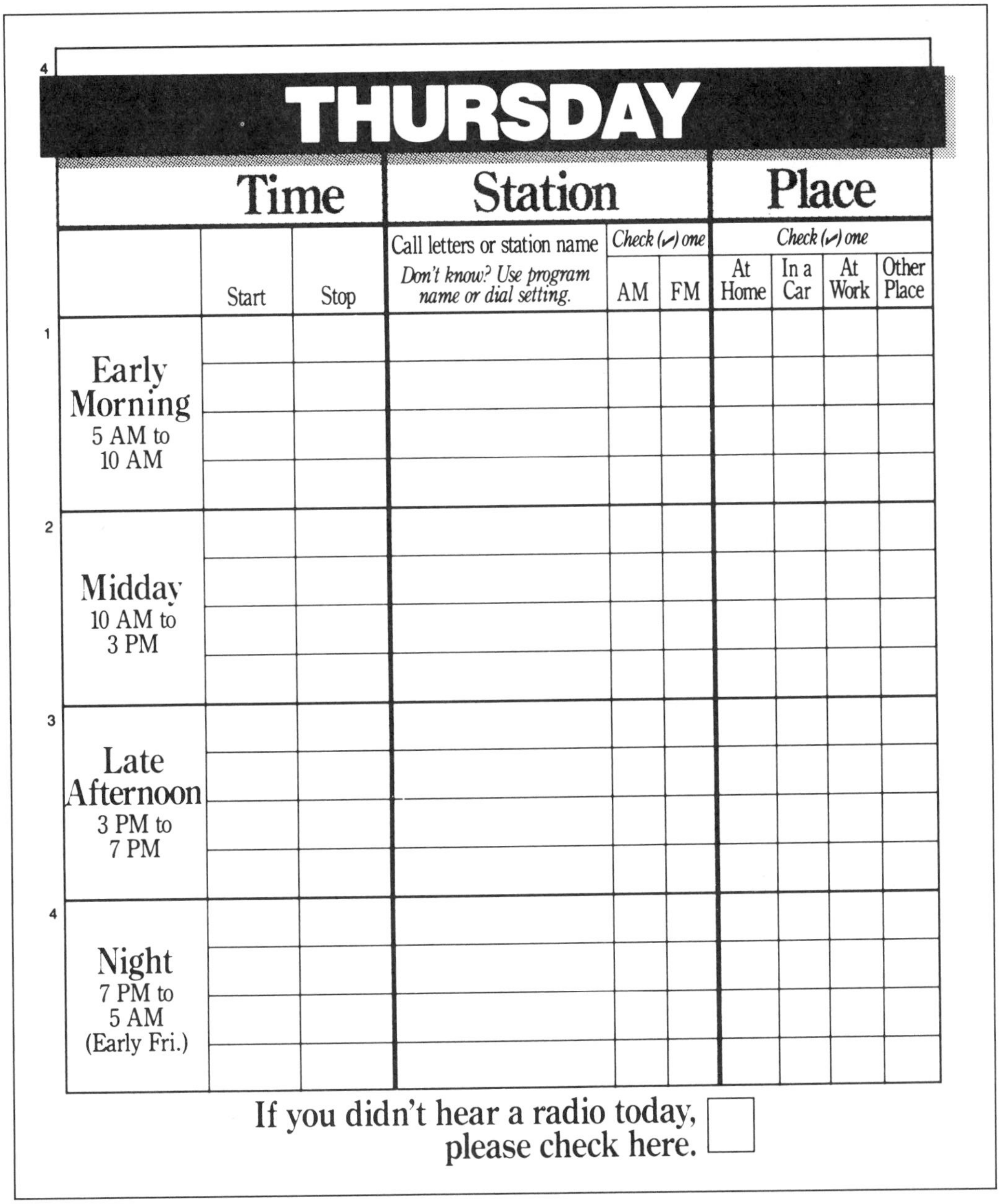

FIGURE 6.5
Diary log sheet. This current sheet includes a checkmark box for "at work," a feature not found in earlier sheets. Courtesy Arbitron.

product consumption and media usage data. In late 1994, Arbitron purchased 50 percent of the Scarborough Research Corporation with plans to offer stations expanded data about their listeners.

Dozens of other research companies throughout the country (among them Coleman Research, Bolton Research Corporation, Mark Kassof and Company, Spectrum Research, Frank N. Magid Associates, DIR, Star, Paragon Research, Shane Media, Hagan Media Research, Gallup Services, Mediabase, TAPSCAN, Rantel Research, and others cited herein) provide broadcasters with a broad range of useful audience information. Many utilize approaches similar to Arbitron (and formerly Birch) to collect data, while others use different methods. "Southeast Media Research offers four research methods: focus groups, telephone stud-

FIGURE 6.6
Diary holder demographic information page and further instructions. Courtesy Arbitron.

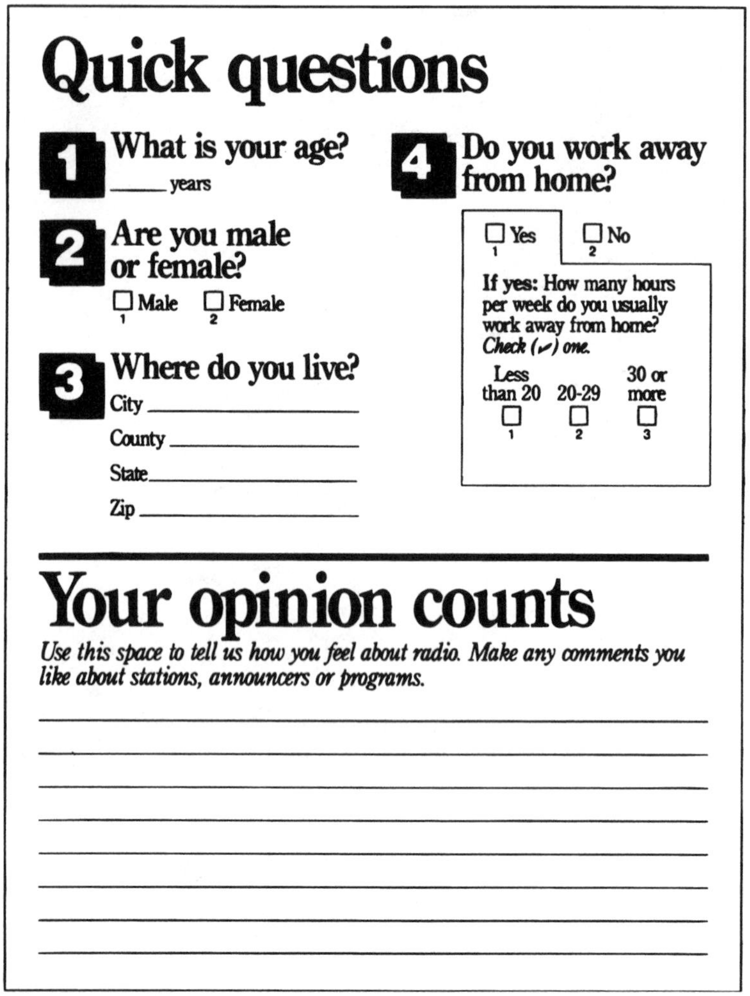

ies, mail intercepts, and music tests," explains Don Hagen, the company's president. Christopher Porter, associate director of Surrey Research, says that his company uses similar techniques.

Meanwhile, audience researcher Dick Warner claims that the telephone recall method is the most commonly used and effective approach to radio audience surveying. "The twenty-four-hour telephone recall interview, in my estimation, yields the most reliable information. Not only that, it is quick and current—important factors in a rapidly moving and hyperdynamic radio marketplace."

On the subject of music testing, Coleman Research's Rebecca Reising notes that her company's approach is unique. "We developed F.A.C.T., short for Fit Acceptance and Compatibility Test. F.A.C.T.'s proprietary research methods and sophisticated databases make it much more powerful, reliable, and useful than old-fashioned music testing methods. As quick and efficient as old tests, F.A.C.T. provides a sophistication of interpretation of music tests not offered by any other research company."

In the hyperactive radio industry arena, both traditional and novel audience survey techniques must ultimately prove themselves by assisting stations in their unrelenting quest for stronger ratings numbers.

QUALITATIVE AND QUANTITATIVE DATA

Since their inception in the 1930s, ratings services primarily have provided broadcasters with information pertaining to the number of listeners of a certain age and gender tuned to a station at a given time. It was on the basis of these quantitative data that stations chose a format and advertisers made a buy.

Due to the explosive growth of the electronic media in recent years, the audience is presented with many more options, and the radio broadcaster, especially in larger markets, must know more about his intended listeners in

Quick questions

1 **What is your age?**
_____ years

2 **Are you male or female?**
☐ Male   ☐ Female
  1   2

3 **Where do you live?**
City _____
County _____
State _____
Zip _____

4 **Do you work away from home?**
☐ Yes   ☐ No
  1   2

If yes: How many hours per week do you usually work away from home? Check (✔) one.

Less than 20	20-29	30 or more
☐	☐	☐
1	2	3

Your opinion counts

Use this space to tell us how you feel about radio. Make any comments you like about stations, announcers or programs.

_____
_____
_____
_____
_____
_____
_____
_____
_____

order to attract and retain them. Subsequently, the need for more detailed information arose. In the 1990s, in-depth research is available to broadcasters from numerous sources. In this age of highly fragmented audiences, advertisers and agencies alike have become less comfortable with buying just numbers and look for audience qualities, notes Surrey's Christopher Porter. "The proliferation of stations has resulted in tremendous audience fragmentation. There are so many specialized formats out there, and many target the same piece of demographic pie. This predicament, if it can be called that, has made amply clear the need for qualitative, as well as quantitative, research. With so many stations doing approximately the same thing, differentiation is of paramount importance."

Today a station shooting for a top spot in the ratings surveys must be concerned with more than simply the age and sex of its target audience. Competitive programming strategies are built around an understanding and appreciation of the lifestyles, values, and behavior of those listeners sought by a station.

IN-HOUSE RESEARCH TECHNIQUES

Research data provided by the major survey companies can be costly. For this reason, and others, stations frequently conduct their own audience studies. Although stations seldom have the professional wherewithal and expertise of the research companies, they can derive useful information through do-it-yourself, in-house telephone, face-to-face, and mail surveys.

Telephone surveying is the most commonly used method of deriving audience data on the station level. It generally is less costly than the other forms of in-house research, and sample selection is less complicated and not as prone to bias. It also is the most expedient method. There are, however, a few things that must be kept in mind when conducting call-out surveys. To begin

with, not everyone has a phone and many numbers are unlisted. People also are wary of phone interviews for fear that the ultimate objective of the caller is to sell something. The public is inundated by phone solicitors (both human and computerized). Finally, extensive interviews are difficult to obtain over the phone. Five to ten minutes usually is the extent to which an interviewee will submit to questioning. Call-out interview seminars and instructional materials are available from a variety of sources, including the telephone company itself.

Ultimately, it appears that computer network and e-mail services will provide another valuable means for those radio stations which survey their audiences. Some stations are already employing computers for call-out research purposes. There are many obvious benefits, interactivity and archiving foremost among them. The face-to-face or personal interview also is a popular research approach at stations, although the cost can be higher than call-out, especially if a vast number of individuals are being surveyed in an auditorium setting. The primary advantages of the in-person interview are that questions can be more substantive and greater time can be spent with the respondees. Of course, more detailed interviews are time consuming and usually require refined interviewing skills, both of which can be cost factors.

Mail surveys can be useful for a host of reasons. To begin with, they eliminate the need to hire and train interviewers. This alone can mean a great deal in terms of money and time. Since there are no interviewers involved, one source of potential bias also is eliminated. Perhaps most important is that individuals questioned through the mail are somewhat more inclined toward candor since they enjoy greater anonymity. The major problem with the mail survey approach stems from the usual low rate of response. One in every five questionnaires mailed may actually find its way back to the station. The length of the questionnaire must be kept relatively short, and the questions succinct and direct. Complex questions

FIGURE 6.7
Arbitron estimates
show where a
station stands in
its market.
Courtesy Arbitron.

AVERAGE SHARE TRENDS

The Trends section provides an indication of individual station performance and the relative standing among stations for periods prior to the most recent survey. The duration reported in this section may include as many as five discrete Arbitron survey rating periods, always including the most recent. Trends may not always reflect actual changes over time due to changes in methodology, station operations, etc.

1▶ The estimates reported are average persons shares in the Metro Survey Area, by individual stations, on the basis of broad demographics in specific dayparts. A share is the percent of all listeners in a demographic group that are listening to a specific station. This percent is calculated by dividing the Average Quarter-Hour Persons to a station by the Average Quarter-Hour Persons to all stations.

2▶ Total listening (Metro Totals) in the market, expressed as a rating, is also reported. This is the sum of all reported stations' Average Quarter-Hour estimates plus those for stations not meeting Minimum Reporting Standards and unidentified listening.

3▶ Reported dayparts are as follows:
Monday-Sunday 6AM-Midnight
Monday-Friday 6AM-10AM
Monday-Friday 10AM-3PM
Monday-Friday 3PM-7PM
Monday-Friday 7PM-Midnight

4▶ Reported demographics are as follows:
Total Persons 12+
Men 18+
Women 18+
Teens 12-17

5▶ A "+" indicates station changed call letters.

6▶ A "**" indicates station not reported for that survey.

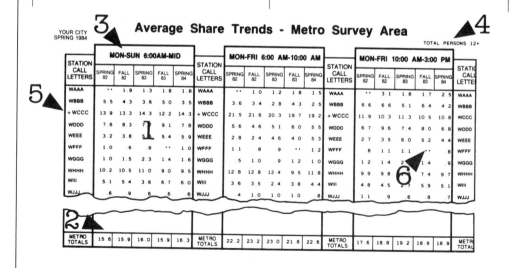

YOUR CITY SPRING 1984

Average Share Trends - Metro Survey Area

TOTAL PERSONS 12+

STATION CALL LETTERS	MON-SUN 6:00AM-MID					STATION CALL LETTERS	MON-FRI 6:00 AM-10:00 AM					STATION CALL LETTERS	MON-FRI 10:00 AM-3:00 PM					STATION CALL LETTERS
	SPRING 82	FALL 82	SPRING 83	FALL 83	SPRING 84		SPRING 82	FALL 82	SPRING 83	FALL 83	SPRING 84		SPRING 82	FALL 82	SPRING 83	FALL 83	SPRING 84	
WAAA	**	1.9	1.3	1.8	1.8	WAAA	**	1.0	1.2	1.8	1.5	WAAA	**	3.1	1.8	1.7	2.5	WAAA
WBBB	5.5	4.3	3.6	5.0	3.5	WBBB	3.6	3.4	2.8	4.3	2.5	WBBB	6.6	6.6	5.1	6.4	4.2	WBBB
+WCCC	13.9	13.3	14.3	12.2	14.3	+WCCC	21.5	21.6	20.3	19.7	19.2	+WCCC	11.9	10.3	11.3	10.5	10.8	WCCC
WDDD	7.6	8.3	7.7	9.1	7.8	WDDD	5.6	4.6	5.1	6.0	5.5	WDDD	6.7	9.6	7.4	8.0	6.8	WDDD
WEEE	3.2	3.8	5.5	5.4	5.9	WEEE	2.8	2.4	4.6	4.0	5.3	WEEE	2.7	3.5	6.0	5.2	4.4	WEEE
WFFF	1.0	.6	.8	**	1.0	WFFF	1.1	.8	.9	**	1.2	WFFF	.8	1.1	1.1	**	.8	WFFF
WGGG	1.0	1.5	2.3	1.4	1.6	WGGG	.5	1.0	.9	1.2	1.0	WGGG	1.2	1.4	2.1	1.4	.9	WGGG
WHHH	10.2	10.5	11.0	9.0	9.5	WHHH	12.8	12.8	12.4	9.5	11.8	WHHH	9.9	9.8	7.3	7.4	9.7	WHHH
WIII	5.1	5.4	3.6	6.7	6.0	WIII	3.6	3.5	2.4	3.8	4.4	WIII	4.8	4.5	2.7	5.9	5.1	WIII
WJJJ	.6	.9	.6	.6	.8	WJJJ	.4	1.0	1.0	1.0	.8	WJJJ	1.1	.9	.6	.8	.7	WJJJ
METRO TOTALS	15.6	15.9	16.0	15.9	16.3	METRO TOTALS	22.2	23.2	23.0	21.8	22.6	METRO TOTALS	17.6	18.8	19.2	18.9	18.9	METRO TOTAL

Average Quarter-Hour Listening Estimates

PHILADELPHIA

SUNDAY
3 00PM-7 00PM

AVERAGE PERSONS—TOTAL SURVEY AREA, IN HUNDREDS												STATION CALL LETTERS
TOT. PERS. 12+	MEN					WOMEN					TNS. 12-17	
	18-24	25-34	35-44	45-54	55-64	18-24	25-34	35-44	45-54	55-64		
392		13	27	27	27	3	6	30	21	65	6	KYW
-1												WBCB
625	15	38	42	59	106	3	31	47	23	66	9	WCAU
562	59	16	41	5	26	65	78	53	8		178	WCAU FM
100		23				4	8	8	9			WDAS
437	107	38	7	25	13	48	107	11	13	15	53	WDAS FM
358		7	19	41	28	3	28	4	58	47		WEAZ
108	11	6	4	35		5	20	8	12		3	WFIL
12					3							WFLN
166			13	12	14	3	18	10	28	9		WFLN FM
178			13	15	14	3	18	10	28	9		TOTAL
109		10			8		5	14	33	26	13	WHAT
52		1				18					33	WIFI
269	14	99	31			28	74	1	6		16	WIOQ
172		4	12	4	20		15	24	16	40	27	WIP
96		2	5	19		13	10	30	17			WKSZ
427	18	94	11	34		29	100	33	38	14	30	WMGK
423	95	93	3	2	6	73	75	15			61	WMMR
551		15	13	70	65	16	28	39	73	41		WPEN
3			1								2	WSNI
146		28	21			23	49				25	WSNI FM
149		28	22			23	49				27	TOTAL
32	11					9	4				3	WSSJ
585	116	116	31		12	69	79	43	5	7	107	WUSL
19									19			WVCH
270	15			26		5	28	27	13	55	1	WWDB
97	10	4	8	11		2	15	14	20	4	2	WWSH
432	88	86		2	10	124	40	3	9	5	65	WYSP
40				17			17			4	2	WZZD
122				40	30		2				32	WJBR FM
113	12	20				54					27	WPST
52	12	11		2		16			8		3	WSTW
24			18		6							WTTM
55									17	13		WOR

AVERAGE PERSONS—METRO SURVEY AREA, IN HUNDREDS												STATION CALL LETTERS
TOT. PERS. 12+	MEN					WOMEN					TNS. 12-17	
	18-24	25-34	35-44	45-54	55-64	18-24	25-34	35-44	45-54	55-64		
350		13	22	14	24	3	6	26	11	62	6	KYW
-1												WBCB
570	15	38	42	50	86	3	31	47	17	46	9	WCAU
420	59	16	27	5	26	59	51	28			126	WCAU FM
91		23				4	8	8				WDAS
343	32	38	7	25	13	48	88	11	13	15	53	WDAS FM
239		7	2	5	23	3		4	34	42		WEAZ
100	11	6	4	29		5	18	8	12		3	WFIL
12					3							WFLN
144			13	12	14	3	5	10	28	9		WFLN FM
156			13	15	14	3	5	10	28	9		TOTAL
109		10			8		5	14	33	26	13	WHAT
23		1				14					8	WIFI
263	14	94	31			28	74		6		16	WIOQ
144		4	12	4	20		13	24			30	27	WIP
76		2	5	10		9	10	30	10			WKSZ
374	18	83	11	27		29	70	28	38	14	30	WMGK
352	74	88	3	2	6	71	47	15			46	WMMR
533		15	13	70	47	16	28	39	73	41		WPEN
3			1								2	WSNI
146		28	21			23	49				25	WSNI FM
149		28	22			23	49				27	TOTAL
32	11					9	4				3	WSSJ
466	75	87	31		12	51	79	38	5	7	81	WUSL
-1												WVCH
237	15			26		5	28	27		55	1	WWDB
68	10	4	8	11		2	15	7	5	4	2	WWSH
289	63	52		2	10	75	8	3	9	5	62	WYSP
40				17			17				2	WZZD
18								2			3	WJBR FM
67	12	20				20					15	WPST
18					2	16						WSTW
24			18		6							WTTM
13											13	WOR

SHARES—METRO SURVEY AREA												STATION CALL LETTERS
TOT. PERS. 12+ %	MEN					WOMEN					TNS. 12-17 %	
	18-24 %	25-34 %	35-44 %	45-54 %	55-64 %	18-24 %	25-34 %	35-44 %	45-54 %	55-64 %		
5.7		1.9	6.7	3.9	6.5	.6	8	6.3	3.6	16.2	1.1	KYW
												WBCB
9.3	3.1	5.5	12.8	13.9	23.3	.6	4.4	11.4	5.6	12.0	1.7	WCAU
6.9	12.2	2.3	8.2	1.4	7.0	11.3	7.2	6.8			23.4	WCAU FM
1.5			3.4				.8	1.1	1.9			WDAS
5.6	6.6	5.5	2.1	7.0	3.5	9.2	12.4	2.7	4.2	3.9	9.9	WDAS FM
3.9			1.0		.6	1.4	6.2	6		1.0	11.1	11.0	WEAZ
1.6	2.3	9	1.2	8.1		1.0	2.5	1.9	3.9		.6	WFIL
.2					8							WFLN
2.4			4.0	3.3	3.8	6	7	2.4	9.2	2.3		WFLN FM
2.6			4.0	4.1	3.8	6	7	2.4	9.2	2.3		TOTAL
1.8		1.5			2.2		.7	3.4	10.8	6.8	2.4	WHAT
.4		.1				2.7					1.5	WIFI
4.3	2.9	13.7	9.4			5.4	10.4		2.0		3.0	WIOQ
2.4		6	3.6	1.1	5.4		1.8	5.8		7.8	5.0	WIP
1.2		.3	1.5	2.8		1.7	1.4	7.2	3.3			WKSZ
6.1	3.7	12.1	3.3	7.5		5.5	9.9	6.8	12.4	3.7	5.6	WMGK
5.8	15.3	12.8	9	6	1.6	13.6	6.6	3.6			8.6	WMMR
8.7		2.2	4.0	19.5	12.7	3.1	3.9	9.4	23.9	10.7		WPEN
			3								4	WSNI
2.4		4.1	6.4			4.4	6.9				4.6	WSNI FM
2.4		4.1	6.7			4.4	6.9				5.0	TOTAL
.5	2.3					1.7	.6				6	WSSJ
7.6	15.5	12.7	9.4		3.3	9.8	11.1	9.2	1.6	1.8	15.1	WUSL
												WVCH
3.9	3.1			7.2		1.0	3.9	6.5		14.4	2	WWDB
1.1	2.1	6	2.4	3.1		4	2.1	1.7	1.6	1.0	4	WWSH
4.7	13.0	7.6		6	2.7	14.3	1.1	7	2.9	1.3	11.5	WYSP
.7			4.7				2.4			1.0	4	WZZD
.3				3.5			5			8		WJBR FM
1.1	2.5	2.9				3.8					2.8	WPST
.3			6			3.1						WSTW
4		5.5		1.6			2.4					WTTM
.2										3.4		WOR

TOTAL LISTENING IN METRO SURVEY AREA	6102	484	686	329	359	369	523	710	414	306	383	538

Footnote Symbols: (*) means audience estimates adjusted for actual broadcast schedule.

ARBITRON RATINGS

FIGURE 6.8
Sample page from an Arbitron *Radio Report.* Courtesy Arbitron.

create resistance and may result in the survey being ignored or discarded.

Large and major market outlets usually employ someone to direct research and survey efforts. This person works closely with upper management and department heads, especially the program director and sales manager. These two areas require data on which to base programming and marketing decisions. At smaller outlets, area directors generally are responsible for conducting surveys relevant to their department's needs. A case in point would be the PD who plans a phone survey during a special broadcast to help ascertain whether it should become a permanent program offering. To accomplish this task, the programmer enlists the aid of a secretary and two interns from a local college. Calls are made, and data are collected and analyzed.

The objective of a survey must be clear from the start, and the methodology used to acquire data should be as uncomplicated as possible. Do-it-yourself surveys are limited in nature, and overly ambitious goals and expectations are seldom realized. However, in-house research can produce valuable information that can give a station a competitive edge. Today, no radio station can operate in a detached way and expect to prosper.

Every station has numerous sources of

FIGURE 6.9
Radio station promotional pieces often reflect research and survey data provided by Arbitron and other companies. Courtesy WBMX and WFMT.

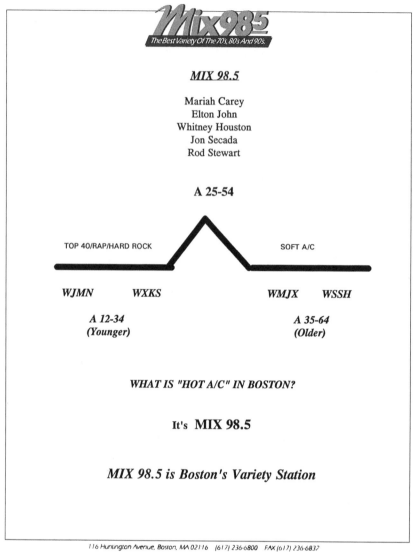

information available to it. Directories containing all manner of data, such as population statistics and demographics, manufacturing and retailing trends, and so on, are available at the public library, city hall, chamber of commerce, and various business associations. The American Marketing Association and American Research Foundation also possess information designed to guide stations with their in-house survey efforts.

RESEARCH DEFICITS

Although broadcasters refer deferentially to the ratings surveys as the "Book" or "Bible," the stats they contain are audience-listening "estimates," no more, and, it is hoped, no less. Since their inception, research companies have been criticized for the methods they employ in collecting audience listening figures. The most prevalent complaint has had to do with the selection of samples. Critics have charged that they invariably are limited and exclusionary. Questions have persisted as to whether those surveyed are truly representative of an area's total listenership. Can one percent of the radio universe accurately reflect general listening habits? The research companies defend their tactics and have established a strong case for their methodology.

WFMT Listener Profile

FIGURE 6.9
continued

AGE		EDUCATION		INCOME		OCCUPATION	
18-34	21.0%	Some H.S.	--	$34,999 or less	22.8%	Not Employed	34.5%
35-54	38.5%	H.S. Grad.	20.9%	$35-49,999	29.8%	Blue Collar	6.7%
55+	40.4%	Some College	12.2%	$50-74,999	19.3%	Proprietor/Mgr.	29.7%
Avg. Age	49.6	College Grad.	40.2%	$75,000+	28.0%	Clerical	7.3%
		Adv. Degree	26.7%			Prof./Tech.	19.9%

General Population Profile

AGE		EDUCATION		INCOME		OCCUPATION	
18-34	35.3%	Some H.S.	7.9%	$34,999 or less	34.0%	Not Employed	33.6%
35-54	38.5%	H.S. Grad.	34.8%	$35-49,999	30.7%	Blue Collar	17.2%
55+	26.2%	Some College	23.4%	$50-74,999	21.8%	Proprietor/Mgr.	18.2%
Avg. Age	43.2	College Grad.	24.1%	$75,000+	13.4%	Clerical	15.1%
		Adv. Degree	9.7%			Prof./Tech.	10.4%

Source: The *Media Audit*, Nov-Dec 1994, Chicago, Illinois. International Demographics, Inc. Projected persons age 18+

WFMT98.7
CHICAGO

In the 1970s, ratings companies were criticized for neglecting minorities in their surveys. In efforts to rectify this deficiency, both Arbitron and Birch established special sampling procedures. The incidence of nontelephone households among Blacks and Hispanics tends to be higher. The survey companies also had to deal with the problem of measuring Spanish-speaking people. Arbitron found that using the personal-retrieval technique significantly increased the response rate in the Spanish community, especially when bilingual interviewers were used. The personal-retrieval technique did not work as well with Blacks, since it was difficult to recruit interviewers to work in many of the sample areas. Thus, Arbitron used a telephone retrieval procedure that involved call-backs to selected households over a seven-day period to document listening habits. In essence, the interviewer filled out the diaries for those being surveyed. In 1982, Arbitron implemented Differential Survey Treatment (DST), a technique designed to increase the response rate among Blacks. The survey company provides incentives over the customary

fifty cents to a dollar to certain Black households. Up to five dollars is paid to some respondents. DST employs follow-up calls to retrieve diaries.

Birch/Scarborough Research employed special sampling procedures and bilingual interviewers to collect data from the Hispanic population. According to the company, its samples yielded a high response rate among Blacks. Thus Birch did not use other special sampling controls. Ethnic listening reports containing average quarter-hour and cume estimates for Hispanics, Blacks, and others were available from the company in a format similar to that of its Capsule Market Report.

Both survey companies employed additional procedures to survey other nontelephone households, especially in markets that have a large student or transient population. In the late 1970s, a Boston station targeting young people complained that Arbitron failed to acknowledge the existence of over 200,000 college students who did not have personal phone listings. The station, which was rated among the top five in the market at the time, contended that a comprehensive survey of the city's lis-

THE MARKET

RADIO ADVERTISING VS. OTHER MEDIA

According to figures from Robert Coen, Senior Vice President, Director of Forecasting at McCann Erickson, radio presently accounts for $8.8 billion in annual revenue. Of this $8.8 billion, $2.1 billion is attributed to "national" sales, while $6.7 billion is considered "local" sales. The combined total translates into approximately 7% of the total $130 billion advertising marketplace.

By contrast, television (broadcast and cable) accounts for $28.3 billion in annual revenue, approximately 22% of the total marketplace, print (newspaper and magazine) accounts for $40.5 billion, or 31% of the market, direct mail accounts for $23.2 billion, or 18% of the market, and other media account for $29.6 billion or 23%.

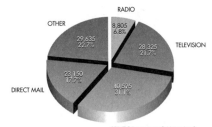

TOTAL ADVERTISING BY MEDIA
1990 PROJECTIONS BY McCANN-ERICKSON
TOTAL DOLLARS IN MILLIONS

CONSUMER INFLUENCE VS. SHARE OF MARKET

A look at each medium's average consumer influence time versus share of total advertising revenue shows a startling imbalance for radio in the total marketplace. Despite its 36% of consumer influence time, radio only has 7% of total advertising revenue.

The other media consumer influence shares vs. share of total marketplace are significantly more in line. Television has a 49% consumer influence time, controlling 22% of the advertising marketplace, Print (newspapers and magazines) has 15% consumer influence time, controlling 31% of the marketplace.

SHARE OF TIME SPENT WITH MEDIA VS.
SHARE OF TOTAL ADVERTISING DOLLARS

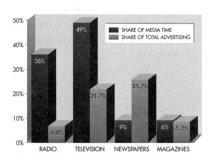

FIGURE 6.10
The Interep Radio Store Research Division conducts ongoing research designed to enhance radio's take of the national advertising budget. Courtesy Interep Radio Stores.

tening audience would bear out the fact that they were, in fact, number one.

Similar complaints of skewed or inconclusive surveys persist today, but the procedures and methods used by the major radio audience research companies, although far from perfect, are more effective than ever. Christopher Porter says the greatest misconception about research data is that they are absolutes etched in granite. "The greatest fallacy is that research findings are gospel. This goes not only for the quantitative studies but for focus groups as well. Regardless of the methodology, any findings should be used as a 'gut adjuster,' rather than a 'gut replacer.' Sampling error is often ignored in a quantitative study, even in an Arbitron report. When we report that 25 percent of a sample feels some way about something, or when a station with a 4.1 beats one with a 3.8 in a book, most station managers and PDs take all these statistics at face value."

Rip Ridgeway, former vice president of Arbitron Ratings, believes that stations place too much emphasis on survey results. "I think that station hierarchy puts too much credence on the ratings estimates. They're an indicator, a sort of report card on a station's performance. They're not the absolute end all. To jump at the next numbers and make sweeping changes based on them generally is a big mistake."

Radio executive Lorna Ozman concurs with both Porter and Ridgeway and warns that research should help direct rather than dictate what a station does. "I use research, rather than letting it use me. The thing to remember is that no methodology is without a significant margin of error. To treat the results of a survey as gospel is dangerous. I rely on research to provide me with the black and white answers and depend on myself to make determinations on the gray areas. Research never provided a radio station with the glitter to make it sparkle."

Station general manager Richard Bremkamp also expresses concern over what he perceives as an almost obsessive emphasis on survey statistics. "The concern for numbers gets out of hand. There are some really good-sounding stations out there that don't do good book, but the money is in the numbers. Kurt Vonnegut talks about 'the Universal Will to Become' in his books. In radio that can be expanded to the Universal Will to Become Number One. This is good if it means the best, but that's not always what it means today."

The proliferation of data services has drawn criticism from broadcasters who feel that they are being oversurveyed and overresearched. When Arbitron introduced its computerized monthly ratings service (Arbitrends), the chairman of its own radio advisory council opposed the venture on the grounds that it would cause more confusion and create more work for broadcasters. He further contended that the monthly service would encourage short-term buying by advertisers. Similar criticism was lodged against Birch/Scarborough's own computerized service, BirchPlus. However, both services experienced steady growth.

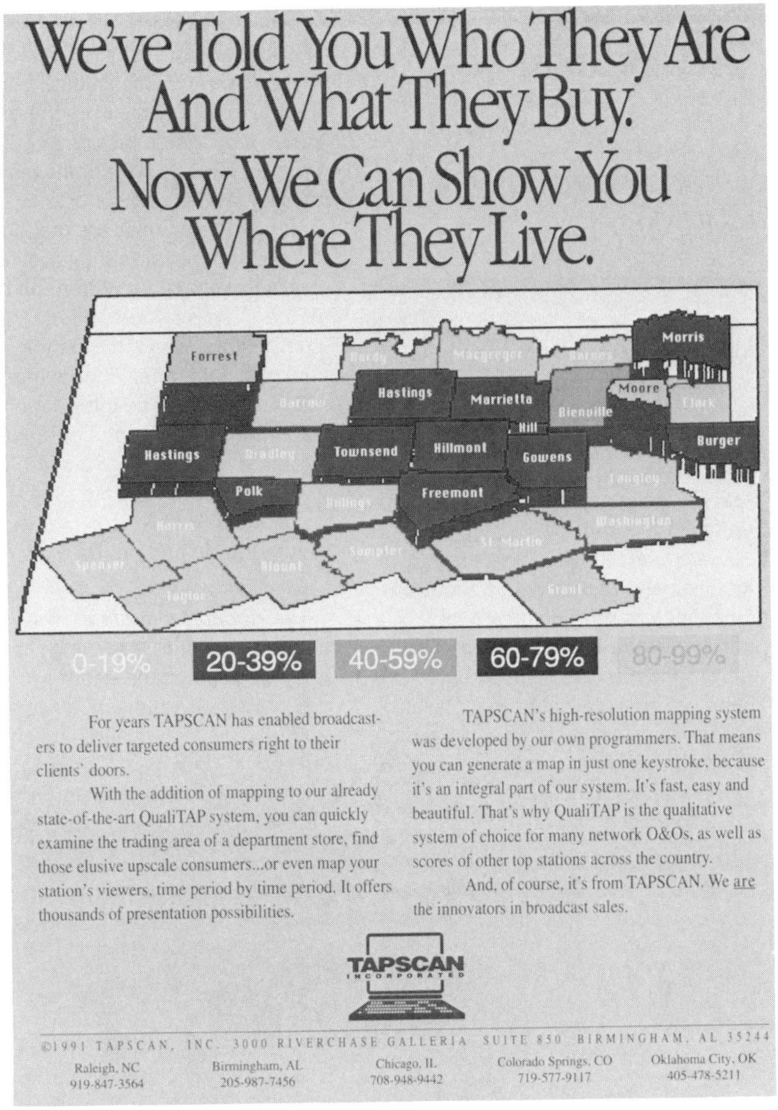

HOW AGENCIES BUY RADIO

The primacy of numbers perhaps is best illustrated through a discussion of how advertising agencies place money on radio stations. It is the media buyer's job to effectively and efficiently invest the advertiser's money, in other words, to reach the most listeners with the budget allotted for radio use.

According to media buyer Lynne Price, the most commonly employed method determines the cost per point (CPP) on a given station. Lynne explains the procedure: "A media buyer is given a budget and a gross rating point (GRP) goal. Our job is to buy to our GRP goal, without going over budget, against a predetermined target audience, i.e., adults 25 to 54, teens, men 18 to 34, etc. Our CPP is derived by taking the total budget and dividing by the GRP goal, or total number of rating points we would like to amass against our target audience. Now, using the CPP as a guideline, we take the cost per spot on a given station, and divide by the rating it has to see how close to the total CPP the station is. This is where the negotiation comes in. If the station is way off, you can threaten not to place advertising until they come closer to what you want to spend."

FIGURE 6.11
The more a station knows about its audience, the more effectively it can program itself. Courtesy TAPSCAN.

Ted Bolton

What a Research Company Does

Bolton is a quantitative research company—our main objective is to get the opinions of radio listeners on virtually anything having to do with the sound of a given station (or future station). There are several methods we use to get the info:

Perceptual studies. These are in-depth surveys that gather opinions on issues such as music preferences, station personalities, competitors, etc. All respondents are included based on their age, listening habits (e.g., favorite station), favorite music types, ethnicity, county of residence, and anything else of importance to the station (client). These surveys run 15-20 minutes.

Respondents are selected at random. We also do tracking studies, which include respondents from the original survey and which measure changes in opinion (usually six months out from the original survey).

Perceptual studies are the foundation of research, because they provide overall market information: perceptions of the client's station and its competitors, which the station uses to make programming, marketing, and sales decisions. The information gathered also aids us in designing a research program for the station.

Music and program testing. There are several commonly used methods for testing music and program elements.

1. *Auditorium testing.* This is the industry standard for testing music that may be aired on the station. Respondents are screened according to station listenership, and paid an incentive—between $35 and $50, depending on market size. The typical test involves 100–150 respondents; they are usually split into two groups. Each group gathers at a hotel, where they listen on speakers to 350–400 hooks (5–10 seconds of a song, the most memorable part). In total, 700-800 hooks will be tested. The respondents score each hook, using a 1–5 (or similar) scale. They also note if they are familiar with the song, and give it a burn score if they're tired of hearing the song.

Bolton does "Personalized Music Tests" (PMT) instead of auditorium tests. Respondents in a PMT come to a facility at a time of their choice, and they test the hooks on Walkman-style cassette players with headphones. We've found that we have a better turn-out than the auditorium tests, and that the results are better because each respondent hears the hooks the same way (in an auditorium, respondents are at different distances from the speakers) without distractions from other listeners. Auditorium tests are still the industry standard, however.

2. *Perceptual analyzer tests.* In both methods described above, respondents typically score hooks with paper and pencil. The data then must be coded and tabulated before the client sees the results. A recent development, perceptual analyzer tests, gives a client instant information. Respondents give their scores on dials or keypads, which are hooked into a computer. As the scores are given, the computer produces an instant, continuous EKG-like graph that shows averages for different groups—a station's "core" listeners and its competitors' listeners, for instance. We've found this method (we call the test "Bolt-Scan") to be most useful in testing music in the original order that it actually aired, thereby showing us both popular songs that keep listeners tuned in, and the stuff that causes tune-out. It works the same way for testing morning shows, comedy bits, and so on.

3. *Call-out music research.* This is a staple of radio programming. Respondents of a specific age and listening group—typically core listeners in a tight age range—are called at home and asked to score 25–30 hooks. The test is obviously short, and can be conducted biweekly or even weekly. This

FIGURE 6.12

information is most useful for trending the familiarity, popularity, and burn-out of songs (usually new songs) over time. Call-out is a task often assigned to station interns, but a number of research companies offer the service as well.

4. *Focus groups and listener panels.* Focus groups are, of course, used for all kinds of consumer product testing, and radio is no exception. Again, respondents are screened for age and listenership. Chosen participants are paid a small incentive, and come to a focus group facility or hotel room in groups of 8–12. Station staff observe while the moderator asks the respondents about their likes and dislikes of station attributes, including music preferences, personalities, competing stations, etc.

Listener panels are more informal and are usually done in-house by the station. Respondents are recruited on the air or from the station's database. Again, respondents are asked for their opinions about the station.

5. *Statistics.* Although some complicated statistical methods are sometimes used, the vast majority of data analysis involves "descriptive" statistics: frequency and average scores. Frequency is a computation of how many people fit into a category or choose a given response to a question. Averages (mean, median, mode) are measures of average or typical performance. For most research reports, these basic statistics are enough.

Some more sophisticated statistics are used to look for relationships and differences between groups: correlation ANOVA (analysis of variance), t-tests, chi-square, and cluster and factor analysis. For instance, we often use cluster analysis to separate respondents into distinct groups, such as "Modern Rock" and "Classic Rock" lovers. This helps a station to determine which artists are unique elements in a specific audience's tastes.

New development. Two new areas are developing that may show the direction for the research industry:

1. *Computer-assisted and fully automated interviewing.* Live interviewers conduct virtually all interviews at this point. But computers can be used to do part of an interview—a live interviewer screens the respondent, and then lets the computer ask the questions—or the computer can even do the whole interview. This technology is largely untested, but could be useful as people become more resistant to spending 20-30 minutes or more on the phone. Computer interviewing allows respondents to pause tests, and all the responses are instantly coded, because responses are given on the telephone keypad. Resistance to technology and issues of control over responses will have settled before computer-conducted interviews are widespread.

2. *Integrating research and marketing programs.* At this point, research and marketing are kept separate. Research must be kept "pure"—as free from bias as possible—in order to claim that it is "representative of the marketplace." Marketing, conversely—especially telemarketing—is sales-driven and meant to be directive. In order to increase listenership, though, a hybrid of research and telemarketing—"teleresearch"—is being developed which simultaneously gathers unbiased information (through research calls) and encourages listening (through telemarketing calls). All information gathered is then put into a database, which the station can use for marketing and market intelligence.

Why do research? Not every radio programmer is a convert to the gospel of research. It is expensive, and stations are always on a tight budget. Many programmers have done just fine using experience and their gut to get them through. Additionally, as people become more guarded with their privacy, both in terms of personal information and home invasion with the telephone, research becomes a more difficult and expensive proposition.

Yet the market intelligence research provides is vital to the stations. As deregulation approaches, and station values skyrocket, gut feelings alone won't be enough to keep a multi-million dollar station on top. Research provides information that can make the gut feeling more of a sure thing, or tell you that your gut is all wrong. But it is a tool, much as a computer or a tape recorder is. Even the best research is useless if it's not studied and properly applied. And shoddy research can do a lot more harm than good.

Bolton Research Director Doug Keith assembled the preceding material.

The other method used to justify station buys is cost per thousand (CPM). Using this technique, the buyer determines the cost of reaching 1,000 people at a given station. The CPM of one station is then compared with that of another's to ascertain efficiency. In order to find out a station's CPM, the buyer must know the station's average quarter-hour audience (AQH persons) estimate in the daypart targeted and the cost of a commercial during that time frame. The computation below will provide the station's CPM:

$$\frac{\$30 \text{ for } 60 \text{ seconds}}{25 \text{ } (000) \text{ AQH}} = \$1.20 \text{ CPM}$$

By dividing the number of people reached into the cost of the commercial, the cost per thousand is deduced. Thus, the lower the CPM, the more efficient the buy. Of course, this assumes that the station selected delivers the

FIGURE 6.13 Research companies analyze a station's intended target audience. Courtesy International Demographics.

target audience sought. Again, this is the responsibility of the individual buying media for an agency. It should be apparent by now that many things are taken into consideration before airtime is purchased.

CAREERS IN RESEARCH

The number of media research companies has grown rapidly since the late 1960s. Today over 150 research houses nationwide offer audience measurement and survey data to the electronic media and allied fields. Job opportunities in research have increased proportionately. Persons wanting to work in the research area need sound educational backgrounds, says Surrey's Porter. "College is essential. An individual attempting to enter the field today without formal training is at a serious disadvantage. In fact, a master's degree is a good idea." Dr. Rob Balon, president of Balon and Associates, agrees with Porter. "Entering the research field

today requires substantial preparation. College research courses are where to start."

Researcher Ed Noonan is of the same opinion. "It is a very competitive and demanding profession. Formal training is very important. I'd advise anyone planning a career in broadcast research to get a degree in communications or some related field and heavy-up on courses in research methodology and analysis, statistics, marketing, and computers. Certain business courses are very useful, too."

To Don Hagen of Southeast Media Research, a strong knowledge of media is a key criterion when hiring. "One of the things that I look for in a job candidate is a college background in electronic media. That's the starting point. You have to know more today than ever before. Audience research has become a complex science."

As might be expected, research directors also place considerable value on experience. "The job prospect who offers some experience in the research area, as well as a diploma, is particularly attractive," notes researcher Dick Warner. Ed Noonan concurs. "Actual experience in the field, even if it is gained in a summer or part-time job, is a big plus." Christopher Porter advises aspiring researchers to work in radio to get a firsthand feel for the medium. "A hands-on knowledge of the broadcast industry is invaluable, if not vital, in this profession."

In the category of personal attributes, Dick Warner puts inquisitiveness at the top of the list. "An inquiring mind is essential. The job of the researcher is to find and collect facts and information. Curiosity is basic to the researcher's personality." Don Hagen adds objectivity and perceptiveness to the list, while Dwight Douglas of Burkhart/Douglas and Associates emphasizes interactive skills. "People skills are essential, since selling and servicing research clients are as important as the research itself."

While not everyone is suited for a career in audience research, those who are find the work intellectually stimulating and financially rewarding.

FIGURE 6.14
Research data
provides stations
with a
comprehensive
profile of their
audience. Courtesy
The Media Audit.

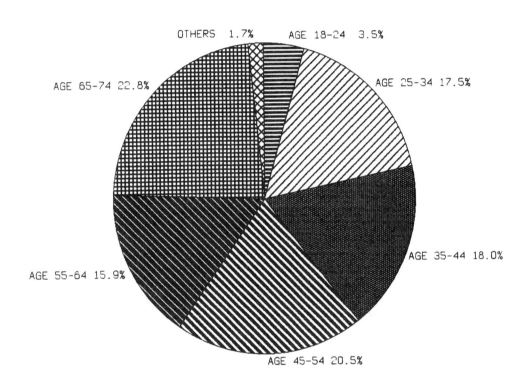

THE MEDIA AUDIT

MEDIA PROFILE REPORT

* CHICAGO * NOV-DEC 1994 *

ADULTS AGE 18 +

AUDIENCE ANALYSIS FOR : **WFMT-FM**

THE MEDIA AUDIT   7-DAY CUME PERSONS :   242,700

AGE PROFILE

OTHERS 1.7%   AGE 18-24 3.5%

AGE 65-74 22.8%   AGE 25-34 17.5%

AGE 55-64 15.9%   AGE 35-44 18.0%

AGE 45-54 20.5%

MEDIA AUDIENCE ANALYSIS IS BASED ON   64 RESPONDENTS IN THE AUDIENCE OF THE MEDIA.

THE FUTURE OF RESEARCH IN RADIO

Most experts agree that the role of research in radio will continue to grow. They base their predictions on the ever-increasing fragmentation and niching of the listening audience, which makes the job of targeting and positioning a more complex one. "The field of broadcast research has grown considerably in the last decade, and there is every reason to suspect that the growth will continue. As demographic targets and formats splinter, there will be an increasingly greater need to know. Much of the gut feel that has propelled

FIGURE 6.15
Call-out music
research helps a
station determine
what and what not
to program. In this
example, callers
were played bits of
songs over the
phone and asked to
rank them.
Courtesy Balon
Associates.

'HATE IT' - TOP 50

	RESPONSES
MICKEY - TONI BASIL	24
GUITAR MAN - ELVIS PRESLEY	19
I'LL TUMBLE FOR YA - CULTURE CLUB	18
UNDERCOVER OF THE NIGHT - ROLLING STONES	14
RIO - DURAN DURAN	13
LITTLE RED CORVETTE - PRINCE	13
PLAY THAT FUNKY MUSIC - WILD CHERRY	12
PLEASE COME TO BOSTON - DAVE LOGGINS	12
AFTER THE LOVING - ENGELBERT HUMPERDINCK	12
COME ON EILEEN - DEXY'S MIDNIGHT RUNNERS	12
A WOMAN NEEDS LOVE - RAY PARKER	12
THE TIDE IS HIGH - BLONDIE	11
I WRITE THE SONGS - BARRY MANILOW	11
CARS - GARY NUMAN	11
CRUMBLIN' DOWN - JOHN COUGAR	11
ANY DAY NOW - RONNIE MILSAP	11
SUDDENLY LAST SUMMER - MOTELS	11
NIGHT FEVER - BEE GEES	10
JACK AND DIANE - JOHN COUGAR	10
SQUEEZE BOX - THE WHO	10
HELLO MY LOVER - ERNIE KADOE	10
MORNING TRAIN - SHEENA EASTON	10
MINUTE BY MINUTE - DOOBIE BROTHERS	10
YOU MAKE MY DREAMS COME TRUE - HALL AND OATES	10
THE ONE THAT YOU LOVE - AIR SUPPLY	10
FUNKY TOWN - LIPPS, INC.	10
ANOTHER ONE BITES THE DUST - QUEEN	10
EMOTIONAL RESCUE - ROLLING STONES	10
DR. HECKYL AND MR. JIVE - MEN AT WORK	10
POISON ARROW - ABC	10
WE TOO - LITTLE RIVER BAND	10
ONE THING LEADS TO ANOTHER - FIXX	9
KEY LARGO - BERTIE HIGGINS	9
LOVIN' YOU - MINNIE RIPERTON	9
WHAT ARE WE DOING IN LOVE - DOTTIE WEST AND KENNY ROGERS	9
TRY AGAIN - CHAMPAGNE	9
PHYSICAL - OLIVIA NEWTON-JOHN	9
TELL IT LIKE IT IS - AARON NEVILLE	8
YOU DON'T MESS AROUND WITH JIM - JIM CROCE	8
I CAN'T GO FOR THAT - HALL AND OATES	8
TOO MUCH HEAVEN - BEE GEES	8
'65 LOVE AFFAIR - PAUL DAVIS	8
SOMEWHERE DOWN THE ROAD - BARRY MANILOW	8
PRIVATE EYES - HALL AND OATES	8
IT'S RAINING - IRMA THOMAS	8
SLOW HAND - POINTER SISTERS	8
I LOVE THE NIGHT LIFE - ALICIA BRIDGES	8
SOLITARY MAN - NEIL DIAMOND	8
YOU NEEDED ME - ANNE MURRAY	8
LOVE TAKES TIME - ORLEANS	8

radio programming will give way to objective research that is based on a plan," contends Dwight Douglas.

Christopher Porter sees the fragmentation and niching as creating a greater demand for research. "With the inevitability of more competition in already overcrowded markets, the need to stay abreast of market developments is critical. Yes, the role of research will continue to grow."

While the role of research in the programming of large market stations is significant, the expense involved will continue to limit many smaller stations to in-house methods, claims Ed Noonan. "Professional research services can be very costly. This will keep research to a minimum in lesser markets, although there will be more movement there than in the past. Call-out research will continue to be a mainstay for the small station."

"Cost-effective ways to perform and utilize sophisticated psychographic data have made the computer standard equipment at most stations, small and large alike. Research is becoming a way of life everywhere, and computers are an integral part of the information age. Computers will encourage more do-it-yourself research at stations, as well," contends WGAO station manager Vic Michaels.

Dick Warner contends that technological advances also will improve the nature and quality of research. "We'll see more improvement in methodology and a greater diversity of applicable data as the result of high-tech innovations. I think the field of research will take a quantum leap in the 1990s. It has in the 1970s and 1980s, but the size of the leap will be greater in the last decade of this century."

Today it is common for stations to budget 5 to 10 percent of their annual income to research, and it is probable that this figure will increase, says Christopher Porter. "As it evolves, it is likely that the marketplace will demand that more funds be allocated for research purposes. Research may not guarantee success, but it's not getting any easier to be successful without it."

INTRODUCTION

Every year at Benchmark, we mail out 7-day listening diaries so we can learn more about the overall process. This summer, we mailed out diaries and then actually called those who returned diaries and set up interviews with them. We exposed them to some of the leading position lines and diary manipulators in the business today.

The results of these interviews, along with our own professional judgements, are presented in this report. Depending on your perspective, the results may surprise and even shock you. But the listeners, in their own inarticulate fashion, are making a point. And they're saying that they don't understand it when we speak in radio jargon to them, or when we use hackneyed, tired, old positioning cliches. They understand when we speak to them in **their** language, using their terms, and reinforcing the station's benefits that **they** perceive.

We have spent the last year refining techniques that enable stations to talk to listeners in a *productive* fashion; i.e., so they'll **remember** they listened to your station. The basis for this is our Benchmark Perceptual/Strategic Study. Please call me or Marjorie Myers-Drasnin at 1-800-274-5164 so that we can discuss a plan to help you achieve a maximum level of communication with your listener.

Rob Balon, CEO
The Benchmark Company
Austin, Texas

Research has been a part of radio broadcasting since its modest beginnings in the 1920s, and it appears that it will play an even greater role in the operations of stations as the next century unfolds.

CHAPTER HIGHLIGHTS

1. Beginning in the late 1920s, surveys were conducted to determine the most popular stations and programs with various audience groupings. Early surveys (and their methods) included C. E. Hooper, Inc. (telephone), Cooperative Analysis of Broadcasting (telephone), and The Pulse (in-person). In 1968 Radio's All Dimension Audience Research (telephone to 6,000 households) began to provide information for

FIGURE 6.16 An introductory statement in a report prepared by Rob Balon for The Benchmark Company that contains findings stemming from research about the way radio stations position themselves. Courtesy The Benchmark Company.

FIGURE 6.17
Services available from one research company. Courtesy The Benchmark Company.

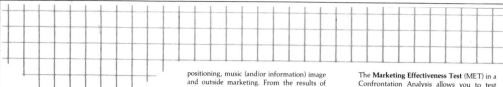

THE BENCHMARK STUDY™
Perceptual/Strategic

The Benchmark Study is a custom telephone perceptual study of your marketplace conducted from our international interviewing center in Austin. This is essential once per year in every competitive market and is recommended at six month intervals for new formats and highly volatile, changing, competitive situations. A benchmark is anything a listener uses to reconstruct radio listening in an unaided recall situation (such as an Arbitron diary or Birch interview). The higher the level of unaided recall, the higher the ratings. You receive detailed descriptions of your station's (and competitors') benchmarks as well as strategic recommendations for maximizing benchmark opportunities. Don't let budget constraints keep you from investing in needed research. Benchmark has devised an innovation which allows these studies to generate direct revenue for your station. This can cover your research investment and even become an additional revenue center.

The **Marketing Effectiveness Test** (MET) is a part of The Benchmark Study. This examines listener understanding of your station's handle, positioning, music (and/or information) image and outside marketing. From the results of your station's study, we develop detailed strategies to put this research into action.

FORMAT PREFERENCE STUDIES

Format Preference Studies actually use a variety of methods depending on the market situation. The goals are to identify a potentially profitable niche in the market, to design a product for that niche that serves the listener's needs and to create a marketing plan that will establish and nurture the new format. In testing for a music niche, we use a system of playing *actual audio music mixes* down the phone line instead of relying on verbal descriptions of the artists.

THE CONFRONTATION
ANALYSIS™ Evaluation/Tactical

Confrontation Analysis is an auditorium research technique which has replaced focus groups as the primary method of gathering *evaluative* information. It takes its name from the process in which listeners "confront" taped audio and video segments of programming and advertising. This "confrontation" triggers open-end evaluative responses that would be impossible to obtain in a telephone survey. Yet, unlike focus groups, Confrontation Analysis eliminates bias of group interaction, moderator bias and solves the problem of small sample size. You receive reliable evaluations on the effectiveness of various tactical approaches.

The **Marketing Effectiveness Test** (MET) in a Confrontation Analysis allows you to test audience receptiveness to new and alternative marketing tactics, advertising creative and positioning niches. The tactical MET ensures that you are effectively communicating with the diarykeeper.

FREQUENCY BASED
MUSIC TEST™ Cluster Analysis

Available in the form of call-out research and *auditorium* testing, the Frequency Based Music Test is a proven technique which effectively measures burnout as well as answering the question of how often listeners *really* want to hear a record. As with all Benchmark research services, all the work is done in-house with no "farming out." From recruiting to data processing, we deliver you a professional, high quality music report.

CUSTOM RESEARCH
AND CONSULTATION

With over ten years of radio research experience, The Benchmark Company will design custom research to fit your exact needs. From agency/advertiser studies to vulnerability testing and advertising pre-tests, The Benchmark Company has the experience to help you solve problems and take advantage of opportunities.

FIGURE 6.18
The electronic music testing graph shows where listening drops off. In this test the preferences of Adult Contemporary and Easy Listening listeners were gauged as certain tunes were played. Courtesy FMR Associates.

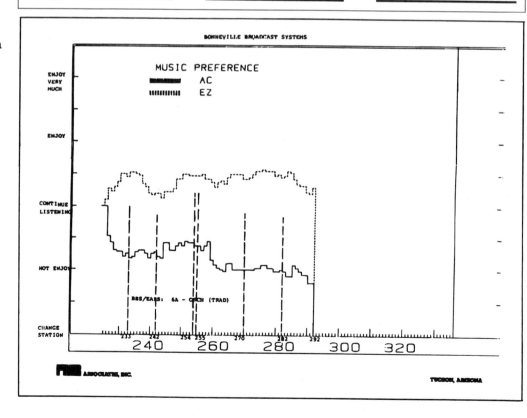

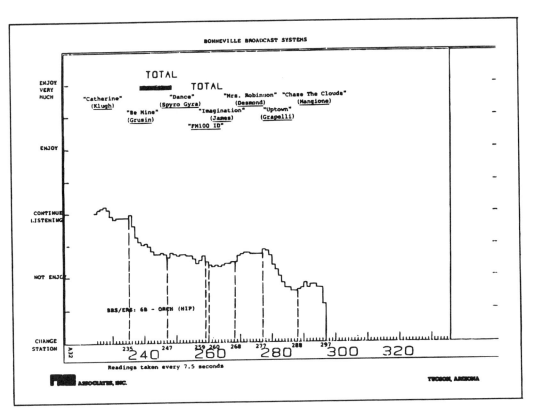

FIGURE 6.19
The electronic
music testing
graph shows
audience reaction
to specific songs.
Even the ID is
tested. Courtesy
FMR Associates.

FIGURE 6.20
Research firms
provide stations
with in-depth data
regarding audience
perceptions.
Courtesy FMR
Associates.

Oldies Morning Show Example 2

ELEMENT	DESCRIPTION	TIME	AVERAGE TIME-POINT SCORE			
			Total	Reg	Occas	Non
1.	Station I.D.	:00- :08	3.0	3.0	3.0	3.0
2.	Song	:08- :32	3.4	3.6	3.4	3.2
3.	Station I.D./Time/Weather/Song Intro	:32- :42	3.5	3.7	3.6	3.4
4.	Song	:42- :58	3.5	3.8	3.6	3.3
5.	Station I.D./Song Intro	:58-1:10	3.5	3.7	3.4	3.4
6.	Song	1:10-1:28	3.5	4.0	3.2	3.3
7.	Time/Song I.D.	1:28-1:36	3.5	4.0	3.4	3.2
8.	Spot	1:36-1:54	3.2	3.6	3.1	2.8
9.	Calendar of Events	1:54-3:00	2.8	2.7	2.8	2.7
10.	Commercials	3:00-3:16	2.3	2.4	2.2	2.3
11.	Time/Station I.D.	3:16-3:24	2.6	2.6	2.5	2.7
12.	Traffic Report/Station I.D.	3:24-3:38	3.2	3.1	3.3	3.3
13.	Song	3:38-4:10	3.6	4.1	3.4	3.5
14.	Station I.D./Contest	4:10-4:26	3.6	4.0	3.4	3.4
15.	Song	4:26-4:48	3.4	4.0	3.3	3.1
16.	Station I.D./Time	4:48-5:04	3.9	4.0	3.8	3.9
17.	Song	5:04-5:22	4.2	4.3	4.0	4.2
	TOTAL SEGMENT AVERAGE		3.4	3.6	3.4	3.3

Key:
5 - Most Positive (Listen Closely)
4 - Enjoy
3 - Neutral (Continue Listening)
2 - Dislike (Turn Down/"Mentally" Tune-out)
1 - Most Negative (Turn-Off/Switch Station)

Guideline:	
3.7 or better	High-testing
3.3 - 3.6	Mid-to-High
3.0 - 3.2	Listenable
2.6 - 2.9	Low-Testing
2.5 or below	Negative-Testing

FIGURE 6.21
Some broadcasters
contend that there
are inherent
failures in across-
the-board GRP/CPP
buying by agencies.
RAB publishes this
report to caution
against the pitfalls
of buying formulas.
Courtesy RAB.

networks. The current leader among local market audience surveys is Arbitron (week-long diary).

2. In 1963 the Broadcast Rating Council was established to monitor, audit, and accredit ratings companies. In 1982 it was renamed the Electronic Media Planning Council to reflect its involvement with cable television ratings.

3. Arbitron measures listenership in the Metro Survey Area (MSA), that is, the city or urban center, and the Total Survey Area (TSA), which covers the surrounding communities.

4. A station's primary listening locations are called Areas of Dominant Influence (ADI).

5. The Arbitron daily diary logs time tuned to a station; station call letters or program name; whether AM or FM; where listening occurred; and the listener's age, sex, and area of residence.

6. From the late 1970s to the early 1990s, Birch/Scarborough gathered data by calling equal numbers of male and female listeners aged 12 and over. Clients were offered seven different report formats, including a computerized data retrieval system. The company went out of business on December 31, 1991.

7. With today's highly fragmented audiences, advertisers and agencies are less comfortable buying just ratings numbers and look for audience qualities. Programmers must consider, not only the age and sex of the target audience, but also their lifestyles, values, and behavior.

8. Station in-house surveys utilize telephone, computer, face-to-face, and mail methods.

9. In response to complaints about "missed" audiences, the major survey companies adjusted their survey techniques to assure inclusion of minorities and nontelephone households. Today's survey results are more accurate.

10. Ratings data should only direct, not dictate, what a station does.

11. Media buyers for agencies use station ratings to determine the most cost-effective buy for their clients. Two methods they use are the cost per rating point (CPP) and the cost per thousand (CPM).

12. Although the significant increase in the number of broadcast research companies (over 150 nationwide) has created a growing job market, a college education is necessary. Courses in communications, research methods, statistics, marketing, computers, and business are useful. Beneficial personal traits include inquisitiveness, objectivity, perceptiveness, and interpersonal skills.

FIGURE 6.22
APPENDIX 6A:
RAB'S Radio
Research Glossary.

RADIO RESEARCH GLOSSARY

RAB Guide To Key Terms And Formulas

A wealth of terminology has come to describe the many dimensions of Radio audience research. As new buying and planning techniques are developed, new terms are added to the vocabulary and even established terms take on new meanings. To help you understand and use today's language of Radio audience research, RAB has prepared this brief guide.

FIGURE 6.22
continued

AUDIENCE RESEARCH TERMS

AVERAGE QUARTER-HOUR AUDIENCE (AQH PERSONS)

An average of the number of people listening for at least five minutes in each quarter-hour over a specified period of time. In modern Radio average quarter-hour measurement should be considered a measure of Total Time Spent Listening (see below).

AVERAGE QUARTER-HOUR RATING (AQH RATING)

Average Quarter-Hour Audience expressed as a percentage of the population being measured. *

$$AQH \ PERSONS \div POPULATION = AQH \ RATING$$

SHARE OF AUDIENCE (SHARE)

The percentage of those listening to Radio in the AQH that are listening to a particular station. *

$$AQH \ PERSONS \ / \ ONE \ STATION \div AQH \ PERSONS \ / \ RADIO = SHARE$$

Because the AQH actually reflects Total Time Spent Listening, Share Of Audience is the share of Total Time Spent Listening to Radio.

CUMULATIVE AUDIENCE (CUME PERSONS)

Also called Unduplicated Audience, it is the number of different people listening for at least five minutes during a specified period of time. Cume Audience is the potential group that can be exposed to advertising on a radio station, just as readership is the potential exposure group for a magazine or newspaper.

CUMULATIVE RATING (CUME RATING)

Cumulative Audience expressed as a percentage of the population being measured. *

$$CUME \ PERSONS \div POPULATION = CUME \ RATING$$

TOTAL TIME SPENT LISTENING (TTSL)

The number of quarter-hours of listening to Radio or to a station by the population group being measured. *

$$AQH \ PERSONS \times QUARTER\text{-}HOURS \ IN \ TIME \ PERIOD = TTSL \ \ (IN \ QUARTER \ HOURS)$$

AVERAGE TIME SPENT LISTENING (TSL)

The time spent listening by the average person who listens to Radio or to a station.

$$TTSL \div CUME \ PERSONS = TSL$$

Average Time Spent Listening is an indicator of audience availability to advertising messages. The more time spent listening, the greater opportunity for exposure and ability to develop frequency.

AUDIENCE TURNOVER (T/O)

The number of times the Average Quarter-Hour Audience is replaced by new listeners in a specified period of time. Audience Turnover is also the number of announcements required to reach approximately 50% of the station's Cumulative Audience in the time period.

$$CUME \ PERSONS \div AQH \ PERSONS = T/O$$

* The population being measured can be all people or any demographic group.

FIGURE 6.22
continued

SCHEDULE MEASUREMENT TERMS

REACH

The number of different people who are exposed to a schedule of announcements, i.e., those listening during a quarter-hour when announcements are aired.

Reach can also be expressed as a Rating (percentage of the population being measured):

$$PERSONS\ REACHED \div POPULATION = REACH\ RATING$$

GROSS IMPRESSIONS

The total number of exposures to a schedule of announcements. Not a measure of the number of different people exposed to a commercial.

$$AQH\ PERSONS \times NUMBER\ OF\ ANNOUNCEMENTS = GROSS\ IMPRESSIONS$$

FREQUENCY

The average number of times the audience reached by an advertising schedule (those listening during a quarter-hour when an announcement is aired) is exposed to a commercial.

$$GROSS\ IMPRESSIONS \div REACH = FREQUENCY$$

GROSS RATING POINTS (GRPs)

Gross Impressions expressed as a percentage of the population being measured.* One Rating Point equals one percent of the population.

$$GROSS\ IMPRESSIONS \div POPULATION = GRPs$$

It can also be derived by combining AQH Ratings:

$$AQH\ RATING \times NUMBER\ OF\ ANNOUNCEMENTS = GRPs$$

COST PER THOUSAND (CPM)

The basic term to express Radio's unit cost. It establishes 1000 as the basic unit for comparing Radio values. Most frequently used to compare the cost of 1000 Gross Impressions on different stations, it can also be used to compare the cost of reaching 1000 people.

$$\frac{SCHEDULE\ COST}{GROSS\ IMPRESSIONS\ (IN\ THOUSANDS)} = CPM$$

COST PER RATING POINT (CPP)

An expression of Radio's unit cost using a Rating Point, which is one percent of the population being measured*. Cost Per Rating Point is often used for planning Radio in conjunction with GRPs.

$$\frac{SCHEDULE\ COST}{GROSS\ RATING\ POINTS} = CPP$$

* The population being measured can be all people or any demographic group.

FIGURE 6.22
continued

REACH / FREQUENCY EVALUATIONS

REACH / FREQUENCY FORMULAS

There are three factors in any Reach/Frequency formula: 1) Reach, 2) Frequency and 3) GRPs. Their relationship is expressed in these formulas, with any two factors predicting the third:

$$\text{REACH} \times \text{FREQUENCY} = \text{GRPs}$$
$$\text{GRPs} \div \text{FREQUENCY} = \text{REACH}$$
$$\text{GRPs} \div \text{REACH} = \text{FREQUENCY}$$

For example, if 100 GRPs are purchased and the advertiser has determined a 4 Frequency is necessary, the Reach will be 25. These formulas make Radio planning extremely flexible.

EFFECTIVE FREQUENCY

The minimum level of frequency—number of exposures—determined to be effective in achieving the goals of an advertising campaign (e.g., awareness, recall, sales, etc.). This level will vary with individual products or services and the marketing objectives of the campaign.

FREQUENCY DISTRIBUTION

A tabulation separating those reached by a schedule, according to their minimum levels of exposure: 2 or more times, 3 or more times, 4 or more times, etc.

EFFECTIVE REACH

The number of people reached by a schedule at the pre-determined level of Effective Frequency.

EFFECTIVE RATING POINTS (ERPs)

Effective Reach expressed as a percentage of the population being measured. *

$$\text{EFFECTIVE REACH} \div \text{POPULATION} = \text{ERPs}$$

* The population being measured can be all people or any demographic group.

For further information please contact RAB Research.

Radio Advertising Bureau, Inc.
485 Lexington Avenue, New York, N.Y. 10017 • (212) 599-6666

FIGURE 6.23
APPENDIX 6B:
ARBITRON'S
Glossary of Terms.

XII. GLOSSARY OF TERMS

ADVANCE RATINGS/An Arbitron Ratings Radio Special Service's service that provides a client, via telephone, with selected estimates which will appear in his market report (RMR) as soon as the report has been approved for printing.

AGE/SEX POPULATIONS/Estimates of population, broken out by various age/sex groups within a county.

AM-FM TOTAL/A figure shown in market reports for AM-FM affiliates in time periods when they are simulcast.

ARBITRENDS/An Arbitron Ratings service, introduced in 1984 and available through Radio Special Services. Delivers averages of tabulated Radio audience listening estimates directly to clients' microcomputers in two types of reports: (1) Rolling Average Report, containing averages of listening estimates from three consecutive Arbitron Radio survey months; and (2) Quarterly Report, containing estimates from a three-month Arbitron survey.

ARBITRON INFORMATION ON DEMAND (AID)/An Arbitron Ratings Radio Special Services information service for direct access clients (via terminals) and indirect access clients (via AID division of Radio Special Services Department). Provides audience estimates and Reach and Frequency information, based on the same diaries that are used in the processing of the Radio Market Reports (RMRs).

ARBITRON SURVEY WEEK NUMBER/All fifty-two weeks in a year are assigned a number from 01-53 consecutively, beginning with the week in which January 1 falls. Appears on the diary label and serves as a quality check to ensure that a diary is placed in the correct week of a survey.

AREA OF DOMINANT INFLUENCE (ADI)/An exclusive geographic area, defined by Arbitron Television, consisting of sampling units in which the home-market television stations receive a preponderance of viewing. Every county in the United States (excluding Alaska and Hawaii) is allocated exclusively to one ADI.

A-SALE TAPE/An Arbitron Ratings Radio Special Services data tape of ADI estimates for one or more of the top fifty ADI markets available to agency and station clients that subscribe to the RMRs. Also known as "ASL" or "ADI" tape.

ASCRIPTION/A statistical technique that allocates radio listening proportionate to each conflicting station's diaries as calculated on a county basis using up to four surveys' TALO from the previous year, excluding the most recently completed survey for those markets with back-to-back surveys. Diary credit is randomly assigned automatically to a station based on its share of total diaries in the county.

AUDIENCE/A group of households, or a group of individuals, that are counted in a radio audience according to any one of several alternative criteria.

AVERAGE QUARTER-HOUR PERSONS/The estimated number of persons who listened at home and away to a station for a minimum of five minutes within a given quarter-hour. The estimate is based on the average of the reported listening in the total number of quarter-hours the station was on the air during a reported time period. This estimate is shown for the MSA, TSA and, where applicable, the ADI.

AVERAGE QUARTER-HOUR RATING/The Average Quarter-Hour Persons estimate expressed as a percentage of the universe. This estimate is shown in the MSA and, where applicable, the ADI.

AVERAGE QUARTER-HOUR SHARE/The Average Quarter-Hour estimate for a given station expressed as a percentage of the Average Quarter-Hour Persons estimate for the total listening in the MSA within a given time period. This estimate is shown only in the MSA.

AWAY FROM HOME LISTENING/Estimates of listening for which the diarykeeper indicated listening was done away from home, either "in a car" or "some other place."

CLIENT TAPE/A magnetic tape containing the same data as the Arbitron Ratings reports, sent to clients who subscribe to the printed report.

CONDENSED RADIO MARKET/Generally a small to middle-sized radio market; most are surveyed only once, in the Spring. The Metro and TSA sample objectives are considerably less than those for Standard Radio Markets and an abbreviated version of the Standard Radio Market Report is produced.

CONFLICT/Two or more stations using the same or similar slogan/program/personality/sports identification in the same county and qualifying under Arbitron's "One Percent TALO" criteria.

CONSOLIDATED METROPOLITAN STATISTICAL AREA (CMSA)/As defined by the U. S. Government's Office of Management and Budget; a grouping of closely related Primary Metropolitan Statistical Areas.

COUNTY SLOGAN EDIT FILE/A county-by-county listing of stations whose signals penetrate a county. Denotes all One Percent TALO qualifying stations. Includes each station's call letters, slogan ID, city and county of license, exact frequency and network affiliation(s). An internal document used to process diary entries.

FIGURE 6.23
continued

CUME PERSONS/The estimated number of different persons who listened at home and away to a station for a minimum of five minutes within a given daypart. (Cume estimates may also be referred to as "cumulative," "unduplicated," or "reach" estimates.) This estimate is shown in the MSA, TSA and ADI.

CUME RATING/The estimated number of Cume Persons expressed as a percentage of the universe. This estimate is shown for the MSA only.

DAYPART/The days of a week and the portion of those days for which listening estimates are calculated (e.g., Monday-Sunday 6AM-Midnight, Monday-Friday 6AM-10AM).

DEMOGRAPHICS/Statistical identification of human populations according to sex, age, race, income, etc.

DISCRETE DEMOGRAPHICS/Uncombined or non-overlapping sex/age groupings for listening estimates (e.g., men and/or women 18-24, 25-34, 35-44, etc.) as opposed to "target" group demographics (e.g., men and/or women 18+, 18-34, 18-49, 25-49).

DUAL CITY OF IDENTIFICATION/A multi-city identification, with the city of license required to be named first in all multi-city identification announcements, according to FCC guidelines (see section 73.1201 (B) (2), amended October 19, 1983). Also known as Home Market Guidelines, at Arbitron.

EFFECTIVE SAMPLE BASE (ESB)/An estimate of the size of simple random samples (in which all diaries have equal values) that would be required to provide the same degree of statistical reliability as the sample actually used to produce the estimates in a report.

ELECTRONIC MEDIA RATING COUNCIL (EMRC)/An organization that accredits broadcast ratings services; performs annual audits of the compliance of a service with certain minimum standards.

ETHNIC CONTROLS/Arbitron Ratings placement and weighting techniques used in certain sampling units to establish proper representation of the black and/or Hispanic populations in the Metros of qualifying Arbitron Radio Markets.

EXCLUSIVE CUME LISTENING/The estimated number of Cume Persons who listened to one and only one station within a given daypart.

EXPANDED SAMPLE FRAME (ESF)/A universe that consists of unlisted telephone households — households which do not appear in the current or available telephone directories because either (a) they have requested their telephone number not to be listed or (b) they are households where the assigned telephone number is not listed in the directory because of the date of installation and the date of telephone directory publication.

FACILITY FORM/(See Station Information Form.)

FLIP/A computerized edit procedure that assigns aberrated call letters to legal call letters, or the AM designation of a set of call letters may be changed to an FM designation, e.g., WODC-AM flips to WOBC-AM and WOBC-FM flips to WOBC-AM.

GROUP QUARTERS/Residences of all persons not living in nuclear households. The population in group quarters includes, for example, persons living in college dormitories, homes for the aged, military barracks, rooming houses, hospitals and institutions.

HOME MARKET GUIDELINES/The criteria by which a radio station with multi-city identification can be reported Home to an Arbitron Radio Metro Area. Also known as Dual City of Identification.

HOME NUMBER/A unique four-digit number assigned to each household within a county being sampled.

IN-TAB/The number of usable diaries actually tabulated in producing an Arbitron Ratings report.

LISTED SAMPLE/Names, addresses, and telephone numbers of selected potential diary-keepers derived from telephone directories.

LOCAL MARKET REPORT (LMR)/A syndicated report for a designated market; also known as SRMR (Standard Radio Market Report) and RMR (Radio Market Report).

MARKET TOTALS/The estimated number of persons in the market who listened to reported stations, as well as to stations that did not meet the Minimum Reporting Standards, and/or to unidentified stations.

MENTION/The number of different diaries in which a station is mentioned once with at least five minutes of listening, in a quarter-hour, (does not indicate all the entries to a station in one diary); appears in county and station TALO.

METROPOLITAN STATISTICAL AREA (MSA)/ As defined by the U.S. Government's Office of Management and Budget; a free-standing metropolitan area, surrounded by nonmetropolitan counties and not closely associated with other metropolitan areas.

METRO TOTALS AND ADI TOTALS (Total Listening in Metro Survey Area or Total Listening in the ADI)/The Metro and ADI total estimates include estimates of listening to reported stations as well as to stations that did not meet the Minimum Reporting Standards plus estimates of listening to unidentified stations.

NETWORK AFFILIATE/A broadcasting station, usually independently owned, in contractual agreement with a network in which the station grants the network an option on specific time periods for the broadcast of network-originated programs.

FIGURE 6.23
continued

NEW ENGLAND COUNTY METROPOLITAN AREA (NECMA)/As defined by the U.S. Government's Office of Management and Budget; New England MSA or PMSA definition adjusted to a whole county definition.

ONE PERCENT (1%) TALO RULE/An Arbitron radio procedure that establishes a cutoff point for resolving conflicts. The cutoff is one percent of the *previous year's* TALO by county, by station. All "potential" conflicting stations are analyzed to determine whether they qualify for conflict resolution. If only one of the two or more stations "potentially" in conflict receives one percent or more of the mentions in that county, then that station will receive credit for the contested entries in that county. However, if two or more of the stations "potentially" in conflict receive one percent or more of the total mentions in that county, each is considered in conflict. Ascription procedures are then instituted to determine proper listening credit.

PREMIUM/A token cash payment most often mailed with the diaries; serves as an inducement for a diarykeeper to participate in the survey and to return the diary to Arbitron. A premium is sent for each person twelve years of age and older in a household. The amount of the premium may vary.

PRIMARY METROPOLITAN STATISTICAL AREA (PMSA)/As defined by the U.S. Government's Office of Management and Budget; a metropolitan area that is closely related to another.

RATING/(See Average Quarter-Hour Rating and Cume Rating.)

REACH (Station)/Each county in which it has been determined by Arbitron Ratings that the signal for a specific radio station may be received.

R-SALE TAPE/An Arbitron Radio Special Services data tape of Metro and TSA estimates from one or more of the RMRs; available to clients who subscribe to the RMRs; also known as "RSL" or "Market Report data tape."

SHARE/The percentage of individuals listening to radio, who are listening to a specific station at a particular time.

SLOGAN/An on-air identifier used in place of or in conjunction with a station's call letters or exact frequency.

STATION INFORMATION FORM/A computer-generated form that lists essential station information including: power (day and night), frequency, sign-on/sign-off times, simulcasting (if any), slogan ID, network affiliates and national representative (if applicable). The Station Information Form is forwarded for verification to the applicable station prior to the survey period.

STATION INFORMATION PACKET/A set of forms mailed by Arbitron Ratings to a radio station approximately fifty days prior to each survey; allows station to change its slogan ID, sign-on/sign-off times, and make routine programming changes; included are forms for: Station Information, Programming Schedule Information, and Sports Programming.

STATION REACH FILE/A county-by-county file of stations that can be received in a county. This file is based on previous diary history and is updated with recent diary information as well as changes in power/antenna height; replaces the subjective review of all diary mentions within a county each survey to determine whether or not the mentions are "logical." Also, a station-by-station file of total counties reached.

TARGET DEMOGRAPHICS/Audience groupings containing multiple discrete demographics (e.g., men and/or women 18+, 18-34, 18-49, 25-49, etc.) as opposed to discrete demographics (e.g., men and/or women 18-24, 25-34, 35-44, etc.).

TECHNICAL DIFFICULTIES (TD)/Time period(s) of five or more consecutive minutes, in a quarter-hour, during the survey period in which a station listed in an Arbitron Ratings market report notified Arbitron Ratings in writing of technical difficulties including, but not limited to, times it was off the air or operating at reduced power.

TIME SPENT LISTENING (TSL)/An estimate of the amount of time the average person spends listening to a radio or to a station during a specified time period.

$$\frac{\text{Quarter-Hours in Time Period} \quad \text{X} \quad \text{Average Quarter-Hours Persons Audience}}{\text{Cume Audience}} = \text{TSL}$$

TOTAL AUDIENCE LISTENING OUTPUT (TALO)/The number of diaries in which a station is "mentioned" in (a) a market, (b) a county, or (c) another designated geographic area; a county-by-county printout showing the stations that are mentioned in the in-tab diaries and the number of mentions for each station; can be used to rank stations, to calculate weekly cumes and raw bases.

UNCOMBINED LISTENING ESTIMATES/(See Discrete Demographics.)

UNIVERSE/The estimated total number of persons in the sex/age group and geographic area being reported.

UNLISTED SAMPLE/(See Expanded Sample Frame.)

FIGURE 6.23
continued

UNUSABLE DIARIES/Returned diaries determined to be unusable according to established Arbitron Ratings Radio Edit procedures.

UUUU/Unidentified; listening that could not be interpreted as belonging to a specific station.

FREQUENTLY USED ABBREVIATIONS

ADI	Area of Dominant Influence
AID	Arbitron Information on Demand
AQH	Average Quarter-Hour
CMSA	Consolidated Metropolitan Statistical Area
CRMR	Condensed Radio Market Report
CSB	Client Service Bulletin
DST	Differential Survey Treatment
EMRC	Electronic Media Rating Council
ESB	Effective Sample Base
ESF	Expanded Sample Frame
HDBA	High Density Black Area
HDHA	High Density Hispanic Area
LMR	Local Market Report
MMAC	MetroMail Advertising Company
MRS	Minimum Reporting Standards
MSA	Metro Survey Area
	Metropolitan Statistical Area
MSI	Market Statistics, Inc.
NECMA	New England County Metropolitan Area
PMSA	Primary Metropolitan Statistical Area
PPDV	Persons Per Diary Value
PPH	Persons Per Household
PSA	Primary Signal Area
PUR	Persons Using Radio
QM	Quarterly Measurement
RMR	Radio Market Report
SRDS	Standard Rate and Data Service, Inc.
SRMR	Standard Radio Market Report
SU	Sampling Unit
TALO	Total Audience Listening Output
TAR	Trading Area Report
TD	Technical Difficulty
TSA	Total Survey Area
TSL	Time Spent Listening

SUGGESTED FURTHER READING

Arbitron Company. *Research Guidelines for Programming Decision Makers.* Beltsville, Md.: Arbitron Company, 1977.

Aspen Handbook on the Media—1977–79 Edition: A Selective Guide to Research, Organizations, and Publications in Communications. New York: Praeger, 1977.

Balon, Robert E. *Radio In the '90's.* Washington: NAB Publishing, 1990.

Bartos, Rena. *The Moving Target, What Every Marketer Should Know about Women.* New York: The Free Press, 1982.

Broadcast Advertising Reports. New York: Broadcast Advertising Research, periodically.

Broadcasting Yearbook. Washington, D.C.: Broadcasting Publishing, 1935 to date, annually.

Browne, Bortz, and Coddington (consultants). *Radio Today—And Tomorrow.* Washington, D.C.: NAB, 1982.

Buzzard, Karen. *Electronic Media Ratings.* Boston: Focal Press, 1992.

Chappell, Matthew N., and Hooper, C. E. *Radio Audience Measurement.* New York: Stephen Day Press, 1944.

Compaine, Benjamin, et al. *Who Owns the Media? Confrontation of Ownership in the Mass Communication Industry,* 2nd ed. White Plains, N.Y.: Knowledge Industry Publications, 1982.

Duncan, James. *American Radio.* Kalamazoo, Mich.: Author, twice yearly.

———. *Radio in the United States: 1976–82. A Statistical History.* Kalamazoo, Mich.: the author, 1983.

Electronic Industries Association. *Electronic MarketData Book.* Washington, D.C.: EIA, annually.

Fletcher, James E., ed. *Handbook of Radio and Television Broadcasting: Research Procedures in Audience, Programming, and Revenues.* New York: Van Nostrand Reinhold, 1981.

Jamieson, Kathleen Hall, and Campbell, Karlyn Kohrs. *The Interplay of Influence: Mass Media and Their Publics in News, Advertising, and Politics.* Belmont, Calif.: Wadsworth Publishing, 1983.

Lazarsfeld, Paul F., and Kendall, Patrick. *Radio Listening in America.* Englewood Cliffs, N.J.: Prentice-Hall, 1948.

M Street Radio Directory, 1993 edition, Boston: M Street Corporation and Focal Press.

National Association of Broadcasters. *Audience Research Sourcebook.* Washington, D.C.: NAB Publishing, 1991.

National Association of Broadcasters. *Radio Financial Report.* Washington, D.C.: NAB, 1955 to date, annually.

Radio Facts. New York: Radio Advertising Bureau, published annually.

Sterling, Christopher H. *Electronic Media: A Guide to Trends in Broadcasting and Newer Technologies, 1902–1983.* New York: Praeger, 1984.

Webster, James G., and Lichty, Lawrence W. *Ratings Analysis: Theory and Practice.* Hillsdale, N.J.: Lawrence Erlbaum Associates, 1991.

Wimmer, Roger D., and Dominick, Joseph. *Mass Media Research: An Introduction,* 4th ed. Belmont, Calif.: Wadsworth Publishing, 1994.

7 Promotion

PAST AND PURPOSE

In 1959 author Norman Mailer published a book of essays entitled *Advertisements for Myself.* The title could be used to describe a text on radio promotion. Of course, Mailer's book is not about radio promotion, which is a form of self-advertisement. "WXXX—The station without equal," "For the best in music and news tune WXXX," or "You're tuned to the music giant—WXXX" certainly illustrate this point. Why must stations practice self-glorification? The answer is simple: to keep the listener interested and tuned. The highly fragmented radio marketplace has made promotion a basic component of station operations. Five times as many stations vie for the listening audience today as did when television arrived on the scene. It is competition that makes promotion necessary.

Small stations as well as large promote themselves. In the single station market, stations promote to counter the effects of other media, especially the local newspaper, which often is the archrival of small town outlets. Since there is only one station to tune to, promotion serves to maintain listener interest in the medium. In large markets, where three stations may be offering the same format, promotion helps a station differentiate itself from the rest. In this case, the station with the best promotion often wins the ratings war.

Radio recognized the value of promotion early. In the 1920s and 1930s, stations used newspapers and other print media to inform the public of their existence. Remote broadcasts from hotels, theaters, and stores also attracted attention for stations and were a popular form of promotion during the medium's golden age. Promotion-conscious broadcasters of the pretelevision era were just as determined to get the audience to take notice as they are today. From the start, stations used whatever was at hand to capture the public's attention. Call letters were configured in such a way as to convey a particular sentiment or meaning: WEAF/New York Water, Earth, Air, Fire; WOW/Omaha Woodmen Of the World; WIOD/Miami Wonderful Isle Of Dreams. Placards were affixed to vehicles, buildings, and even blimps as a means of heightening the public's awareness.

As ratings assumed greater prominence in the age of specialization, stations became even more cognizant of the need to promote. The relationship between good ratings and effective promotion became more apparent. In the 1950s and 1960s, programming innovators such as Todd Storz and Gordon McLendon used promotions and contests with daring and skill, even a bit of lunacy, to win the attention of listeners.

The growth of radio promotion has paralleled the proliferation of frequencies. "The more stations you have out there, the greater the necessity to promote. Let's face it, a lot of stations are doing about the same thing. A good promotion sets you apart. It gives you greater identity, which means everything when a survey company asks a listener what station he or she tunes. Radio is an advertising medium in and of itself. Promotion makes a station salable. You sell yourself so that you have something to sell advertisers," observes Charlie Morriss, promotion director, KOMP, Las Vegas.

Stations that once confined the bulk of their promotional effort to spring and fall to coincide with rating periods now find it necessary to engage in promotional campaigns on an ongoing basis throughout the year, notes Morriss. "More competition and monthly audience surveys mean that stations have to keep the promotion fire burning continually, the analogy being that if the flame goes out you're likely to go cool in the ratings. So you really have to hype your outlet every opportunity you get. The significance of the role of promotion in contemporary radio cannot be overstated."

The vital role that promotion plays in radio is not likely to diminish as hundreds of more stations enter the airways by the turn of the century. "Promotion has become as much a part of radio as the records and the deejays who spin them, and that's not about to change," contends promotion director John Grube.

PROMOTIONS PRACTICAL AND BIZARRE

The idea behind any promotion is to win listeners. Over the years, stations have used a variety of methods, ranging from the conventional to the outlandish, to accomplish this goal. "If a promotion achieves top-of-the-mind awareness in the listener, it's a winner. Granted,

some strange things have been done to accomplish this," admits Bob Lima, WVMI/WQLD, Biloxi, Mississippi.

Promotions designed to captivate the interest of the radio audience have inspired some pretty bizarre schemes. In the 1950s, Dallas station KLIF placed overturned cars on freeways with a sign on their undersides announcing the arrival of a new deejay, Johnny Rabbitt. It would be hard to calculate the number of deejays who have lived atop flagpoles or in elevators for the sake of a rating point.

In the 1980s the shenanigans continued. To gain the listening public's attention, a California deejay set a world record by sitting in every seat of a major league ballpark that held sixty-five thousand spectators. In the process of the stunt, the publicity-hungry deejay injured his leg. However, he went on to accomplish his goal by garnering national attention for himself and his station. Another station offered to give away a mobile home to contestants who camped out the longest on a platform at the base of a billboard. The challenge turned into a battle of wills as three contestants spent months trying

FIGURE 7.1
Courtesy
Communications
Graphics, Inc.

FIGURE 7.2
Stations promote
their superstars.
Courtesy WRKO.

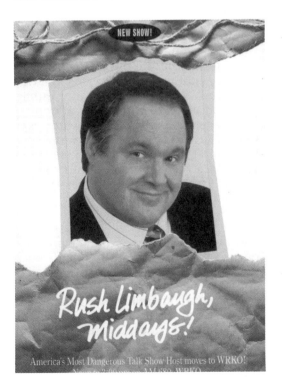

Opposites attract! Wake up to Clapprood and Whitley.
5:30 - 10:00 am on AM 680, WRKO.

FIGURE 7.3
Station promotion
in the early 1920s
with an airborne
antenna. Courtesy
Westinghouse
Electric.

and making a tremendous mess. The station was forced to make a public apology, and it promised full financial restitution."

One of the most infamous examples of a promotion gone bad occurred when a station decided to air-drop dozens of turkeys to a waiting crowd of listeners in a neighborhood shopping center parking lot. Unfortunately, the station discovered too late that turkeys are not adept at flying at heights above thirty feet. Consequently, several cars were damaged and witnesses traumatized as turkeys plunged to the ground. This promotion-turned-nightmare was depicted in an episode of the television sitcom "WKRP in Cincinnati."

The list of glitches is seemingly endless. In the late 1960s a station in central Massachusetts asked listeners to predict how long its air personality could ride a carousel at a local fair. The hardy airman's effort was cut short on day three when motion sickness got the best of him and he vomited on a crowd of spectators and newspaper photographers. A station in California came close to disaster when a promotion that challenged listeners to find a buried treasure resulted in half the community being dug up by overzealous contestants.

These promotions did indeed capture the attention of the public, but in each case the station's image was somewhat tarnished. The axiom that any publicity, good or bad, is better than none at all can get a station into hot water, contends Chuck Davis, promotion director, WSUB/WQGN, New London, Connecticut. "It's great to get lots of exposure for the station, but if it makes the station look foolish, it can work against you."

The vast majority of radio contests and promotions are of a more practical nature and run without too many complications. Big prizes, rather than stunts, tend to draw the most interest and thus are offered by stations able to afford them. In the mid-1980s, WASH-AM, Washington, D.C., and KSSK-AM, Honolulu, both gave a lucky listener a million dollars. Cash prizes always have attracted tremendous response.

to outlast each other. In the end, one of the three was disqualified and the station, in an effort to cease what had become more of an embarrassment than anything else, awarded the two holdouts recreational vehicles.

At this point in the 1990s, radio seems to have assumed a more conservative aspect (perhaps induced by a tottering economy) although the shenanigans have not disappeared entirely. It is difficult to forecast what the rest of the decade holds, but it is a reasonable assumption that wild and off-beat promotions have a permanent place in radio.

Reporter Peg Harney offers testimony that the bizarre still occurs in radio promotions. "As a publicity stunt and also to get people to use the local public library, a station in Ft. Worth, Texas, recently announced it had hidden cash in small denominations in the fiction section. Approximately 800 people descended on the library and proceeded to pull books off the shelves looking for the money. The library had not been notified that the station was going to make the announcement, and it was totally taken by surprise. The librarian said that approximately 4000 books were pulled from the shelves—some of them had pages torn out—and that people were climbing on the bookcases

Valuable prizes other than cash also can boost ratings. For example, Los Angeles station KHTZ-AM experienced a sizable jump in its ratings when it offered listeners a chance to win a $122,000 house. Increased ratings also resulted when KHJ-AM, Los Angeles, gave away a car every day during the month of May.

Promotions that involve prizes, both large and small, spark audience interest, says Rick Peters, vice president of programming, Sconnix Broadcasting. "People love to win something or, at least, feel that they have a shot at winning a prize. That's basic to human nature, I believe. You really don't have to give away two city blocks, either. A listener usually is thrilled and delighted to win a pair of concert tickets."

Although there are numerous examples to support the view that big prizes get big audiences, there is an ample amount to support the contention that low-budget giveaways, involving T-shirts, albums, tickets, posters, dinners, and so forth are very useful in building and maintaining audience interest. In fact, some surveys have revealed that smaller, more personalized prizes may work better for a station than the high-priced items. CDs, concert tickets, and dinners-for-two rank among the most popular contest prizes, according to surveys. Cheaper items usually also mean more numerous or frequent giveaways.

THE PROMOTION DIRECTOR'S JOB

Not all stations employ a full-time promotion director. But most stations designate someone to handle promotional responsibilities. At small outlets, promotional chores are assumed by the program director or even the general manager. Larger stations with bigger operating budgets typically hire an individual or individuals to work exclusively in the area of promotion. "At major market stations, you'll find a promotion department that

Wizard WORLD

THE 106.7 WIZN LISTENER UPDATE   WINTER 1992

Win Win Situation

Carl Tamburo, Mark Waters, Pam Hale, Tom Cawley – these are some of the names you hear announced over and over again when WIZN tells you who's won the latest contests. How do they do it? Are they just plain lucky or do they know something we don't?

We weren't sure either, so we set out to find out what it takes to be jokingly known around the Penthouse as a "Prize Pig." Our suspicion is that these winners are very loyal listeners. That's good for us, and it's been pretty good for them ... a win-win situation all around.

Now don't get angry at these people for taking a share of the prizes ... get even. Surprisingly, these winners are happy to let you in on their strategies. Soon you could be hearing your name on the air a lot more often collecting some of the great booty that WIZN regularly gives away.

Foremost, you need to know when we give prizes away. Each year you can pretty much count on chances to win cash, trips, and good times during our Couching Staff contest (just before the Superbowl);

continued on page 6

Tour the Penthouse

Six months ago the paint was barely dry on the walls at the top floor of 212 Battery Street. Now, let's tour the Wizard's waterfront Penthouse and see how things are shaping up.

Hang a left through the front door. Twenty-seven steps up green

carpeted stairs the walls turn to light purple. Thirteen more and the wall is as deep purple as the carpet. Open the black door and you're greeted eye-level by black and white photos of rockers who've played

continued on page 2

IN THIS ISSUE

NEXT TIME

Words for the Wizard

"... Select Set Thursdays are great, but Monday Mindsets are even better. When can we have a whole day of Mindsets? I'm sure listeners could come up with more than you could play in a week."

Jim Boucher
Burlington

"... myself and quite a few others in my unit at the Chittenden County Correctional Center really, really enjoy Blues for Breakfast ... It's one of the most looked forward to things in the place (other than being released, that is). After all, who would know and feel the blues better than us?"

Larry Jarrett
So. Burlington

106.7 WIZN

FIGURE 7.4
Stations use print to promote. Courtesy WIZN.

---

includes a director and possibly assistants. In middle-sized markets, such as ours, the promotion responsibility is often designated to someone already involved in programming," says WVMI's Bob Lima.

The promotion director's responsibilities are manifold. Essential to the position are a knowledge and understanding of the station's audience. A background in research is important, contends Grube. "Before you can initiate any kind of promotion you must know something about who you're trying to reach. This requires an ability to interpret various research data that you gather through in-house survey efforts or from outside audience research companies, like Birch and

FIGURE 7.5
Giving away "hot" items generates audience interest. Courtesy KNEW-AM.

Arbitron. You don't give away beach balls to 50-year-old men. Ideas must be confined to the cell group you're trying to attract."

Writing and conceptual skills are vital to the job of promotion director, says KOMP's Morriss. "You prepare an awful lot of copy of all types. One moment you're composing press releases about programming changes, and the next you're writing a 30-second promo about the station's expanded news coverage or upcoming remote broadcast from a local mall. A knowledge of English grammar is a must. Bad writing reflects negatively on the station. The job also demands imagination and creativity. You have to be able to come up with an idea and bring it to fruition."

WSUB's Chuck Davis agrees with Morriss and adds that while the promotion person should be able to originate concepts, a certain number of ideas come from the trades and other stations. "When this is the case, and it often is, you have to know how to adapt an idea to suit your own station. Of course, the promotion must reflect your location. Lifestyles vary almost by region. A promotion that's successful at a station in Louisiana may bear no relevance to a station with a similar format in Michigan. On the other hand, with some adjustments, it may work as effectively there. The creativity in this example exists in the adaptation."

Promotion directors must be versatile. A familiarity with graphic art generally is necessary, since the promotion director will be involved in developing station logos and image IDs for advertising in the print media and billboards. The promotion department also participates in the design and preparation of visuals for the sales area.

The acquisition of prize materials through direct purchase and trades is another duty of the promotion person, who also may be called on to help coordinate sales co-op arrangements. "You work closely with the sales manager to arrange tie-ins with sponsors and station promotions," contends Morriss.

Like other radio station department heads, it is the promotion director's responsibility to ensure that the rules

and regulations established by the FCC, relevant to the promotions area, are observed. This will be discussed further in a subsequent section of this chapter.

On a final note, RAB's Lynn Christian observes that "The word *marketing* has become the rallying cry for the 1990s. At a BPME meeting the question was raised as to whether a promotion director should be designated 'marketing director,' and given up-graded status in a station—that is, parity with the program director and sales manager. In the light of the horrendous competition and the need to survive in what has been a very soft market, it makes sense to acknowledge the value and importance of an effective promotion (marketing) director."

WHOM PROMOTION DIRECTORS HIRE

In each section devoted to hiring in preceding and subsequent chapters, college training is listed as a desirable, if not necessary, attribute. This is no less true in the area of radio station promotion. "My advice to an individual interested in becoming a promotion person would be to get as much formal training as possible in marketing, research, graphics, writing, public relations and, of course, broadcasting. The duties of the promotion director, especially at a large station, are diverse," notes Rick Peters of TK Communications.

Charlie Morriss agrees with Peters and adds, "A manager reviewing the credentials of candidates for a promotion position will expect to find a statement about formal training, that is, college. Of course, nothing is a substitute for a solid track record. Experience is golden. This is a very hands-on field. My advice today is to get a good education and along the way pick up a little experience, too."

Familiarity with programming is important, contends Lima, who suggests that prospective promotion people spend some time on the air. "Part-timing it on mike at a station, be it a small commercial outlet or a college facility, gives a person special insight into the nature of the medium that he or she is promoting. Working in sales also is valuable. In the specific skills department, I'd say the promotion job candidate should have an eye for detail, be well organized, and possess exemplary writing skills. It goes without saying that a positive attitude and genuine appreciation of radio are important as well."

Both John Grube and Chuck Davis cite wit and imagination as criteria for the job of promotion. "It helps to be a little wacky and crazy. By that I mean able to conceive of entertaining, fun concepts," says Grube. Davis concurs, "This is a convivial medium. The idea behind any promotion or contest is to attract and amuse the listener. A zany, off-the-wall idea is good, as long as it is based in sound reasoning. Calculated craziness requires common sense and creativity, and both are qualities you need in order to succeed in promotion."

The increasing competition in the radio marketplace has bolstered job opportunities in promotion. Thus the future appears bright for individuals planning careers in this facet of the medium.

TYPES OF PROMOTION

There are two primary categories of station promotion—"on-air" and "off-air." The former will be examined first since it is the most prevalent form of radio promotion. Broadcasters already possess the best possible vehicle to reach listeners, so it should come as no surprise that on-air promotion is the most common means of getting word out on a station. The challenge confronting the promotion director is how to most effectively market the station so as to expand and retain listenership. To this end, a number of promotional devices are employed, beginning with the most obvious—station call letters. "The value of a good set of call letters is inestimable," says KGLD's Bremkamp. "A

good example is the call letters of a station I once managed which have long been associated with the term *rich* and all that it implies: 'Hartford's Rich Music Station—WRCH.'"

Call letters convey the personality of a station. For instance, try connecting these call letters with a format: WHOG, WNWS, WEZI, WODS, WJZZ, WIND, WHTS. If you guessed Country, News, Easy Listening, Oldies, Jazz, Talk, and Hits you were correct. The preceding call letters not only identify their radio stations, but they literally convey the nature or content of the programming offered.

When stations do not possess call letters that create instant recognition, they often couple their frequency with a call letter or two, such as JB-105 (WPJB-FM 105) or KISS-108 (WXKS-FM 108). This also improves the retention factor. Slogans frequently are a part of the on-air ID. "Music Country—WSOC-FM, Charlotte"; "A Touch of Class—WTEB-FM, New Bern"; "Texas Best Rock—KTXQ-FM, Fort Worth" are some examples. Slogans exemplify a station's image. When effective, they capture the mood and flavor of the station and leave a strong impression in the listener's mind. It is standard pro-

FIGURE 7.6 Stations sales packets are designed to correspond with the station's on-air image. Courtesy WMJX-FM.

On-air contests are another way to capture and hold the listener's attention. Contests must be easy to understand (are the rules and requirements of the contest easily understood by the listener?) and possess entertainment value (will nonparticipants be amused even though they are not actually involved?). A contest should engage the interest of all listeners, players and nonplayers alike.

A contest must be designed to enhance a station's overall sound or format. It must fit in, be compatible. Obviously, a mystery sound contest requiring the broadcast of loud or shrill noises would disrupt the tranquility and continuity of an Easy Listening station and result in tune-out.

Successful contests are timely and relevant to the lifestyle of the station's target audience, says Bob Lima. "A contest should offer prizes that truly connect with the listener. An awareness of the needs, desires, and fantasies of the listener will help guide a station. For example, giving away a refrigerator on a hot hit station would not really captivate the 16-year-old tuned. This is obvious, of course. But the point I'm making is that the prizes that are up for grabs should be something the listener really wants to win, or you have apathy."

The importance of creativity already has been stated. Contests that attract the most attention often are the ones that challenge the listener's imagination, contends Morriss. "A contest should have style, should attempt to be different. You can give away what is perfectly suitable for your audience, but you can do it in a way that creates excitement and adds zest to the programming. The goal of any promotion is to set you apart from the other guy. Be daring within reason, but be daring."

On-air promotion is used to inform the audience of what a station has to offer: station personalities, programs, special features and events. Rarely does a quarter hour pass on any station that does not include a promo that highlights some aspect of programming:

gramming policy at many stations to announce the station's call letters and even slogan each time a deejay opens the microphone. This is especially true during ratings sweeps when listeners are asked by survey companies to identify the stations they tune in to. "If your calls stick in the mind of your audience, you've hit a home run. If they don't, you'll go scoreless in the book. You've got to carve them into the listener's gray matter and you start by making IDs and signatures that are as memorable as possible," observes Rick Peters.

It is a common practice for stations to "bookend"—place call letters before and after all breaks between music. For example, "WHJJ. Stay tuned for a complete look at local and national news at the top of the hour on WHJJ." Deejays also are told to graft the station call letters onto all bits of information: "WHJJ Time," "WHJJ Temperature," "WHJJ Weather," and so on. There is a rule in radio that call letters can never be overannounced. The logic behind this is clear. The more a station tells its audience what it is tuned to, the more apt it is to remember, especially during rating periods.

"Tune-in WXXX's News at Noon each week-day for a full hour of . . ."

"Irv McKenna keeps 'Nightalk' in the air midnight to six on the voice of the valley—WXXX. Yes, there's never a dull moment . . ."

"Every Saturday night WXXX turns the clock back to the fifties and sixties to bring you the best of the golden oldies . . ."

"Hear the complete weather forecast on the hour and half hour throughout the day and night on your total service station—WXXX . . ."

On-air promotion is a cost-efficient and effective means of building an audience when done correctly, says John Grube. "There are good on-air promotions and weak or ineffective on-air promotions. The latter can inflict a deep wound, but the former can put a station on the map. As broadcasters, the airtime is there at our disposal, but we sometimes forget just how potent an advertising tool we have."

Radio stations employ "off-air" promotional techniques to reach people not tuned. Billboards are a popular form of outside promotion. To be effective they must be both eye-catching and simple. Only so much can be stated on a billboard, since people generally are in a moving vehicle and only have a limited amount of time to absorb a message. Placement of the billboard also is a key factor. In order to be effective, billboards must be located where they will reach a station's intended audience. Whereas an All-News station would avoid the use of a billboard facing a high school, a rock music outlet may prefer the location.

Bus cards are a good way to reach the public. Cities often have hundreds of buses on the streets each day. Benches and transit shelters also are used by billboard companies to get their client's message across to the population. Outside advertising is an effective and fairly cost-efficient way to promote a radio station, although certain billboards at heavy traffic locations can be extremely expensive to lease.

Newspapers are the most frequent means of off-air promotion. Stations like the reach and targeting that news-papers can provide. In large metro areas, alternative newspapers, such as the *Boston Phoenix*, are very effective in delivering certain listening cells. The *Phoenix* enjoys one of the largest readerships of any independent press in the country. Its huge college-age and young professional audience makes it an ideal promotional medium for stations after those particular demographics. While the readership of the more conventional newspapers traditionally is low among young people, it is high in older adults, making the mainstream publications useful to stations targeting the over-40 crowd.

While newspapers with large circulations provide a great way to reach the population at large, they also can be very costly, although some stations are able to trade airtime for print space. Newspaper ads must be large enough to stand out and overcome the sea of advertisements that often share the same page. Despite some drawbacks, newspapers usually are the first place radio broadcasters consider when planning an off-air promotion.

Television is a costly but effective promotional tool for radio. A primary advantage that television offers is the chance to target the audience that the station is after. An enormous amount of information is available pertaining to television viewership. Thus a station that wants to reach the 18- to 24-year-olds is able to ascertain the programs and features that best draw that particular cell.

The costs of producing or acquiring ready-made promos for television can run high, but most radio broadcasters value the opportunity to actually show the public what they can hear when they tune to their station. WBZ-AM in Boston has used local television extensively to promote its morning personality, Dave Maynard. Ratings for the Westinghouse-owned station have been consistently high, and management points to their television promotion as a contributing factor.

Bumper stickers are manufactured by the millions for distribution by practically every commercial radio station in the country. The primary purpose of

FIGURE 7.8
Newspapers are
popular radio
promotion vehicles.
Courtesy WAAF,
WBCN, WHK, and
WFNX.

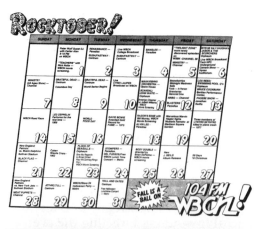

stickers is to increase call letter awareness. Over the years, bumper stickers
have developed into a unique pop-art
form, and hundreds of people actually
collect station decals as a hobby. Some
station bumper stickers are particularly prized for the lifestyle or image
they portray. Youths, in particular, are
fond of displaying their favorite station's call letters. Stations appealing to
older demographics find that their

audiences are somewhat less enthusiastic about bumper stickers.

Stations motivate listeners to display
bumper stickers by tying them in with
on-air promotions:

WXXX WANTS TO GIVE YOU A THOUSAND
DOLLARS. ALL YOU HAVE TO DO IS PUT
AN X-100 BUMPER STICKER ON YOUR
CAR TO BE ELIGIBLE. IT'S THAT SIMPLE.
WHEN YOUR CAR IS SPOTTED BY THE X-

FIGURE 7.9
Bumper stickers
visually convey a
station's sound and
image. Courtesy Mo
Money Associates.

100 ROVING EYE, YOUR LICENSE NUM-
BER WILL BE ANNOUNCED OVER THE
AIR. YOU WILL THEN HAVE THIRTY MIN-
UTES TO CALL THE STATION TO CLAIM
YOUR ONE THOUSAND DOLLARS. . . ."

Hundreds of ways have been invented to entice people to display station call letters. The idea is to get the
station's name out to the public, and
10,000 cars exhibiting a station's
bumper sticker is an effective way to do
that. Says Ed Shane, "Visibility is part
of the answer. Station promotion has
reached new levels of creativity and
intrusion—skateboard jumpers, rolling
radios, inflatables in the shapes of
boots, guitars, frogs."

Thousands of items displaying station call letters and logos are given
away annually by stations. Among the
most common promotional items
handed out by stations are posters, T-
shirts, calendars, key-chains, coffee
mugs, music hit-lists, book covers,
pens, and car litter bags. The list is
vast.

Plastic card promotions have done
well for many stations. Holders are
entitled to a variety of benefits, including discounts at various stores and
valuable prizes. The bearer is told to listen to the station for information as to
where to use the card. In addition,
holders are eligible for special on-air
drawings.

Another particularly effective way to
increase a station's visibility is to sponsor special activities, such as fairs,
sporting events, theme dances, and
concerts. Hartford's Big Band station,
WRCQ-AM, has received significant
attention by presenting an annual
music festival that has attracted over
25,000 spectators each year, plus the
notice of other media, including television and newspapers.

Personal appearances by station personalities are one of the oldest forms of
off-air promotion but are still very effective. Remote broadcasts from malls,
beaches, and the like also aid in getting
the word of the station out to the
public.

One last means of marketing a station is offered by Jay Williams, Jr.,
CEO, Direct Marketing Results. "Promotion and marketing have never been
more critical. In the current economy,
stations have to do everything they can
to draw and hold an audience. Direct
marketing through mail and/or by telephone is a very cost-effective way to target an audience and to keep a station in
front of radio listeners, especially during rating periods. Telepromoting is
becoming more prevalent. Directed or
targeted marketing makes sense
because stations must be more effective
with what they have. The business of
radio is changing. Audiences are frag-
menting, brand loyalties are eroding.

FIGURE 7.10
Plastic cards draw
listeners and
generate business
for participating
sponsors. Courtesy
WIZN.

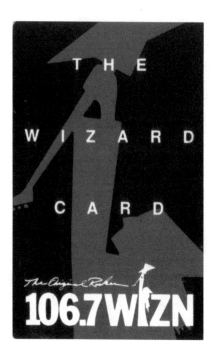

They run the gamut from placing advertisements and coupons on the back of bumper stickers to joining the circus for the day; for example, "WXXX brings the 'Greatest Show on Earth' to town this Friday night, and you go for half price just by mentioning the name of your favorite radio station—WXXX." The ultimate objective of a station/sponsor collaborative is to generate attention in a cost-efficient manner. If a few dollars are made for the station along the way, all the better.

As stated previously, the promotion director also works closely with the station's sales department in the preparation and design of sales promotion materials, which include items such as posters, coverage maps, ratings breakouts, flyers, station profiles, rate cards, and much more.

Mass marketing is losing its impact. Person-to-person or individualized marketing delivers tangible results."

Ed Shane concurs with Williams, adding, "Direct marketing is the wave of the one-to-one future. More direct mail, telemarketing, database management, and computer interaction."

SALES PROMOTION

Promoting a station can be very costly, as much as half a million dollars annually in some metro markets. To help defray the cost of station promotion, advertisers are often recruited. This way both the station and the sponsor stand to benefit. The station gains the financial wherewithal to execute certain promotions that it could not do on its own, and the participating advertiser gains valuable exposure by tying in with special station events. Stations actually can make money and promote themselves simultaneously if a substantial spot schedule is purchased by a client as part of a promotional package.

There are abundant ways to involve advertisers in station promotion efforts.

RESEARCH AND PLANNING

To effectively promote a station, the individual charged with the task must possess a thorough knowledge of the station and its audience. This person must then ascertain the objective of the promotion. Is it the aim of the promotion to increase call letter awareness, introduce a new format/feature/personality, or bolster the station's community service image? Of course, the ultimate goal of any promotion is to enhance listenership.

Effective promotions take into account both internal and external factors. An understanding of the product, consumer, and competition is essential to any marketing effort, including radio. Each of these three areas presents the promotion director with questions that must be addressed before launching a campaign. As stated earlier in this chapter, it is imperative that the promotion or contest fit the station's sound. In other words, be compatible with the format. This accomplished, the next consideration is the relevancy of the promotion to the station's audience. For example, does it fit the listener's lifestyle? Third, is the idea fresh enough in the market to attract and sustain interest?

Observes programming executive Corinne Baldassano, "You must institute ongoing research to make sure your target audience is happy with what it's hearing. You can make adjustments depending on the feedback you get from the research. Those stations that have succeeded have been single-minded in their desire to achieve their goals. They establish a market position and do everything they can to fulfill audience expectations. The other major ingredient for success with promotions is fun. You have to have fun with the promotion while you're doing all the work. If you and your staff have a good time, it is conveyed to the listeners and potential advertisers. It makes a station hard to beat."

Concerning the basic mechanics of the contest, the general rule is that if it takes a long time to explain, it is not appropriate for radio. "Contests that require too much explanation don't work well in our medium. That is not to say that they have to be thin and one-dimensional. On the contrary, radio contests can be imaginative and captivating without being complicated or complex," notes John Grube.

The planning and implementation of certain promotions may require the involvement of consultants who possess the expertise to ensure smooth sailing. Contests can turn into bad dreams if potential problems are not anticipated. Rick Sklar, who served as program director for WABC in New York for nearly twenty years, was responsible for some of the most successful radio promotions ever devised, but not all went without a hitch. In his autobiographical book, *Rocking America* (St. Martin's Press, 1984), Sklar told of the time that he was forced to hire, at great expense, 60 office temporaries for a period of one month to count the more than 170 million ballots received in response to the station's "Principal of the Year" contest. The previous year the station had received a paltry 6 million ballots.

On another occasion, Sklar had over 4 million WABC buttons manufactured as part of a promotion that awarded

up to $25,000 in cash to listeners spotted wearing one. What Sklar did not anticipate was the huge cost involved in shipping several million metal buttons from various points around the country. The station had to come up with thousands of unbudgeted dollars to cover air freight. Of course, both miscalculations were mitigated by the tremendous success of the promotions, which significantly boosted WABC's ratings.

An even more bizarre experience befell Dallas deejay Ron Chapman when he jokingly asked listeners to send $20 without explaining why. The listening faithful, assuming Chapman's request to be a part of a legitimate sta-

FIGURE 7.11 Direct marketing helps increase audience awareness of a radio station. Courtesy DMR.

FIGURE 7.12
Oldies station WHK maintains a high profile by promoting events consistent with its image.

was basically an Easter egg hunt at a local park here in Biloxi attracted over 11,000 people. A local bottler co-sponsored the event and provided many of the thousands of dollars in prizes. The station hyped the event for several weeks over the air, and a little off-air promotion was done. The reason the promotion worked so well is that there actually was a need for a large, well-organized Easter event. We did our homework in selecting and executing this promotion, which turned out to be a big winner."

Charlie Morriss shares an account of a successful promotion at his radio station. "The KOMP jocks recently flew around the Las Vegas skies in several World War II fighter planes owned by Miller Beer while our call letters were written by a skywriter. The effect was stunning. This promotion worked because skywriting is so rare these days and not many people have seen a squadron of vintage warplanes. It also worked because it didn't cost us a penny. It was a trade agreement with Miller. We gave them the exposure, and they gave us the airshow. Of course, a lot of details had to be worked out in advance."

The most effective promotion in recent years at New London's WSUB involves awarding contestants an elaborate night out on the town. Chuck Davis relates: "Our 'Night Out' promotion has been popular for some time. The station gets premium concert tickets through a close alliance with a New York concert promoter. It then finds a sponsor to participate in the giveaway and provides him with counter signs and an entry box so that people may register in his store. The sponsor then becomes part of the promotion and in return purchases an air schedule. Contestants are told to go to the store to register for the 'Night Out,' thus increasing store traffic even more. In addition, the sponsor agrees to absorb the expense of a limousine to transport the winners, who also are treated to a preconcert dinner at a local restaurant that provides the meals in exchange for promotional consideration—a mention on the air in our 'Night Out' promos. In

tion promotion, mailed in nearly a quarter of a million dollars. This left the station (KVIL-FM) with the interesting problem of what to do with the money. "We're flabbergasted," exclaimed Chapman. "We never expected this to happen." The moral to this tale is never underestimate the power of the medium. Plan before implementing.

Careful planning during the developmental phase of a promotion generally will prevent any unpleasant surprises, says Bob Lima. "Practical and hypothetical projections should be made. Radio can fool you by its pulling power. If a promotion catches on, it can exceed all expectations. You've got to be prepared for all contingencies. These are nice problems to have, but you can get egg on your face. Take a good look at the long and short of things before you bolt from the starting gate. Don't be too hasty or quick to execute. Consider all the variables, then proceed with care."

Lima tells of a successful promotion at WVMI that required considerable organization and planning. "We called it 'The Great Easter Egg Hunt.' What

FIGURE 7.13
Promotion with
celebrities.
Courtesy LIVE 105.

FIGURE 7.13
Promotion with celebrities. Courtesy LIVE 105.

the end, the station's cost amounts to a couple of phone calls, a cardboard sign, and a box. Reaction has always been great from all parties. The sponsor likes the tie-in with the promotion. The restaurant is very satisfied with the attention it receives for providing a few dinners, and the concert promoter gets a lot of exposure for the acts that he books simply by giving the station some tickets. It works like a charm. We please our audience and also put a few greenbacks in the till."

The point already has been made that a station can ill-afford not to promote itself in today's highly com-

FIGURE 7.14
From a stations's
media kit. Courtesy
KGIL-AM.

petitive marketplace. Promotions are an integral part of contemporary station operations, and research and planning are what make a promotion a winner.

BUDGETING PROMOTIONS

Included in the planning of a promotion are cost projections. The promotion director's budget may be substantial or all but nonexistent. Stations in small markets often have minuscule budgets compared to their giant metro market counterparts. But then again, the need to promote in a one- or two-station market generally is not as great as it is in multistation markets. To a degree, the promotion a station does is commensurate with the level of competition.

A typical promotion at an average-size station may involve the use of newspapers, plus additional hand-out materials, such as stickers, posters, buttons, and an assortment of other items depending on the nature of the

promotion. Television and billboards may also be utilized. Each of these items will require an expenditure unless some other provision has been made, such as a trade agreement in which airtime is swapped for goods or ad space.

The cost involved in promoting a contest often constitutes the primary expense. When WASH-FM in Washington, D.C., gave away a million dollars, it spent $200,000 to purchase an annuity designed to pay the prize recipient $20,000 a year for 50 years. The station spent nearly an equal amount to promote the big giveaway. Most of the promotional cost resulted from a heavy use of local television.

KHTZ-FM in Los Angeles spent over $300,000 on billboards and television to advertise its dream-house giveaway. The total cost of the promotion approached a half million dollars. The price tag of the house was $122,000. Both of these high-priced contests accomplished their goals—increased ratings. In a metro market, one rating point can mean a million dollars in ad revenue. "A promotion that contributes to a two- or three-point jump in the ratings is well worth the money spent on it," observes Rick Peters.

The promotion director works with the station manager in establishing the promotion budget. From there, it is the promotion director's job to allocate funds for the various contests and promotions run throughout the station's fiscal period. Just as in every other area of a station, computers are becoming a prominent fixture in the promotion department. "If you have a large budget, a microcomputer can make life a lot easier. The idea is to control the budget and not let it control you. Computers can help in that effort," states Marlin R. Taylor, who also contends that large sums of money need not be poured into promotions if a station is on target with its programming. "In 1983, the Malrite organization came to New York and launched Z-100, a contemporary hit-formatted outlet, moving it from 'worst to first' in a matter of months. They did a little advertising and gave away some money. I estimate that their giveaways totaled less dollars than some of their competitors spent on straight advertising. But the station's success was built on three key factors: product, service, and employee incentives. Indeed, they do have a quality product. Second, they are providing a service to their customers or listeners, and, third, the care and feeding of the air staff and support team are obvious at all times. You don't necessarily have to spend a fortune on promotion."

Since promotion directors frequently are expected to arrange trade agreements with merchants as a way to defray costs, a familiarity with and understanding of the station's rate structure is necessary. Trading airtime for use in promotions is less popular at highly rated stations that can demand top dollars for spots. Most stations, however, prefer to exchange available airtime for goods and services needed in a promotion, rather than pay cash.

Contests must not place participants in any danger or jeopardize property. Awarding prizes to the first five people who successfully scale a treacherous mountain or swim a channel filled with alligators certainly would be construed by the FCC as endangering the lives of those involved. Contestants have been injured and stations held liable more than once. In the case of the station in California that ran a treasure hunt resulting in considerable property damage, it incurred the wrath of the public, town officials, and the FCC. In a more tragic example of poor planning a listener was killed during a "find the disc jockey" contest. The station was charged with negligence and sustained a substantial fine.

Stations are expected to disclose the material terms of all contests and promotions conducted. These include the following:

Entering procedures
Eligibility requirements
Deadlines
When or if prizes can be won
Value of prizes

PROMOTIONS AND THE FCC

Although the FCC has dropped most of its rules pertaining to contests and promotions, it does expect that they be conducted with propriety and good judgment. The basic obligation of broadcasters to operate in the public interest remains the primary consideration. Section 73.1216 of the FCC's rules and regulations (as printed in the *Code of Federal Regulations*) outlines the do's and don't's of contest presentations.

Stations are prohibited from running a contest in which contestants are required to pay in order to play. The FCC regards as lottery any contest in which the elements of prize, chance, and consideration exist. In other words, contestants must not have to risk something in order to win.

FIGURE 7.15
Cash giveaways are popular with audiences if not quite as popular from a station's perspective. Courtesy WRC-AM.

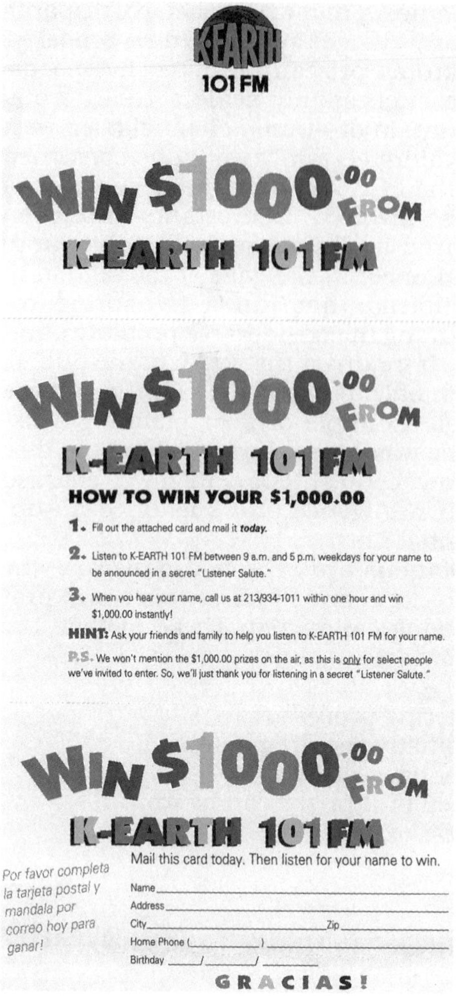

FIGURE 7.16
Stations are
expected to make
contest rules clear
to the public.
Courtesy K-EARTH
101 FM.

Although the FCC does not require that a station keep a contest file, most do. Maintaining all pertinent contest information, including signed prize receipts and releases by winners, can prevent problems should questions or a conflict arise later.

Stations that award prizes valued at $600 or more are expected by law to file a 1099 MISC form with the IRS. This is done strictly for reporting purposes and stations incur no tax liability. However, failure to do so puts a station in conflict with the law.

BROADCAST PROMOTION AND MARKETING EXECUTIVES

The Broadcasters Promotion Association (BPA) was founded in 1956 as a nonprofit organization expressly designed to provide information and services to station promotion directors around the world. In the late 1980s its name was changed to Broadcast Promotion and Marketing Executives (BPME). The objectives of the BPME are as follows:

Increase the effectiveness of broadcast promotion personnel
Improve broadcast promotion methods, research principles, and techniques
Enhance the image and professional status of its members and members of the broadcast promotion profession
Facilitate liaison with allied organizations in broadcasting, promotion, and government
Increase awareness and understanding of broadcast promotion at stations, in the community, and at colleges and universities

The BPME conducts national seminars and workshops on promotion-related subjects. Further information about the organization may be obtained by contacting: Administrative Secretary, BPME Headquarters, 6255 Sunset

Procedure for awarding prizes
Tie-breaking procedures

The public must not be misled concerning the nature of prizes. Specifics must be stated. Implying that a large boat is to be awarded when, in fact, a canoe is the actual prize would constitute misrepresentation, as would suggesting that an evening in the Kontiki Room of the local Holiday Inn is a great escape weekend to the exotic South Seas.

The FCC also stipulates that any changes in contest rules must be promptly conveyed to the public. It makes clear, too, that any rigging of contests, such as determining winners in advance, is a direct violation of the law and can result in a substantial penalty, or even license revocation.

Blvd., Suite 624, Los Angeles, CA 90028.

CHAPTER HIGHLIGHTS

1. To keep listeners interested and tuned, stations actively promote their image and call letters. Small market stations promote themselves to compete for audience with other forms of media. Major market stations use promotion to differentiate themselves from competing stations.

2. Radio recognized the value of promotion early and used print media, remote broadcasts, and billboards to inform the public. Later, ratings surveys proved the importance of effective promotions.

3. Greater competition because of the increasing number of stations and monthly audience surveys means today's stations must promote themselves continually.

4. The most successful (attracting listenership loyalty) promotions involve large cash or merchandise prizes.

5. A successful promotion director possesses knowledge and understanding of the station's audience; a background in research, writing, and conceptual skills; the ability to adapt existing concepts to a particular station; and a familiarity with graphic art. The promotion director is responsible for acquiring prizes through trade or purchase and for compliance with FCC regulations covering promotions.

6. On-air promotions are the most common method used to retain and expand listenership. Such devices as slogans linked to the call letters and contests are common.

7. To "bookend" call letters means to place them at the beginning and conclusion of each break. To "graft" call letters means to include them with all informational announcements.

8. Contests must have clear rules and must provide entertainment for players and nonplayers alike. Successful contests are compatible with the station's sound, offer prizes attractive to the target audience, and challenge the listener's imagination.

9. Off-air promotions are intended to attract new listeners. Popular approaches include billboards, bus cards, newspapers, television, bumper stickers, discount cards, giveaway items embossed with call letters or logo, deejay personal appearances, special activity sponsorship, remote broadcasts, direct mail, faxing, and telemarketing.

10. To offset the sometimes substantial cost of an off-air promotion, stations often collaborate with sponsors to share both the expenses and the attention gained.

11. FCC regulations governing promotions are contained in Section 73.1216. Basically, stations may not operate lotteries, endanger contestants, rig contests, or mislead listeners as to the nature of the prize.

12. The BPME provides information and services to station promotion directors worldwide.

SUGGESTED FURTHER READING

Aaker, David A., and Myers, John G. *Advertising Management*, 2nd ed. Englewood Cliffs, N.J.: Prentice-Hall, 1982.

Bergendorff, Fred L. *Broadcast Advertising and Promotion: A Handbook for Students and Professionals.* New York: Hastings House, 1983.

Dickey, Lew. *The Franchise: Building Radio Brands.* Washington: NAB Publications, 1994.

Eastman, Susan Tyler, and Klein, Robert A. *Promotion and Marketing for Broadcasting and Cable*, 2nd ed. Prospect Heights, Ill.: Waveland Press, 1991.

Gompertz, Rolf. *Promotion and Publicity Handbook for Broadcasters.* Blue Ridge Summit, Pa.: Tab Books, 1977.

Macdonald, Jack. *The Handbook of Radio Publicity and Promotion.* Blue Ridge Summit, Pa.: Tab Books, 1970.

Matelski, Marilyn. *Broadcast Programming and Promotion Work Text.* Boston: Focal Press, 1989.

National Association of Broadcasters. *Best of the Best Promotions*, III. Washington: NAB Publications, 1994.

Nickels, William. *Marketing Communications and Promotion*, 3rd ed. New York: John Wiley and Sons, 1984.

Peck, William A. *Radio Promotion Handbook*. Blue Ridge Summit, Pa.: Tab Books, 1968.

Rhoads, B. Eric, et al., eds. *Programming and Promotions*. West Palm Beach, FL.: Streamline Press, 1995.

Roberts, Ted E. F. *Practical Radio Promotions*. Boston: Focal Press, 1992.

Savage, Bob. *Perry's Broadcast Promotion Sourcebook*. Oak Ridge, Tenn.: Perry Publications, 1982.

Stanley, Richard E. *Promotions*, 2nd ed. Englewood Cliffs, N.J.: Prentice-Hall, 1982.

8 Traffic and Billing

THE AIR SUPPLY

A station sells airtime. That is its inventory. The volume or size of a given station's inventory chiefly depends on the amount of time it allocates for commercial matter. For example, some stations with Easy Listening and Adult Contemporary formats deliberately restrict or limit commercial loads as a method of enhancing overall sound and fostering a "more music–less talk" image. Other outlets simply abide by commercial load stipulations as outlined in their license renewal applications.

A full-time station has over 10,000 minutes to fill each week. This computes to approximately 3,000 minutes for commercials, based on an 18-minute commercial load ceiling per hour. In the eyes of the sales manager, this means anywhere from 3,000 to 6,000 availabilities or slots—assuming that a station sells 60- and 30-second spot units—in which to insert commercial announcements.

From the discussion up until now, it should be apparent that inventory control and accountability at a radio station are no small job. They are, in fact, the primary duty of the person called the traffic manager.

THE TRAFFIC MANAGER

A daily log is prepared by the traffic manager (also referred to as the traffic director). This document is at once a schedule of programming elements (commercials, features, public service announcements) to be aired and a record of what was actually aired. It serves to inform the on-air operator of what to broadcast and at what time, and it provides a record for, among other things, billing purposes.

Let us examine the process involved in logging a commercial for broadcast beginning at the point at which the salesperson writes an order for a spot schedule.

The salesperson writes an order and returns it to the station.
Order is then checked and approved by the sales manager.
Order is typed by the sales secretary.
Copies of formalized order are distributed to: traffic manager, sales manager, billing, salesperson, and client.
Order is placed in the traffic scheduling book or entered into the computer for posting to the log by traffic manager.
Order is logged, commencing on the start date according to the stipulations of the buy.

Although the preceding is both a simplification and generalization of the actual process, it does convey the basic idea. Keep in mind that not all stations operate in exactly the same manner. The actual method for preparing a log will differ, too, from station to station depending on whether it is done manually or by computer. Those outlets using the manual system often simplify the process by preparing a master or semipermanent log containing fixed program elements and even long-term advertisers. Short-term sponsors and other changes will be entered on an ongoing basis. This method significantly reduces typing. The master may be imprinted on plastic or Mylar\tm, and entries can be made and erased according to need. Once the log is prepared, it is then copied and distributed.

FIGURE 8.1
Traffic directors
move a different
kind of traffic.

It is the traffic manager's responsibility to see that an order is logged as specified and that each client is treated fairly and equitably. A sponsor who purchases two spots, five days a week during morning drive, can expect to receive good rotation for maximum reach. It is up to the traffic manager to schedule the client's commercials in as many quarter-hour segments of the daypart as possible. The effectiveness of a spot schedule is reduced if the spots are logged in the same quarter-hour each day. If a spot is logged at 6:45 daily, it is only reaching those people tuned at that hour each day. However, if on one day it is logged at 7:15 and then at 8:45 on another, and so on, it is reaching a different audience each day. It also would be unfair to the advertiser who purchased drivetime to only have spots logged prior to 7:00 a.m., the beginning of the prime audience period.

The traffic manager maintains a record of when a client's spots are aired to help ensure effective rotation. Another concern of the traffic manager is to keep adequate space between accounts of a competitive nature. Running two restaurants back-to-back or within the same spot set likely would result in having to reschedule both at different times at no cost to the client.

It also falls within the traffic manager's purview to make sure copy and production tapes are in on time. Most stations have a policy, often stated in their rate card, requiring that commercial material be on hand at least 48 hours before it is scheduled for broadcast. Carol Bates, traffic manager of WGNG-AM (now WICE-AM), Providence, says that getting copy before the air date can be a problem. "It is not unusual to get a tape or copy a half-hour before it is due to air. We ask that copy be in well in advance, but sometimes it's a matter of minutes. No station is unfamiliar with having to make up spots due to late copy. It's irritating but a reality that you have to deal with."

Jan Hildreth, traffic manager of WARA-AM, Attleboro, Massachusetts,

FIGURE 8.2
Unless the traffic manager keeps a close eye on the air supply, it can be exhausted, placing the station in an oversold situation. Courtesy Marketron.

Marketron

WDMO-AM
EXCLUDED: PRIORITIES:
    DAYS:
MINS. BUMPED PRIOR TO 8/ 7/89
  5.00   RATE CLASS AAA
  3.30   RATE CLASS AA
  4.30   RATE CLASS A
  0.30   RATE CLASS B

OVERSOLD REPORT
SLOTS: 910
AS OF 8/ 7/89

	UNITS AVAILABLE				ORIG-INVEN	BOOKED	BUMPED	PCT SOLD	UNSOLD	PRIORITIES #6	#7
	60	30	10	5							
CLASS AAA											
8/ 7 MON	1	1	1	0	60:20	58:40	2:00	101%	1:40	6:00	9:00
8/ 8 TUE	0	1	1	0	60:20	59:40	1:00	101%	0:40	8:00	7:00
8/ 9 WED	2	1	1	0	60:20	57:40	5:00	104%	2:40	6:30	3:00
8/10 THU	1	0	1	0	60:20	59:10	3:00	104%	1:10	5:30	5:00
8/11 FRI	1	0	1	0	60:20	59:10	3:00	104%	1:10	4:30	5:00
8/12 SAT	2	0	2	0	135:45	133:25	11:00	107%	2:20	23:00	11:00
8/13 SUN	0	0	0	0	0:00	0:00	0:00	0%	0:00	0:00	0:00
TOTAL	7	3	7	0	437:25	427:45	25:00	104%	9:40	53:30	40:00
CLASS AA											
8/ 7 MON	0	0	1	0	60:20	60:10	2:30	104%	0:10	11:00	8:00
8/ 8 TUE	1	0	1	0	60:20	59:10	3:00	104%	1:10	12:30	9:30
8/ 9 WED	1	0	1	0	60:20	59:10	1:00	100%	1:10	14:00	9:30
8/10 THU	0	1	1	0	60:20	59:40	2:00	103%	0:40	14:30	5:30
8/11 FRI	1	0	1	0	60:20	59:10	2:00	102%	1:10	11:30	7:30
8/12 SAT	0	0	0	0	0:00	0:00	0:00	0%	0:00	0:00	0:00
8/13 SUN	19	1	0	1	196:05	176:30	3:00	92%	19:35	26:30	19:30
TOTAL	22	2	5	1	497:45	473:50	13:30	98%	23:55	90:00	59:30
CLASS A											
8/ 7 MON	20	0	1	1	75:25	55:10	1:30	76%	20:15	6:00	12:00
8/ 8 TUE	22	0	1	0	75:25	53:15	0:00	71%	22:10	7:00	11:00
8/ 9 WED	21	0	1	1	75:25	54:10	0:00	72%	21:15	9:00	10:00
8/10 THU	20	1	1	0	75:25	54:45	0:00	73%	20:40	8:30	10:00
8/11 FRI	22	1	1	1	75:25	52:40	0:00	70%	22:45	8:30	10:00
8/12 SAT	10	0	1	0	60:20	50:10	0:00	84%	10:10	4:30	11:00
8/13 SUN	0	0	0	0	0:00	0:00	0:00	0%	0:00	0:00	0:00
TOTAL	115	2	6	3	437:25	320:10	1:30	74%	117:15	43:30	64:00
CLASS B											
8/ 7 MON	51	0	1	1	105:35	54:20	0:00	52%	51:15	2:00	9:00
8/ 8 TUE	52	0	2	0	105:35	53:15	0:30	51%	52:20	2:30	9:00
8/ 9 WED	53	0	1	1	105:35	52:20	0:00	50%	53:15	1:30	9:00
8/10 THU	53	0	2	0	105:35	52:15	1:00	51%	53:20	2:00	8:00
8/11 FRI	47	1	1	1	105:35	57:50	0:00	55%	47:45	7:30	10:00
8/12 SAT	46	1	2	0	105:35	58:45	0:00	56%	46:50	5:30	10:00
8/13 SUN	60	1	1	1	105:35	44:50	0:00	43%	60:45	1:30	9:00
TOTAL	362	3	10	4	739:05	373:35	1:30	51%	365:30	22:30	64:00

says that holiday and political campaign periods can place added pressure on the traffic person. "The fourth quarter is the big money time in radio. The logs usually are jammed, and availabilities are in short supply. The workload in the traffic department doubles. Things also get pretty chaotic around elections. It can become a real test for the nerves. Of course, there's always the late order that arrives at 5 p.m. on Friday that gets the adrenalin going."

There are few station relationships closer than that of the traffic department with programming and sales.

FIGURE 8.3
Stations often project availabilities to avoid being oversold. Computers simplify this task. Courtesy Marketron.

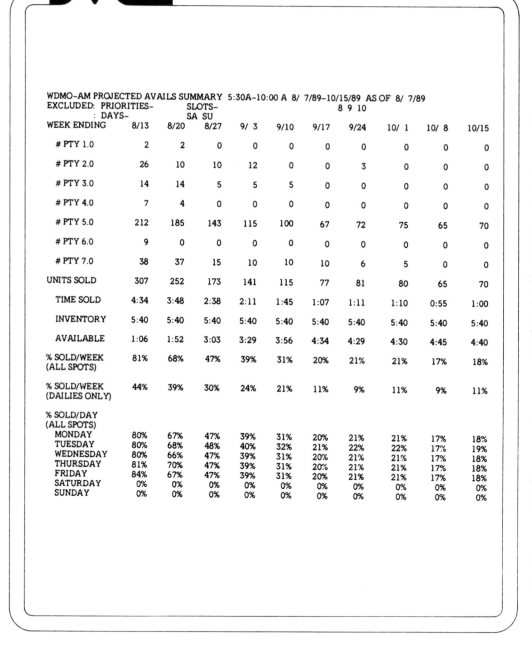

Marketron

WDMO–AM PROJECTED AVAILS SUMMARY 5:30A–10:00 A 8/ 7/89–10/15/89 AS OF 8/ 7/89
EXCLUDED: PRIORITIES– SLOTS– 8 9 10
: DAYS– SA SU

WEEK ENDING	8/13	8/20	8/27	9/ 3	9/10	9/17	9/24	10/ 1	10/ 8	10/15
# PTY 1.0	2	2	0	0	0	0	0	0	0	0
# PTY 2.0	26	10	10	12	0	0	3	0	0	0
# PTY 3.0	14	14	5	5	5	0	0	0	0	0
# PTY 4.0	7	4	0	0	0	0	0	0	0	0
# PTY 5.0	212	185	143	115	100	67	72	75	65	70
# PTY 6.0	9	0	0	0	0	0	0	0	0	0
# PTY 7.0	38	37	15	10	10	10	6	5	0	0
UNITS SOLD	307	252	173	141	115	77	81	80	65	70
TIME SOLD	4:34	3:48	2:38	2:11	1:45	1:07	1:11	1:10	0:55	1:00
INVENTORY	5:40	5:40	5:40	5:40	5:40	5:40	5:40	5:40	5:40	5:40
AVAILABLE	1:06	1:52	3:03	3:29	3:56	4:34	4:29	4:30	4:45	4:40
% SOLD/WEEK (ALL SPOTS)	81%	68%	47%	39%	31%	20%	21%	21%	17%	18%
% SOLD/WEEK (DAILIES ONLY)	44%	39%	30%	24%	21%	11%	9%	11%	9%	11%
% SOLD/DAY (ALL SPOTS)										
MONDAY	80%	67%	47%	39%	31%	20%	21%	21%	17%	18%
TUESDAY	80%	68%	48%	40%	32%	21%	22%	22%	17%	19%
WEDNESDAY	80%	66%	47%	39%	31%	20%	21%	21%	17%	18%
THURSDAY	81%	70%	47%	39%	31%	20%	21%	21%	17%	18%
FRIDAY	84%	67%	47%	39%	31%	20%	21%	21%	17%	18%
SATURDAY	0%	0%	0%	0%	0%	0%	0%	0%	0%	0%
SUNDAY	0%	0%	0%	0%	0%	0%	0%	0%	0%	0%

Programming relies on the traffic manager for the logs that function as scheduling guides for on-air personnel. Sales depends on the traffic department to inform it of existing availabilities and to process orders onto the air. "It is crucial to the operation that traffic have a good relationship with sales and pro-gramming. When it doesn't, things begin to happen. The PD has to let traffic know when something changes; if not, the system breaks down. This is equally true of sales. Traffic is kind of the heart of things. Everything passes through the traffic department. Cooperation is very important," observes Barbara Kafulas, traffic manager, WNRI-AM, Woonsocket, Rhode Island.

THE TRAFFIC MANAGER'S CREDENTIALS

A college degree usually is not a criterion for the job of traffic manager. This is not to imply that skill and training are not necessary. Obviously the demands placed on the traffic manager are formidable, and not everyone is qualified to fill the position. "It takes a special kind of person to effectively handle the job of traffic. Patience, an eye for detail, plus the ability to work under pressure and with other people are just some of the qualities the position requires," notes radio executive Bill Campbell.

Typing or keyboard skills are vital to the job. A familiarity with computers and word processing has become necessary, as most stations have given up the manual system of preparing logs in favor of the computerized method.

Many traffic people are trained in-house and come from the administrative or clerical ranks. It is a position that traditionally has been filled by women. While traffic salaries generally exceed that of purely secretarial positions, this is not an area noted for its high pay. Although the traffic manager is expected to handle many responsibilities, the position generally is perceived as more clerical in nature than managerial.

Traffic managers frequently make the transition into sales or programming. The considerable exposure to those particular areas provides a solid foundation and good springboard for those desiring to make the change.

DIRECTING TRAFFIC

Computers vastly enhance the speed and efficiency of the traffic process.

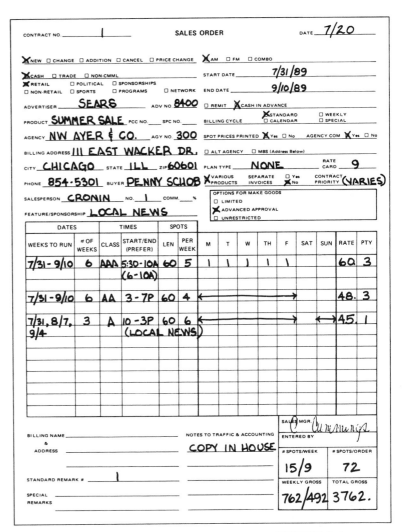

FIGURE 8.4
Handwritten sales order. The station salesperson usually writes the order at the client's business for initial approval. Courtesy Marketron.

Computers store copious amounts of data, retrieve information faster than humanly possible, and schedule and rotate commercials with precision and equanimity, to mention only a few of the features that make the new technology especially adaptive for use in the traffic area.

Computers are an excellent tool for inventory control, contends former broadcast computer consultant Vicki Cliff. "Radio is a commodity not unlike a train carload of perishables, such as tomatoes. Radio sells time, which is progressively spoiling. The economic laws of supply and demand are classically applied to radio. Computers can assist in plotting that supply and demand curve in determining rates to be charged for various dayparts at any

SOFTWARE
SPECIALISTS
INC.

RADIO TRAC

"The Traffic and Accounting Software of the 90's"

RADIO TRAC is one of the most comprehensive traffic and accounting software packages available. It is a completely integrated system, rich in flexibility and user-friendliness.

RADIO TRAC was designed and developed with the benefits of 17 years background in radio traffic management, 25 years in accounting/controller functions, and over 20 years experience in the development of computer software applications.

RADIO TRAC's customizing features allow each station a perfect fit for its particular needs and changing situations without costly additional software programming.

The TRAFFIC module, from initial order entry to the finished program log, is designed for speed, ease of data entry, immediate log format changes, and printed or on-line display of pertinent information.

The ACCOUNTS RECEIVABLE module provides a complete billing process including preparation of the sales journal, invoices, statements and accounts receivable; it also provides advertiser information, a cash receipts entry process, sales projections, a myriad of information screens and reports on all accounts receivable-related functions, and posting to the General Ledger.

The ACCOUNTS PAYABLE module processes and reports accounts payable data including accounts payable invoices, check preparation vendor information, activity reports, and posting to the General Ledger.

The GENERAL LEDGER module gathers posted entries from the Accounts Payable, Accounts Receivable and from the General Ledger systems and ultimately produces an easy to read customized financial statement.

The SALES module provides useful information on screen and in report form on salesperson performance, bulk contract status and advertiser activity.

The MANAGEMENT module supplies useful administrative and personnel information.

FIGURE 8.5
Integrated traffic/accounting systems are much in vogue at radio stations.

given moment. Inventory control is vital to any business. Radio is limited in its availabilities and seasonal in its desirability to the client. In a sold-out state, client value priorities must be weighted to optimize the station's billing. All things being equal, the credit rating of the client should be the deciding factor. Computers can eliminate the human subjectivity in formulating the daily log."

The cost of computerizing traffic has kept a small percentage of stations from converting from the manual system, notes Cliff. "Purchasing a personal computer for the traffic function is okay, but if a station desires to perform several functions, it may find itself out of luck. There are limits as to what can be done with a personal computer. On the other hand, a larger capacity, main-frame computer can be a major expense, although the functions it can perform are extraordinary. An on-line, real-time system can be costly also. Line expenses can really add up. Those station managers considering computerizing must gather all possible data to determine if the system they're considering will cost-justify itself."

In a recent interview in *Radio Ink*, WPOC-FM's Jim Dolan observed that "The move right now is toward putting your sales force in the field armed with laptops and instantaneous on-line access to inventory, availability, and contract information. And the PC-based systems seem to be evolving faster than the minicomputer systems in this regard."

A variety of traffic and billing software are available. Dozens of companies, most notably Marketron, Columbine, Bias Radio, Custom Business Systems, Jefferson Pilot, and The Management, specialize in providing broadcasters with software packages. Prices for computer software vary depending on the nature and content of the program.

Compatible hardware is specified by the software manufacturer. Most software is IBM-PC compatible, however. Companies and consultants specializing in broadcast computerization are listed in *Broadcasting Yearbook*.

According to the NAB, the majority of stations are utilizing computers, especially in the area of traffic, in the 1990s.

According to Jay Williams, Jr., "Traffic and accounting for many affiliated stations are now being done by modem from corporate offices. Even sales is becoming more computer-powered. It won't be long before avails and pricing will be offered to stations, acceptance of an order will be confirmed by a keystroke, and the commercial downloaded on command."

BILLING

At most stations, advertisers are billed for the airtime they have purchased

Traffic Order

Date _____

FIGURE 8.6
Spot schedule
order form.
Courtesy
WMJX-FM.

CODE GROUP

Station		Product No.		
Agency Number		Category		
Advertiser Number		Billing Period		
Contract Number		Invoice Times		
Salesman Number		Agency Commission	%	
Salesman Name				
Advertiser				
To Be Logged As				

Agency Estimate No.	
Rate Card	
Co-Op Invoice	YES NO SORT!
Dealer Invoice	YES NO # OF SHARES
Discount	%
Annct. Type	
Advertiser Avail.Code	
Contract Avail.Code	

BUYER INFORMATION

Name	
Telephone	
Account Type	L R N
X	Y
Z	

CREDIT INFORMATION

Name	
Telephone	
Credit Check	YES NO INITIALS
Credit Limit	
Contract Year	START END

Special Instructions:

Option	Line No.	Group	Type Prd Adv	Minum. Separ.	Broadcast Dates Start	End	Flight Section	Class	Spots/Prm Duration	Annct Length	Time Schedule Start	End	Spots Per Week	Broadcast Pattern Mon Tue Wed Thur Fri Sat Sun	Control	Rate	Plan Unit Madc Charges Live	Video	Audio	Agency Copy Number	Rack Number	Cr. Code M/G For

Agency Name and Address			
	Zip/PC.		

Billing Name and Address			
	Zip/PC.		

	PACKAGE PLAN FREQ. RATE	B'casts Rate	Totals By Rate	B'casts Rate	Totals By Rate

Contract Totals	Gross
	Agency Comm.
	Net
X To Cancel Order Before Start	

PRINTED IN U.S.A.

after a portion or all of it has run. Few stations require that sponsors pay in advance. It is the job of the billing department to notify the advertiser when payment is due. Al Rozanski, former business manager of WMJX-FM, Boston, explains the process involved once a contract has been logged by the traffic department. "We send invoices out twice monthly. Many stations bill weekly, but we find doing it every two weeks cuts down on the paperwork considerably. The first thing my billing person does is check the logs to verify that the client's spots ran. We don't bill them for something that wasn't aired. Occasionally a spot will be missed for one reason or another, say a technical problem. This will be reflected on the log because the on-air person will indicate this fact. Invoices are then generated in triplicate by our computer. We use an IBM System 34 computer and Columbine software. This combination is extremely versatile and efficient. The station retains a copy of the invoice and mails two to the client, who then returns one with the payment. The client also receives an affidavit detailing when spots were aired. If the client requests, we will notarize the invoice. This is generally necessary for clients involved in co-op contracts." The billing procedure at WMJX-FM is representative of that at many stations.

Not all radio stations have a full-time business manager on the payroll. Thus the person who handles billing

FIGURE 8.7
Traffic managers
check logs before
placing them in the
control room.
Courtesy WHJJ-AM.

FIGURE 8.8
Manual and computerized traffic boards permit the traffic department to keep tabs on avails and rotation schedules. Courtesy Marketron.

Marketron

```
WDMO-AM DAILY SPOT REPORT FOR TUE  8/ 8/89  7:00A- 8:00A AS OF  8/ 7/89                    PAGE 1
SLOT    CONTR  LN  ADVERTISER        LEN  LAST  CONTRACT TIMES & DAYS    COMP                  S
TIMES     #    #                        * DATE  FROM       TO MTWTFSS   CODES  RATE PTY        T
7A HOUR
02:00 SUS AVAILS:                0/   0                                                        1
        529    1  ATLANTIC RECO     60   9/10  6:00A  10:00A XXXXX-- 63/0       60  5.0        0

05:00 SUS AVAILS:                1/  10                                                        6
        545    1  FIRESTONE         60   8/13  6:00A  10:00A XXXXXX- 15/0       48  6.0        0

10:00 SUS AVAILS:                1/   0                                                        6
        546    2  AMERICAN MOTO     60   8/13  6:00A  10:00A XXXXXX- 12/0       65  2.0        3

14:00 SUS AVAILS:                0/   0                                                        3
        527    1  CARNATION         60   9/ 2  6:00A  10:00A XXXXX-- 35/90      70  2.0        3

16:00 SUS AVAILS:                0/   0                                                        6
        510    3  HOLIDAY INN       30  11/19  5:30A  10:00A 1111111 42/0       55  5.0        0

22:00 SPO AVAILS:                1/  60                                                        7

23:00 SUS AVAILS:                0/   0                                                        6
          1    1  SEARS             60   9/10  5:30A  10:00A 1111100 82/0       60  3.0        0

28:00 NET AVAILS:                1/   0                                                        8
        856    1  GENERAL MOTOR     60  (NET)                       12/110
        534    1  HOWARD JOHNSO     60   9/10  6:00A  10:00A XXXXXX- 42/0       60  5.0        0

32:00 SUS AVAILS:                0/   0                                                        2
        522    1  ANHEUSER BUSC     60   8/13  7:32A   7:33A 0I00100 21/0       70  1.0        0

35:00 SUS AVAILS:                0/   0                                                        6
        515    2  PRUDENTIAL LI     60   8/27  5:00A  10:00A 2222222 43/0       60  5.0        0
        549    1  AMERICAN AIRL     60   8/27  R/C(9)    AAA XXXXXXX  3/0       60  5.0        0

40:00 SUS AVAILS:                1/   0                                                        6
        489    1  YES ON ONE CO     60   8/20  R/C(9)    AAA XXXXXXX 51/0       58  3.0        0

44:00 SUS AVAILS:                0/   0                                                        5
        501    2  GALLO WINES       60   7/ 1  5:30A  10:00A 1111111 21/0       60  5.0        0

46:00 SUS AVAILS:                1/  60                                                        6

52:00 SUS AVAILS:                1/   5                                                        4

53:00 SUS AVAILS:                0/   0                                                        6
        513    2  HERTZ             30   8/27  5:30A  10:00A 1111111 12/0       60  5.0        0
        523    2  KODAK             30*  8/13  6:00A  10:00A 1010100 23/0       35  5.0        0

57:00 SUS AVAILS:                1/  30                                                        6
        555    1  BIC CORP          60   8/20  6:00A  10:00A 2222220 40/0       48  7.0        0
        524    3  RINGLING BRO      30   8/13  R/C(9)    AAA XXXXXXX  6/0       30  5.0        0

        SOLD    : 7A- 8A=15M  0S ( 7A- 7:30A=7M30S, 7:30A- 8A= 7M30S), # SPOTS= 17
        ALLOWED: 7A- 8A=15M15S
------------------------------------------------------------------------------------------
ST= SLOT TYPE, *= MAKEGOOD, += OVER 9 SPOTS/DAY, X= MULTI-DAY,
SUS= STATION SLOT (SUSTAINING OK), SPO= STATION SLOT (SPONSORSHIP ONLY),
NET= NETWORK SLOT, NCM= NON-COML SLOT
```

commonly is responsible for maintaining the station's financial records or books as well. In this case, the services of a professional accountant may be contracted on a regular periodic basis to perform the more complex bookkeeping tasks and provide consultation on other financial matters.

Accounts that fail to pay when due are turned over to the appropriate salesperson for collections. If this does not result in payment, a station may use the services of a collection agency. Should its attempt also fail, the station likely would write the business off as a loss at tax time.

THE FCC AND TRAFFIC

Program log requirements were eliminated by the FCC in the early 1980s as part of the era's formidable deregulation movement. Before then the FCC expected radio stations to maintain a formal log, which—in addition to program titles, sponsor names, and length of elements—reflected information pertaining to the nature of announcements (commercial material, public service announcement), source of origination (live, recorded, network), and the type of program (entertainment, news, political, religious, other). The log in Figure 8.5 (on p. 224) reflects these requirements. Failing to include this information on the log could have resulted in punitive actions against the station by the FCC.

Although stations no longer must keep a program log under existing rules, some sort of document still is necessary to inform programming personnel of what is scheduled for broadcast and to provide information for both the traffic and billing departments pertaining to their particular functions. A log creates accountability. It is both a programming guide and a document of verification.

Stations are now at liberty to design logs that serve their needs most effectively and efficiently. The WMJX log form in Figure 8.13 (on p. 230) is an example of a log that has been designed to meet all of its station's needs in the most economical and uncomplicated way.

There are no stipulations regarding the length of time that logs must be retained. Before the elimination of the FCC program log regulations, stations were required to retain logs for a minimum of two years. Today most stations still hold onto logs for that amount of time for the sake of accountability.

CHAPTER HIGHLIGHTS

1. Each commercial slot on a station is called an *availability*. Avai-

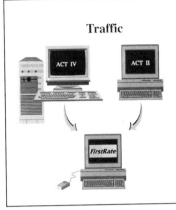

Traffic

The Marketron Act II and IV radio traffic systems supply spot data to a program called FirstRate. This program analyzes and categorizes this data in a completely different way than anything before. The goal is to present a picture of information that will offer greater assistance in manipulating inventory and controlling price. The end objective of FirstRate is to provide the sales staff with the ability to produce a schedule that reflects pricing based upon demand. The computer uses (3) elements to do this. First, it knows how much inventory is sold out into the future. Secondly, it knows what the sales managers

inventory objectives are and last it has a grid rate card developed by the manager to query. This has been developed for each daypart and creates a price based upon the variables of demand. As demand changes, spot prices change, and the severity of these changes is in the control of the manager.

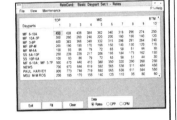

The data transfer to the FirstRate system is extensive. If someone had to do this job manually it would take hours each day. Because of this interface to traffic, we decided to use all the pertinent spot data available in order to create management reports that are new and unique to this assignment.

labilities constitute a station's salable inventory.

2. The traffic manager (or traffic director) controls and is accountable for the broadcast time inventory.

3. The traffic manager prepares a log to inform the deejays of what to broadcast and at what time.

4. The traffic manager is also responsible for ensuring that an ad order is logged as specified, that a record of when each client's spots are aired is maintained, and that copy and production tapes are in on time.

5. Programming relies on the traffic manager for the logs that function as scheduling guides for on-air personnel; the sales department depends upon the traffic manager to inform them of existing availabilities and to process orders onto the air.

6. Although most traffic people are trained in-house and are drawn from the administrative or clerical ranks, they must possess patience, an eye for

FIGURE 8.9
The traffic department provides key information to salespeople through special computer programs that manage inventory. Courtesy Marketron.

FIGURE 8.10
Spots that are not aired when scheduled must be rescheduled ("made good") according to the conditions of the purchase order. Courtesy Marketron.

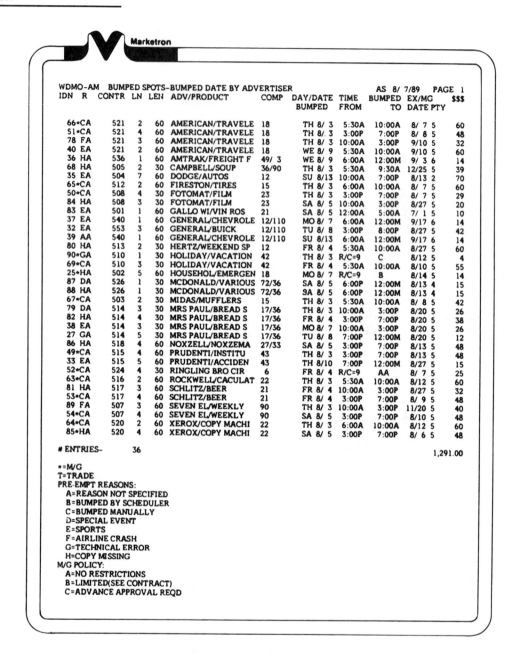

detail, the ability to work under pressure, and keyboarding skills.

7. Most traffic departments have been computerized to enhance speed and efficiency. Therefore, traffic managers must be computer-knowledgeable.

8. Based on the spots aired, as recorded and verified by the traffic department, the billing department sends invoices weekly or biweekly to each client. Invoices are notarized for clients with co-op contracts.

9. Since the FCC eliminated program log requirements in the early 1980s, stations have been able to design logs that inform programming personnel of what is scheduled for broadcast and that provide necessary information for the traffic and billing departments.

FIGURE 8.11
Computerized log
form. Courtesy
WMJX-FM.

ACTUAL TIME	SCHEDULED TIME	LENGTH	PROGRAM TITLES — ANNOUNCEMENTS	CART NUMBER	ANNCT TYPE	PROG. S	T	CONTRACT & LINE NUMBER
ON	OPERATOR		OFF	ON	OPERATOR			OFF

COLUMBINE SYSTEMS, INC.   CSI-498   MADE IN U.S.A.   3-83   79741-1

FIGURE 8.12
Computers were
first installed in
station traffic
departments.
Courtesy WMJX-
FM.

229

FIGURE 8.13
Computerized log.
Courtesy
Marketron.

	PROGRAM LOG		PAGE 8
Marketron			AM TUE 8/ 8/89
			PACIFIC DAYLIGHT TIME

ACTUAL SCHEDULED TIME & LENGTH	PROGRAM OR SPONSOR		COPY	SOURCE & TYPE	
7:00:00– 8:00:00	THE MORNING SHOW		LIVE	L	E
7:00:00– 7:05:00	NEWS ON THE HOUR		LIVE	L	N
02:00 60	ATLANTIC RECORDS/FATS DOMINO		CART #459		CM
05:00 60	FIRESTONE/RADIALS		. . . .		CM
06:00 10	PSA/CONSERVE ENERGY		LIVE		PSA
10:00 60	AMERICAN MOTORS/PACER	(ADJ)	CART #878		CM
7:13:00– 7:16:00	SPORTS REPORT		LIVE	L	S
5	SPONSOR BB		OPEN		
14:00 60	CARNATION/YOGURT		CART #567		CM
5	SPONSOR BB		CLOSE		
16:00 30	HOLIDAY INN/VACATION		LIVE		CM
7:20:00– 7:23:00	LOCAL NEWS			L	N
22:00 60					
23:00 60	SEARS/ALLSTATE		CART #304		CM
28:00 60	(NET SPOT) GM/BUICK	(DB)	CART #5678		CM
29:00 60	HOWARD JOHNSON/GET AWAY WEEKEND		CART #8966		CM
7:30:00– 7:33:00	HEADLINES & NEWS UPDATES		LIVE	L	N
32:00 60	ANHEUSER BUSCH/BUDWEISER	(FP)	CART #523		CM
35:00 60	PRUDENTIAL LIFE/ACCIDENT INS		CART #784		CM
36:00 60	AMERICAN AIRLINES/KANSAS CITY		CART #201		CM
40:00 60	YES ON ONE COMMITTEE/POLITICAL		CART #682		CM
	LIVE DISCLAIMER		4		
7:43:00– 7:46:00	BUSINESS NEWS			L	N
44:00 60	GALLO WINES/VIN ROSE		CART #246 LT		CM
46:00 60					
52:00 10	TEMPERATURE & TIME		LIVE		OA
52:00 5					
53:00 30	HERTZ/WEEKEND SPECIAL		CART #778		CM
53:30 30	KODAK/FILM	*MG	CART #668		CM
57:00 60	BIC CORP/RAZORS	*TR	CART #345		CM
58:00 30	RINGLING BRO CIRCUS		CART #990		CM
58:30 30					
59:55 5	STATION ID		LIVE		ID

TIME ON	LOGKEEPER'S SIGNATURE	TIME OFF	TIME ON	LOGKEEPER'S SIGNATURE	TIME OFF

FIGURE 8.14
Standard form used for clients involved in co-op agreements. Courtesy WMJX-FM.

ANA/RAB FORM FOR SCRIPT (IF TAPE IS USED, PREPARE SCRIPT FROM TAPE)

wmyx 106 fm
The _magic_ is the music. ®

ANA/RAB RADIO "TEAR-SHEET"
FORM AT BOTTOM OF SCRIPT PERMITS KNOWING HOW MANY TIMES THIS SCRIPT RAN, AT WHAT COST.

Client:	For:

Begin:	End:	Date:

STATION DOCUMENTATION STATEMENT APPROVED BY THE CO-OPERATIVE ADVERTISING COMMITTEE OF THE ASSOCIATION OF NATIONAL ADVERTISERS

This announcement was broadcast_____times, as entered in the station's program log. The times this announcement was broadcast were billed to this station's client at our invoice(s) number/dated _____at his earned rate of:

$_____each for_____announcements, for a total of $_____
$_____each for_____announcements, for a total of $_____
$_____each for_____announcements, for a total of $_____

_____
Signature of station official

_____   _____   _____
(Notarize above)   Typed name and title   Station

FIGURE 8.15
Example of station
invoice sent to an
ad agency. Notice
that the
commission is
deducted from the
total gross.
Courtesy
Marketron.

		INVOICE	INVOICE AND AFFIDAVIT NO.
			81001

DATE   8/ 6/89

PAGE   1

TO: NW AYER & CO   (300)
111 EAST WACKER DR
CHICAGO IL 60601

ACCOUNT: SEARS   (8400)
1

SALESMAN: J CRONIN
BILLING INQUIRIES CONTACT NANCY KRUEGER (415) 854-2767

TERMS:

DAY	DATE	CLASS	LENGTH	RATE CARD	ACTUAL TIME		RATES
MO	7/31	AAA	60		7:32A TOOLS		60.00
MO	7/31	A	60		10:22A PONG TV GAME		45.00
TU	8/ 1	A	60		2:22P ALLSTATE (A99)		45.00
TU	8/ 1	AA	60		5:53P BATTERIES		48.00
WE	8/ 2	AAA	60		9:02A TOOLS		60.00
WE	8/ 2	A	60		1:22P PONG TV GAME		45.00
WE	8/ 2	AA	60		6:53P WALLPAPER		48.00
TH	8/ 3	AAA	60		6:23A PONG TV GAME		60.00
TH	8/ 3	A	60		12:22P TIRES		45.00
TH	8/ 3	AA	60		4:02P BATTERIES		48.00
FR	8/ 4	AAA	60		7:29A PONG TV GAME		60.00
FR	8/ 4	A	60		11:22A PAINT		45.00
SU	8/ 6	A	60		10.22A PONG TV GAME		45.00
TU	8/ 1	AAA	60		5:30A TIRES	MISSED	N/C
						M/G TO RUN 8/ 8/89	
MO	7/31	AA	60		3:00P BATTERIES	MISSED	N/C

NOTARIZED AFFIDAVITS REQUIRED

TOTAL SPOTS   13

TOTAL GROSS   654.00
LESS AGENCY COMMISSION   98.10

PAY THIS AMOUNT   555.90

THIS RADIO STATION WARRANTS THAT THE PROGRAM/ANNOUNCEMENTS INDICATED ABOVE WERE BROADCAST IN ACCORDANCE WITH OFFICIAL STATION LOG.
ALL TIMES ARE APPROXIMATE WITHIN 15 MINUTES AND ARE WITHIN THE TIME CLASSIFICATION ORDERED

SUGGESTED FURTHER READING

Diamond, Susan Z. *Records Management: A Practical Guide.* New York: AMACOM, 1983.

Doyle, Dennis M. *Efficient Accounting and Record Keeping.* New York: David McKay and Company, 1977.

Heighton, Elizabeth J., and Cunningham, Don R. *Advertising in the Broadcast and Cable Media,* 2nd ed. Belmont, Calif.: Wadsworth Publishing, 1984.

Keith, Michael C. *Selling Radio Direct.* Boston: Focal Press, 1992.

Murphy, Jonne. *Handbook of Radio Advertising.* Radnor, Pa.: Chilton, 1980.

Slater, Jeffrey. *Simplifying Accounting Language.* Dubuque, Iowa: Kendall-Hall Publishing, 1975.

Warner, Charles, and Buchman, Joseph. *Broadcasting and Cable Selling,* 2nd ed. Belmont, Calif.: Wadsworth Publishing, 1993.

Zeigler, Sherilyn K., and Howard, Herbert H. *Broadcast Advertising: A Comprehensive Working Textbook,* 2nd ed. Columbus, Ohio: Grid Publishing, 1984.

9 Production

A SPOT RETROSPECTIVE

A typical radio station will produce thousands of commercials, public service announcements, and promos annually. Initially, commercials were broadcast live, due to a lack of recording technology. In the 1920s, most paid announcements consisted of lengthy speeches on the virtues of a particular product or service. Perhaps the most representative of the commercials of the period was the very first ever to be broadcast, which lasted over ten minutes and was announced by a representative of a Queens, New York, real estate firm. Aired live over WEAF in 1922, by today's standards the message would sound more like a classroom lecture than a broadcast advertisement. Certainly no snappy jingle or ear-catching sound effects accompanied the episodic announcement.

Most commercial messages resembled the first until 1926. On Christmas Eve of that year the radio jingle was introduced, when four singers gathered for a musical tribute to Wheaties cereal. It was not for several years, however, that singing commercials were commonplace. For the most part, commercial production during the medium's first decade was relatively mundane. The reason was twofold. The government had resisted the idea of blatant or direct commercialism from the start, which fostered a low-key approach to advertising, and the medium was just in the process of evolving and therefore lacked the technical and creative wherewithal to present a more sophisticated spot.

Things changed by 1930, however. The austere, no-frills pitch, occasionally accompanied by a piano but more often done a cappella, was gradually replaced by the dialogue spot that used drama or comedy to sell its product. A great deal of imagination and creativity went into the writing and production of commercials, which were presented live throughout the 1930s. The production demands of some commercials equaled and even exceeded those of the programs they interrupted. Orchestras, actors, and lavishly constructed sound effects commonly were required to sell a chocolate-flavored syrup or a muscle liniment. By the late 1930s, certain commercials had become as famous as the favorite programs of the day. Commercials had achieved the status of pop art.

Still, the early radio station production room was primitive by today's standards. Sound effects were mostly improvised show by show, commercial by commercial, in some cases using the actual objects sounds were identified with. Glass was shattered, guns fired, and furniture overturned as the studio's on-air light flashed. Before World War II, few sound effects were available on records. It was just as rare for a station to broadcast prerecorded commercials, although 78 rpm and wire recordings were used by certain major advertisers. The creation of vinyl discs in the 1940s inspired more widespread use of electrical transcriptions for radio advertising purposes. Today sound effects are taken from specially designed LPs.

The live spot was the mainstay at most stations into the 1950s, when two innovations brought about a greater reliance on the prerecorded message. Magnetic recording tape and 33 LPs revolutionized radio production methods. Recording tape brought about the greatest transformation and, ironically, was the product of Nazi scientists who developed acetate recorders and tape for espionage purposes.

The adoption of magnetic tape by radio stations was costlier and thus occurred at a slower pace than 33 rpm, which essentially required a turntable modification.

Throughout the 1950s, advertising agencies grew to rely on LPs. By 1960, magnetic tape recorders were a familiar piece of studio equipment. More and more commercials were prerecorded. Some stations, especially those automated, did away with live announcements entirely, preferring to tape everything to avoid on-air mistakes.

Commercials themselves became more sophisticated sounding since practically anything could be accomplished on tape. Perhaps no individual in the 1960s more effectively demonstrated the unique nature of radio as an advertising medium than did Stan Freberg. Through skillful writing and the clever use of sound effects, Freberg transformed Lake Michigan into a basin of hot chocolate crowned by a 700-foot-high mountain of whipped cream, and no one doubted the feat.

Today the sounds of millions of skillfully prepared commercials trek through

FIGURE 9.1
Pioneer announcer H. W. Arlin in the early 1920s. Courtesy Westinghouse Electric.

FIGURE 9.2
KDKA's pre–World War II control center. Courtesy Westinghouse Electric.

the ether and into the minds of practically every man, woman, and child in America. Good writing and production are what make the medium so successful.

FORMATTED SPOTS

In the 1950s the medium took to formatting in order to survive and prosper. Today listeners are offered myriad sounds from which to choose. There is something for practically every taste. Stations concentrate their efforts on delivering a specific format, which may be defined as Adult Contemporary, Country, Easy Listening, or any one of a dozen others. As you will recall from the discussion in Chapter 3, each format has its own distinctive sound, which is accomplished through a careful selection and arrangement of compatible program elements. To this end, commercials also must reflect a station's format. What follows is a reference listing containing the observations of several radio producers concerning the effect that five key formats have on the production of commercials, from the copy stage through final mixdown.

ADULT CONTEMPORARY

Copy	Delivery	Mixdown
"The background and lifestyle of the station's audience must always be kept in mind when preparing copy. If the spot doesn't relate, the listener won't connect. Then you don't sell anything. Irrelevant copy may even result in audience tune-out." —Peter Fenstermacher, production director, WTIC-AM, Hartford. "Talk to people. Use short sentences, words with three or less syllables, and use common speech patterns. Remain on the level of your audience. Keep in mind who the A/C format is designed to attract and write from that perspective." —Bill Towery, production director, WFYR-FM, Chicago.	"Announcer delivery on A/C stations is not too punched-up or subdued. No screaming announcers on spots. Nothing too exaggerated. Not hyped, but not sleepy either. Somewhere in between is where you want to be in this format." —Jim Murphy, general manager, WCGY-FM, Lawrence, Massachusetts. "The best way to reach the A/C listener in a spot is by being yourself, natural. The announcer should strive for realism. A one-on-one delivery does the job here. It fits the overall flow and pacing of the programming. Warmth and friendliness are important qualities in A/C spots." —Mike Scalzi, program director, WHBQ-AM, Memphis.	"The production elements in an A/C spot should be congruous with the overall sound. For example, bed music should not deviate greatly from what the listener is accustomed to hearing on the station. For instance, A/C stations would not use music by 'AC/DC' behind a commercial. Using hard rock as bed music would be like breaking format, and the result would be audience dissatisfaction and probable tune-out. You can't be too abrasive." —Mike Nutzger, program director, WGAR-AM, Cleveland. "For the most part we stick with beds that are bright, pleasant, and adult, nothing with a real pronounced beat." —Jim Murphy, WCGY-AM.

CONTEMPORARY HIT

Copy	Delivery	Mixdown
"In this format the copywriter must be aware of current jargon and make an attempt to incorporate it in copy, if possible. Of course, it shouldn't be overdone. A copywriter can use colloquialisms to good advantage. You have to keep the teen audience in mind when creating copy. The key is speaking the language of the young without sounding pretentious." —Jim Cook, program director, WJET-AM, Erie, Pennsylvania.	"This is a high-intensity format. You've got to convey energy and enthusiasm, although it is not necessary to scream at the listener." —Dave Taylor, program director, KAAY-AM, Little Rock, Arkansas. "A wide variety of announcer deliveries work in hit radio. Voice schlock is not necessary, though. High energy and honesty work." —Jim Cook, WJET-AM.	"Hit radio is very production oriented. Spots can be pretty elaborate. In this format, use whatever hooks the listener. What it comes down to is selecting compatible Top 40 production values and mixing them with imagination and skill." —Mark Adams, program editor, WKLR-FM, Toledo, Ohio. "When it comes to choosing music beds you have to use contemporary stuff. Hokey, dated music detracts from the total air sound. Commercials have to be in synch with the station's program sound or they cause product damage." —Jim Cook, WJET-AM.

EASY LISTENING

Copy

"You've got to work with the flow of words in EL. No interjections and harsh repetition. Lead the listener gently from point to point. Use more vowel sounds, with only limited use of consonants." —Enzo DeDominicis, vice president, WRCH/WRCQ, Hartford, Connecticut.

"Write with intelligence and marry the copy to the EL sound. Avoid ultra hip and trendy jargon and stay away from cuteness. Maturity is the buzzword." —Phyllis Moore, general manager, WEZI-FM, Memphis.

Delivery

"Avoid sounding stilted. The one-on-one approach works better than the stiff shirt, even in a conservative format." —Dick Ellis.

"Warmth and friendliness with authority and credibility. Delivery must fit the mood of the format. In EL the mood is set, so working with it is the best way to keep the listener with you." —Enzo De Dominicis, WRCH/WRCQ.

Mixdown

"No pronounced beats in beds. Match the flow. Light jazz is okay, as in Adult Contemporary, and even semi-classical. We keep a complete library of music appropriate for spot production within our format." —Phyllis Moore, WEZI-FM.

"Busy spots, those with several production elements, are fairly uncommon. Two-voicers are rare, too." —Enzo DeDominicis, WRCH/WRCQ.

COUNTRY

Copy

"Write intelligently. Don't insult anyone. The Country audience doesn't exclusively consist of 'good ol' boys' in pickup trucks. A lot of hip business people and professionals enjoy Country radio." —Steve Brelsford, program director, WQHK-AM, Ft. Wayne, Indiana.

"Here, whatever it takes to get the job done, we do. Hard-sell, soft-sell, you name it. We just want to give the client a good piece of copy. Creativity is the key objective. Humor can be very effective in Country." —Gary Tolman, program director, KAST-AM, Astoria, Oregon.

Delivery

"Much of Country is up-tempo, so delivery has to conform to the velocity of the music to achieve flow." —Roger Heinrick, program director, KCLS-AM, Flagstaff, Arizona.

"We don't talk 'country' or 'twang' here. Southern accents just aren't necessary. Announcer delivery is kept pretty free of regionalisms." —J.H. Johnson, program director, KRNR-AM, Roseburg, Oregon.

Mixdown

"Country music beds are not used in much abundance. We use a host of sounds, including pop, light rock, and others. Here we avoid the 'rural' sound in our spot beds. It breaks the sameness and keeps things fresh." —Jarrett Day, KSO-AM, Des Moines, Iowa.

"The Country format doesn't really impose many strictures on the use of effects. Use what sells the spot. Discretion should be employed, however." —Steve Brelsford, WQHK-AM.

ALBUM-ORIENTED ROCK

Copy

"One-on-one writing is important so that the listener feels the message is just for him or her. Stiff or stuffy writing doesn't work in AOR." —Mel Myers, program director, KSNE-FM, Tulsa, Oklahoma.

"The perceptions of the AOR listener must be echoed in the choice of wording used. As a copywriter, you should know how your audience thinks and speaks. You write for lifestyle in this format." —Don Davis, program director, WWDC-FM, Washington, D.C.

Delivery

"The one-on-one style fits well. No real hyped or burn-out deliveries are acceptable." —Dennis Constantine, program director, KBCO-FM, Boulder, Colorado.

"AOR's on-air approach is less aggressive than most contemporary formats, and this same relaxed approach is necessary. Delivery should be natural, hip, and conversational, rather than contrived or forced." —Don Davis, WWDC-FM.

Mixdown

"Beds need to be generic and up-tempo. We find fusion and Jazz work best. Familiar music is avoided, except on concert spots. Edits using up-tempo riffs, drumbeats, etc., are effective." —Dennis Constantine, KBCO-FM.

"Because of the attention paid to audio quality by AOR listeners, production values in commercials are important, too. The mixdown must meet the highest standards. Scratchy records, second or third generation dubbing, and improper levels defeat the purpose of production." —Don Davis, WWDC-FM.

THE PRODUCTION ROOM

Generally speaking, metro market stations employ a full-time production person (known variously as production director, production manager, or production chief). This individual's primary duties are to record voice-tracks and mix commercials and PSAs (public service announcements). Other duties involve the maintenance of the bed and sound effects library and the mixdown of promotional material and special programs, such as public affairs features, interviews, and documentaries.

Stations that do not have a slot for a full-time production person divide work among the on-air staff. In this case, the program director often oversees production responsibilities, or a deejay may be assigned several hours of production duties each day and be called the production director.

At most medium and small outlets, on-air personnel take part in the production process. Production may include the simple dubbing of an agency spot onto a cartridge, a mixdown that requires a single bed (background music) under a 30-second voicer, or a multi-element mixdown of a 60-second two-voicer with sound effects and several bed transitions. Station production can run from the mundane and tedious (the dubbing of 15 60-second Preparation H spots onto several different cartridges) to the exciting and challenging (a commercial without words conveyed through a confluence of sounds).

Most production directors are recruited from the on-air ranks, having acquired the necessary studio dexterity and know-how to meet the demands of the position. In addition to the broad range of mixdown skills required by the job, a solid knowledge of editing is essential. The production director routinely is called on to make rudimentary splices or perform more complex editing chores, such as the rearrangement of elements in a 60-

second concert promo. Editing is covered in more detail later in this chapter.

The production director works closely with many people, but perhaps most closely with the program director. It is expected that the person responsible for production have a complete understanding of the station's programming philosophy and objective. This is necessary since commercials constitute an element of programming and therefore must fit in. A production person must be able to determine when an incoming commercial clashes with the station's image. When a question exists as to the spot's appropriateness, the program director will be called on to make the final judgment, since it is he or she who is ultimately responsible for what gets on the air. In the final analysis, station production is a product of programming. In most broadcast organizations, the production director answers to the program director. It is a logical arrangement given the relationship of the two areas.

The production director also works closely with the station copywriter. Their combined efforts make or break a commercial. The copywriter conceives of the concept, and the producer brings it to fruition. The traffic department also is in close and constant contact with production, since it is one of its primary responsibilities to see that copy gets processed and placed in the on-air studio where it is scheduled for broadcast.

THE STUDIOS

A radio station has two kinds of studios: on-air and production. Both share basic design features and have comparable equipment. For ease of movement and accessibility, audio equipment commonly is set up in a U shape within which the operator or producer is seated.

The standard equipment found in radio studios includes an audio console

FIGURE 9.3
This cutting-edge
multitrack digital
studio is in
contrast with the
analog studio in the
next figures.
Courtesy
Wheatstone.

(commonly referred to as the "board"), reel-to-reel tape machines, cartridge (cart) machines, cassette decks, turntables (less and less), compact disk players, and a patch panel.

Audio Console

The audio console is the centerpiece, the very heart of the radio station. Dozens of manufacturers produce audio consoles, and although design characteristics vary, the basic components remain relatively constant. Consoles come in all different sizes and shapes and may be monaural, stereo, or multitrack, but all contain inputs that permit audio energy to enter the console, outputs through which audio energy is fed to other locations, VU meters that measure the amount or level of sound, pots (faders) that control gain or the quantity of sound, monitor gains that control instudio volume, and master gains for the purpose of controlling general output levels.

Since the late 1960s, the manufacture of consoles equipped with linear faders has surpassed those with rotary faders. While *slide* (another term used) faders perform the same function as the more traditional pots, they are easier to read and handle.

Cue Mode. A low-power amplifier is built into the console so that the operator may hear audio from various sources without it actually being distributed to other points. The purpose of this is to facilitate the setup of certain sound elements, such as records and tapes, for eventual introduction into the mixdown sequence.

Reel-to-Reel Tape Machines

The reel-to-reel machine is the production studio workhorse. All editing is done on the reel-to-reel machine, since it is especially designed for that purpose.

Reel-to-reel machines have three magnetic heads whose purpose it is to record sound (convert electrical energy into magnetic energy), play back sound

FIGURE 9.4
Although each production studio is unique, the basics of layout are fairly consistent from station to station. For the sake of ease and accessibility, most studios are developed in a U shape or a variation thereof. Courtesy WEGX and WHJJ/WHJY.

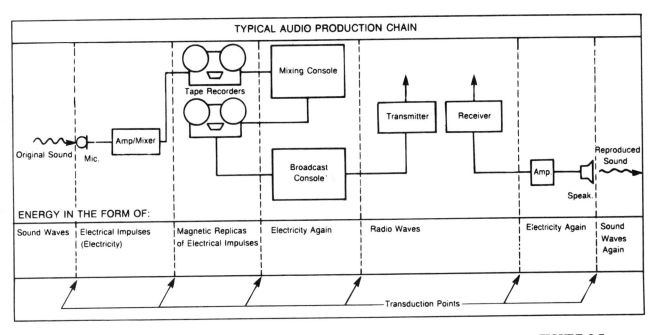

FIGURE 9.5
Transduction
points in a typical
audio chain.

(convert magnetic energy into electrical energy), and erase magnetic impressions from recording tape.

Most reel-to-reel machines are capable of recording at two speeds, although some models offer three. The tape speeds most commonly available on broadcast quality, state-of-the-art reel-to-reel machines are 3¾, 7½, and 15 IPS (inches per second). It is the middle speed, 7½, that is used most frequently by broadcasters. The reason is a practical one. While some stations may possess machines with 3¾ or 15 IPS tape speeds, all have reel-to-reels with 7½ IPS. Therefore, stations and agencies making dubs for distribution do so at 7½ IPS. In-house recording often is accomplished at high speed (15 IPS) because the sound quality is better and editing is easier.

Reel-to-reel tape machines are available in monaural, stereo, and multitrack. The last allows for overdubbing or sound-on-sound recording, which gives the producer greater control over the mix of sound elements. Multitrack recorders come in four, eight, 16, 24, and 36 channel formats. They have become increasingly popular since the 1970s, although they can be extremely

costly—prohibitively so for many small market stations.

Cart Machines

Cartridge tape machines came into use in the late 1950s and drastically simplified the recording and playback process. A continuous loop of quarter-inch magnetic tape within a plastic container (cart) is used to record everything from commercials and promos to music and short public affairs features.

A magnetic pulse is impressed against the tape surface during the recording process, which permits the cart to recue on its own. Carts come in various lengths—10, 40, 70 seconds—depending on need. They consist of a hub (around which tape is loosely

FIGURE 9.6
Multichannel
board. Courtesy
Auditronics.

FIGURE 9.7
Audio console with linear faders, popularly referred to as "slide" board. Courtesy Auditronics.

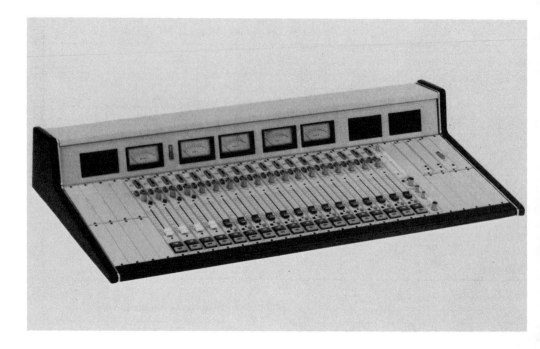

wound), guides, and pressure pads, which keep the tape against the heads of the machine. When the cart machine is activated, a pinch roller presses against a capstan to move the tape.

MINI-DISC MACHINES

At many stations, analog cart machines are being replaced by random access mini-disc (record/playback) technology. These new-age cart machines allow producers to digitally

archive up to 74 minutes of audio on 2.5 inch reusable discs.

Says station manager Vic Michaels, "It's a great replacement for carts, because it's faster, programmable, visual, and competitively priced." Companies like Sony, Harris, Denon, and Otari report heavy sales of the mini-disc machine, which would suggest that the days of the old analog cart deck may be numbered—another victim of the computer age.

Among other things, mini-disc machines offer instant start (no hesitation or drag common in its analog predecessor), back cueing, track selection, end

FIGURE 9.8
Eight-mixer ("pot") monaural console. Courtesy Broadcast Electronics.

marking, automatic fade-in, visual ID and cueing, digital editing, and so forth.

Cassette Tape Machines

Cassette machines are popular because of their ease of handling. The small tape cassettes used can be inserted and removed from the machines without the customary rethreading and rewinding that is required by reel-to-reel machines. Adopted by broadcasters in the mid-1960s, cassette recorders employ $1/8$-inch tape that is moved at 1% IPS. Cassette tapes can hold up to three hours of material. While not as integral to the production mixdown as the cart or reel-to-reel machine, cassettes are nonetheless an important and necessary piece of studio equipment. On-air cassette machines are generally used to play music, features, and actualities, as well as for aircheck purposes.

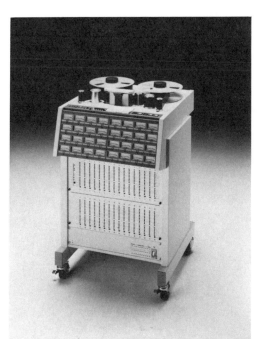

FIGURE 9.9
Cue speaker within console permits producer to set up elements for audio processing. Courtesy LPB.

FIGURE 9.10
(FAR LEFT)
Pot in cue mode. Courtesy Broadcast Electronics.

FIGURE 9.11
(FAR LEFT)
Magnetic head configuration— Erase, Record, and Playback.

FIGURE 9.12
(NEAR LEFT)
Multitrack reel-to-reel recording is widespread. This is a 32-track recorder, using 2-inch audio tape. Courtesy Otari.

FIGURE 9.13
(RIGHT AND FAR
RIGHT)
Standard broadcast
cartridges (carts)
play a key role in
the production
process at most
stations.

FIGURE 9.14
(RIGHT)
Platform-type
magnetic tape
eraser, commonly
referred to as a
bulk eraser.

FIGURE 9.15
(RIGHT)
Cartridge tape
builder. Rather
than do this
themselves, the
majority of stations
simply purchase
new carts or send
existing ones out to
be refilled.
Courtesy Broadcast
Electronics.

FIGURE 9.16
(FAR RIGHT)
Triple-deck
cartridge machine.
Courtesy Broadcast
Electronics.

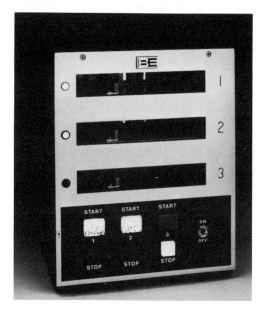

Cassette tape recorders come in stereo and monaural and are available in everything from pocket-sized portables to permanent rack-mounted models.

Audio Tape

Magnetic tape is chemically treated with either polyester or acetate, primarily for the sake of preservation. Polyester-backed tape generally is preferred because of its greater durability. It is less affected by extreme temperatures and humidity. Audio tape is produced in varying thicknesses: .5, 1.0, and 1.5 millimeters (mm). The thinner the tape, the more that can be contained on a reel. For instance, a 7-inch reel of .5 mm holds 2,400 feet of tape, whereas the same size reel of 1.5 mm only holds half as much, 1,200 feet. The industry standard is 1.5-mm tape because it is stronger and more dependable.

The width of audio tape also varies. The most common tape found in a production studio is 1/4-inch in width. This tape is used on standard mono and stereo reel-to-reel and cartridge machines. Multitrack recorders use anywhere from 1/2- to 2-inch-wide tape, depending on the number of tracks a machine possesses.

Head Cleaning. The oxide particles from the magnetic tape leave a residue on tape heads. Dust and dirt also accumulate on heads. This can result in a deterioration of sound quality. Therefore, it is necessary to clean a tape machine's heads frequently. This is generally accomplished with cotton swabs and a liquid head cleaner. Many stations use isopropyl alcohol because it is effective and inexpensive.

The procedure for cleaning heads is simple. Moisten the cotton swab with the head cleaner and run it across each head. When the swab becomes slightly discolored, use a fresh one. It is best to use a new swab for each head. Light pressure may be applied to the heads to ensure that surface particles are removed.

FIGURE 9.17
Mini-disc machines make inroads into studios. Courtesy Denon.

FIGURE 9.18
Equipment rack containing cassette tape machine, cartridge machines, and computer terminal. Rack mounted for convenience and ease of handling. Courtesy WMJX-FM.

FIGURE 9.19
Dirty heads can create numerous problems, not the least of which is a deterioration of sound quality.

FIGURE 9.20
CD cart players are popular in both the on-air studio and the mixing room. Courtesy Denon.

Compact Discs

Compact disc players entered the radio production studio in the 1980s. Although CD players are a more integral part of the on-air studio, their value as a piece of production equipment increases daily. Observes Skip Pizzi: "By far the largest acceptance of digital audio to date has been CD hardware, to the point where it is estimated that over half of the radio stations in the USA use CD to some extent. In major markets, this figure rises steeply. Many of these stations program music exclusively from CD, or nearly so. The practice of providing promotional copies of new releases on CD by record companies (following an earlier period of general reluctance to do so) has fueled a recent surge here. Second and third generation professional CD players are also aiding in the process of acceptance, which shows no signs of abating."

Unlike analog discs, which require actual physical contact with a recording, CD players employ a laser beam to read encoded data at a rate of 4.3218 million bits per second.

A compact disc is 4.7 inches wide and 1.2 mm thick, and players are quite light and compact as well. This feature alone makes them attractive to broad-casters. But what makes a CD player most appealing to broadcasters is its superior sound. Compact disc players offer, among other features, far greater dynamic range than standard turntables and a lower signal-to-noise ratio. They also eliminate the need for physical contact during cueing, and wowing and distortion are virtually gone.

Because digital discs are specially coated, they are much more resistant to damage than are analog discs. This is not to suggest that CDs are impervious. They are not. In fact, the majority of CD-related problems stem from the discs themselves and not the players. Despite initial claims of the invincibility of the digital disc, experience has shown that mishandling of discs is courting disaster. CDs cannot be mistreated—used as Frisbees or placemats for peanut butter sandwiches—and still be expected to work like new. The simple fact is that while compact discs are more resistant to damage than analog discs, they can be harmed.

While a turntable tone arm moves from the outside of a disc inward as the plate revolves in a clockwise direction at 33 or 45 rpm, a CD reads a disc from its core outward, moving from 500 rpm on the inside to 200 rpm on the outer edge of the disc.

Most CD players feature a variety of effect options, which can be of particular use to a production mix. Accessing cuts on a CD player is quick and simple, though excerpting segments from a track for inclusion in a mixdown can be somewhat less expedient than doing so on a conventional analog machine. Nonetheless, CD players are a major addition to the production studio. In the 1990s, compact discs are a wonderful source for bed music and sound effects.

Working with a CD unit is anything but complicated. Press a button and a tray ejects (on top-loaded models a door pops open). A disc is placed into the tray, and the press of the same button returns the tray and disc into the player. The operator then selects the track to be played and presses the appropriately numbered button. The audio rolls.

FIGURE 9.21
In the 1990s record cueing goes on more in the production room than it does in the control room.

Recordable CDs are making inroads into the production room, but since they can only be encoded once, it is probable that mini-disc cart technology will inhibit their wider acceptance in the radio studio environment.

Digital Audio Tape (DAT)

Digital audio technology was first introduced in the 1970s, and twenty years later it is having a substantial impact on the radio industry. Today a growing number of stations have DAT (digital audio tape) recorders. A total departure from analog recording, digital involves the conversion of audio signals into coded pulses—numbers—that are read by a computer sound processor. Digital recording formats quantify numeric values rather than replicate waves. Simply put, in digital, sounds are quantified.

DAT recording is very appealing to broadcasters because it offers a greater dynamic range than analog, and it is nearly noise-free; the flutter (caused by variation in speed of tape transport) and wow common to analog are gone, and there is virtually no loss in quality during dubbing. In addition to improved audio quality, digital audio systems can provide several other enhancements that are of specific value to the radio broadcaster. These include playback speed accuracy, stereo phase stability, fast random access, high storage densities, improved portability, and great cost-effectiveness relative to broadcast-quality analog systems, observes audio expert Skip Pizzi.

During recording, data are conveyed diagonally (helically) from the machine's head to the magnetic tape surface rather than horizontally—as is the case in analog recording.

Digital audio tape differs from analog tape in a number of ways. It is much thinner—0.5 mm is the qualitative equivalent of 1.5 mm tape used with analog machines. More tracks may be written on a smaller surface using digital open-reel tape. For instance, on a 1/4-inch digital tape up to twelve tracks may be recorded, and 1/2-inch supports 28 tracks. Forty tracks may be recorded onto 1-inch-wide digital tape.

Turntables

Even today few audio studios are without a turntable. This is especially true of the production room, even though many stations have phased out regular on-air use of turntables in favor of carted music and compact disks. Nonetheless, turntables usually are present in the control room for backup purposes and for special features that use LPs.

Broadcast turntables are designed for cueing purposes. Turntables are covered with felt to allow records to move freely when being cued, and the tone arm is carefully balanced to prevent record and stylus damage during this process.

Two methods of record cueing are used. One involves a technique called *dead-rolling*, in which an LP is cued and then activated from the turntable's stop position. The other, known as *slip-*

FIGURE 9.22
Lexicon multieffects processor with MIDI function. Courtesy Lexicon.

cueing, requires that the record be held in place and released as the turntable rotates. Using the former technique, an LP must be backtracked approximately one-eighth of a turn from the start of the audio and a 45 rpm one-fourth of a turn. This permits the turntable to achieve its proper speed before the sound is reached. When a turntable does not achieve its proper speed before engaging the recording, distortion will occur in the form of a "wow."

Compressors, Equalizers, and Audio Processing

"There are three domains of audio," says producer Ty Ford. "They are amplitude, frequency, and time." Some stations alter amplitude to create the illusion of being louder without actually changing level. This is called *compressing the signal.* Compressors are used by

production people to enhance loudness as well as to eliminate or cut out ambient noise, thus focusing on specifics of mix. Compression is often used as a method of getting listeners to take greater notice of a piece of production and as a remedy to certain problems.

Equalizers (EQs) work the frequency domain of audio by boosting and/or cutting lows (Hertz \obHz\cb range) and highs (Kilohertz \obkHz\cb range). Equalizers allow producers to correct problems as well as to create parity between different elements of production. They are also useful in creating special effects. EQs are available inboard (part of the audio console) and out-of-board (stand-alone unit) and as part of certain integrated audio effects processors.

Most audio processors (also called *effects processors* or simply *boxes*) are time domain devices. Stations use these digital boxes to create a wide range of effects such as reverb, echo, flange, and so on.

In the last few years, radio stations have become increasingly interested in what audio processors have to offer their mixes. Today these boxes are a familiar, if not integral, item in production rooms at the majority of stations. Their value in the creation of commercials, PSAs, promos, and features is inestimable.

The use of samplers and synthesizers is on the rise in radio production rooms too. Samplers let a production person load a studio audio source (recorder, live mike) into its built-in microprocessor and then manipulate the digitized data with the aid of a musical keyboard to create a multitude of effects. Samplers employ magnetic microfloppies and are wired to an audio console so that the sounds they produce may be integrated into mixdown. Samplers are also found in certain audio effects processors with MIDI (Musical Instrument Digital Interface). A sample is a digital recording of a small bit of sound.

"A lot of musical instrument (MI) gear is being introduced into the radio production studio. Synths, samplers, and sequencers are pretty commonplace today," notes Ty Ford.

FIGURE 9.23
The station's patch panel extends the reach and capacity of the control and production room consoles.

FIGURE 9.24
The look of microphones in the 1920s. Courtesy Jim Steele.

Patch Panel

A patch panel consists of rows of inputs and outputs connected to various external sources—studios, equipment, remote locations, network lines, and so forth. Patch panels essentially are routing devices that allow for items not directly wired into an audio console to become a part of a broadcast or production mixdown.

Microphones

Microphones are designed with different pickup patterns. Omnidirectional microphones are sensitive to sound from all directions (360 degrees), whereas bidirectional microphones pick up sound from two directions (180 degrees). The unidirectional microphone draws sound from only one path (90 degrees), and, because of its highly directed field of receptivity, extraneous sounds are not amplified. This feature has made the unidirectional microphone popular in both the control and production studios, where generally one person is at work at a time. Most studio consoles possess two or more microphone inputs so that additional voices can be accommodated when the need arises.

Omnidirectional and bidirectional microphones often are used when more than one voice is involved. For instance, an omnidirectional may be used for the broadcast or recording of a roundtable discussion, and the bidirectional during a one-on-one interview.

Announcers must be aware of a microphone's directional features. Proper positioning in relation to a microphone is important. Being outside the path of a microphone's pickup (off-mike) affects sound quality. At the same time, being too close to a microphone can result in

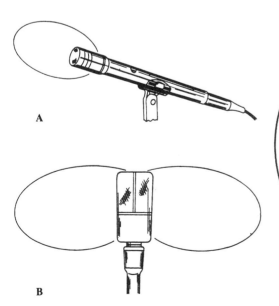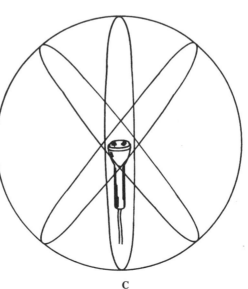

FIGURE 9.25
Microphone pickup patterns: (A) unidirectional, (B) bidirectional, (C) omnidirectional.

FIGURE 9.26
Sound pressure
level (SPL) chart
depicting volume of
different sounds in
relation to human
aural perception.

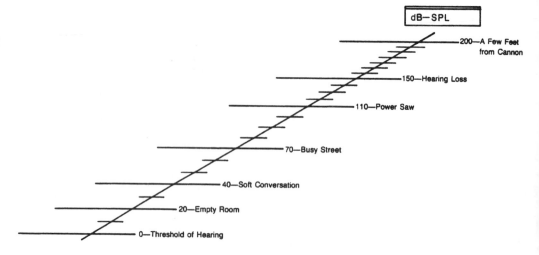

dB–SPL

200—A Few Feet from Cannon

150—Hearing Loss

110—Power Saw

70—Busy Street

40—Soft Conversation

20—Empty Room

0—Threshold of Hearing

distortion, known as popping and blasting. Keeping a hand's length away from a microphone will usually prevent this from occurring. Windscreens and blast filters may be attached to a microphone to help reduce distortion.

EDITING

Tape editing can range from a simple repair to a complicated rearrangement of sound elements. In either case, the ability to splice tape is essential. Figure 9.29 (on pp. 253–256) illustrates splicing steps.

Not all editing is done with a razor and splicing tape. Today the conventional approach to editing tape is losing ground to "nondestructive" digital and multitrack methods.

Multitrack editing is another form of razorless or electronic editing. In multitrack editing, a production ingredient is added to (or eliminated from) a separate track. Mini-disc machines, as previously stated, provide the producer with another tapeless and nondestructive method for editing sound ingredients.

Digital audio workstations, which rely on computer technology, are currently used in a growing number of radio production studios. This tapeless approach involves loading audio into a RAM or hard disc and making edits via a monitor (with the aid of a mouse, a keyboard, or a console). While this technology is fairly

costly (prices are coming down), more and more stations are moving toward the tapeless studio. Workstations are the studio of the future.

About the age-old razor edit approach, Ed Shane comments "Razor blades are about as useful in the production room as they are for shaving in this Norelco age. I recently reviewed software that turns a 386 or 486 computer into a digital workstation for just a few hundred dollars."

COPYWRITING

Poet Stephen Vincent Benet, who wrote for radio during its heyday, called the medium *the theater of the mind*. Indeed, the person who tunes into radio gets no visual aids but must manufacture images on his or her own to accompany the words and sounds broadcast. The station employee who prepares written material is called a copywriter. What he or she writes primarily consists of commercials, promos, and PSAs, with the emphasis on the first of the three.

Not all stations employ a full-time copywriter. This is especially true in small markets where economics dictate that the salesperson write for his own account. Deejays also are called on to pen commercials. At stations with bigger operating budgets, a full-time copy-

FIGURE 9.27
Digital audio puts
the next generation
studio in a box. A
world of production
sound at your
fingertips. Courtesy
Scott.

Move Up from Carts to Touchscreen Digital Audio

Play Any Audio at a Touch

Nothing else makes radio as fast or easy as having all your spots, sounders and sweepers start with your fingertip–*always on-line and ready* to play from hard disk. And *nothing else* makes your station sound as good or as exciting as touchscreen digital and creative talent with the *new Scott Studio System!*

Here's how it works: Six buttons on the left of the 17" computer touchscreen play what's on your program log. Scheduled spots, promos, PSAs and live copy come in automatically from your Scott System Production Bank and your traffic and copy computers. You see legible labels for everything, showing full names, intro times, lengths, endings, announcer initials, outcues, posts, years, tempos and trivia. Your jocks can rearrange anything easily by touching arrows (at mid-screen), or opening windows with the entire day's log and lists of all your recordings.

On the right, 18 "hot keys" start *unscheduled* jingles, sounders, effects, comedy or promos *on the spur of the moment.* You get 26 sets of 18 user-defined instant audio "hot keys" for your jocks' different needs.

Large digital timers automatically count down intro times, and flash 60-, 45-, and 30-seconds before end warnings. You also get clear count-downs the last 15 seconds of each event.

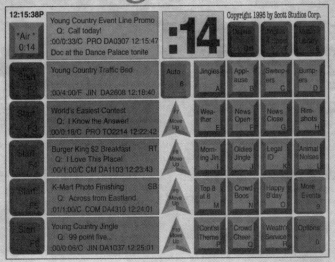

The Scott Studio System is your *best* way to make the move to digital audio and eliminate troublesome carts. Each button on the touchscreen plays whatever you want instantly. All scheduled spots, jingles, promos and scripts come in from your traffic and copy computers.

---

writer often will handle the bulk of the writing chores.

Copywriters must possess a complete understanding of the unique nature of the medium, a familiarity with the audience for which the commercial message is intended, and knowledge of the product being promoted. A station's format will influence the style of writing in a commercial; thus, the copywriter also must be thoroughly acquainted with the station's particular programming approach. Commercials must be compatible with the station's sound. For instance, copy written for Easy Listening usually is more conservative in tone than that written for Contemporary Hit stations, and so on.

ROCKING OUT EVERY FRIDAY AND SATURDAY NIGHT. AT TJ'S THERE'S NEVER A COVER OR MINIMUM, JUST A GOOD TIME. SUNDAY IDAHO'S MONARCHS OF ROCKABILLY, JOBEE LANE, RAISE THE ROOF AT TJ'S. YOU BETTER BE READY TO SHAKE IT, BECAUSE NOBODY STANDS STILL WHEN JOBEE LANE ROCKS. THURSDAY IS HALF-PRICE NIGHT, AND LADIES ALWAYS GET THEIR FIRST DRINK FREE AT BOISE'S NUMBER ONE CLUB FOR FUN AND MUSIC. TAKE MAIN TO MARK STREET, AND LOOK FOR THE HOUSE THAT ROCKS, TJ'S ROCKHOUSE. (SFX: Stinger out)

WXXX

"Home of the Hits"

(SFX: Bed in)

TJ'S ROCKHOUSE, MARK STREET, DOWNTOWN BOISE, PRESENTS CLEO AND THE GANG

WYYY

"Soothing Sounds"

ELEGANT DINING IS JUST A SCENIC RIDE AWAY. (SFX: Bed in and under) THE CRITICALLY

FIGURE 9.28 Production studio rack containing a variety of special equipment—harmonizer, compressor, equalizer, and reverb unit—designed to enhance the sound of commercials, PSAs, and promos. Courtesy WMJX-FM.

ACCLAIMED VISCOUNT (VY-COUNT) INN IN CEDAR GLENN OFFERS PATRONS AN EXQUISITE MENU IN A SETTING WITHOUT EQUAL. THE VISCOUNT'S 18TH CENTURY CHARM WILL MAKE YOUR EVENING OUT ONE TO REMEMBER. JAMISON LONGLEY OF THE WISCONSIN REGISTER GIVES THE VISCOUNT A FOUR STAR RATING FOR SERVICE, CUISINE, AND ATMOSPHERE. THE VISCOUNT (SFX: Royal fanfare) WILL SATISFY YOUR ROYAL TASTES. CALL 675-2180 FOR RESERVATIONS.

TAKE ROUTE 17 NORTH TO THE VISCOUNT INN, 31 STONY LANE, CEDAR GLENN.

There are some basic rules pertaining to the mechanics of copy preparation that should be observed. First, copy is typed in UPPER CASE and double-spaced for ease of reading. Next, left and right margins are set at one inch. Sound effects are noted in parentheses at that point in the copy where they are to occur. Proper punctuation and grammar are vital, too. A comma in the

FIGURE 9.29A
Mark tape to be cut
against playback
head using a light-
colored grease
pencil.

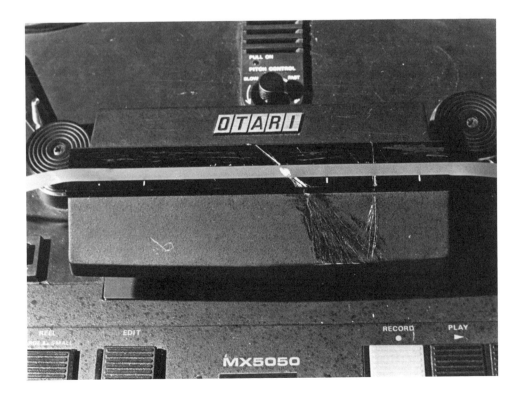

FIGURE 9.29B
Place magnetic tape
in tracks of cutting
block.

FIGURE 9.29C
Make a diagonal cut
with a single-edge
razor. Use a sharp
razor to avoid an
uneven cut.

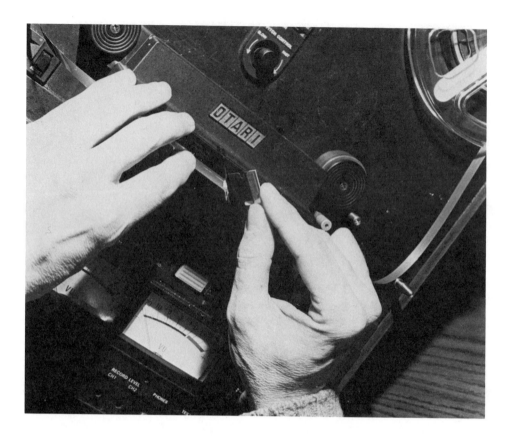

FIGURE 9.29D
Butt ends snugly
together. There
should be no gap or
overlap.

FIGURE 9.29E
Place a ¾-inch-long
piece of splicing
tape against the
nonrecording
surface of the
magnetic tape.

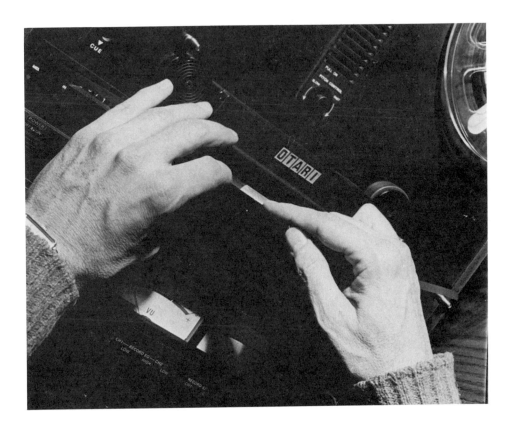

FIGURE 9.29F
Remove the air
bubbles trapped
between the
splicing and
magnetic tapes by
gently rubbing the
surface. Tape must
adhere completely
to ensure
permanent grip.

FIGURE 9.29G
Splicing tape must
be properly aligned.
Protruding splicing
tape can adhere to
surfaces and
results in tape
breakage. Adhesive
also gums up tape
heads.

FIGURE 9.29H
A too-small splice
may come apart,
and too much
decreases the
tape's suppleness
and flexibility. Be
sure to carefully
remove tape from
the tracks of the
cutting block. An
abrupt tug may
cause the tape to
break.

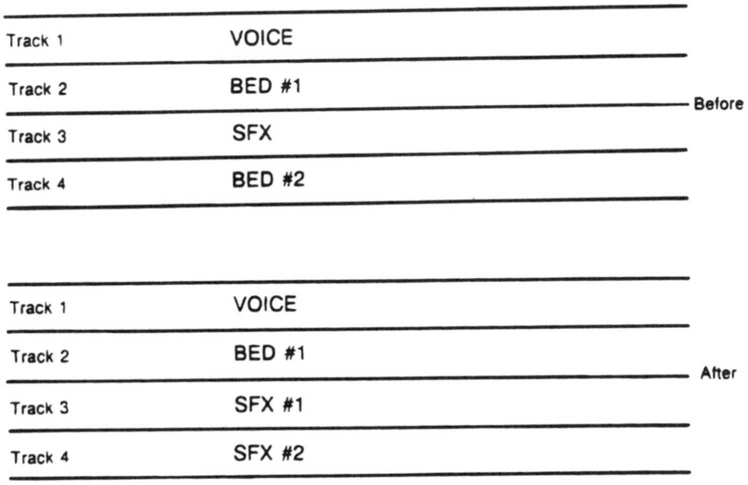

wrong place can throw off the meaning of an entire sentence. Be mindful, also, that commercials are designed to be heard and not read. Keep sentence structure as uncomplicated as possible. Maintaining a conversational style will make the client's message more accessible.

Timing a piece of copy is relatively simple. There are a couple of methods: one involves counting words, and the other counting lines. Using the first approach, 25 words would constitute 10 seconds, 65 words, 30 seconds, and 125 words, one minute. Counting lines is an easier and quicker way of timing copy. This method is based on the assumption that it takes on the average 3 seconds to read one line of copy from margin to margin. Therefore, nine to ten lines of copy would time out to 30 seconds, and 18 to 20 lines to one minute. Of course, production elements such as sound effects and beds must be included as part of the count and deducted accordingly. For example, six seconds worth of sound effects in a 30-second commercial would shorten the amount of actual copy by two lines.

Since everything written in radio is intended to be read aloud, it is important that words with unusual or uncommon pronunciations be given special attention. Phonetic spelling is used to convey the way a word is pronounced. For instance: "DINNER AT THE FO'C'SLE (FOKE-SIL) RESTAURANT IN LAITONE (LAY-TON) SHORES IS A SEA ADVENTURE." Incorrect pronunciation has resulted in more than one canceled account. The copywriter must make certain that the announcer assigned to voice-track a commercial is fully aware of any particulars in the copy. In other words, when in doubt spell it out.

Excessive numbers and complex directions are to be avoided in radio copy. Numbers, such as an address or telephone number, should be repeated and directions should be as simple as possible. The use of landmarks ("ACROSS FROM CITY HALL...") can reduce confusion. Listeners seldom are in a position to write down something at the exact moment they hear it. Copy

FIGURE 9.30 Editing a multitrack involves adding or deleting tracks. Here BED 2 is replaced by SFX 2 on track 4.

should communicate, not confuse or frustrate.

Of course, the purpose of any piece of copy is to sell the client's product. Creativity plays an important role. The radio writer has the world of the imagination to work with and is only limited by the boundaries of his own.

ANNOUNCING TIPS

Over 40,000 men and women in this country make their living as radio announcers. In few other professions is the salary range so broad. A beginning announcer may make little more than minimum wage, while a seasoned professional in a major market may earn a salary in the six-figure range.

While announcer salaries can be very good in smaller markets, the financial rewards tend to be far greater at metro market stations, which can afford to pay more. Of course, competition for positions is keener and expectations are higher. "You have to pay your dues in this profession. No one walks out of a classroom and into WNBC. It's usually a long and winding road. It takes time to develop the on-air skills that the big stations want. It's hard work to become really good, but you can make an enormous amount of money, or at least a very comfortable income, when you do," says radio personality Mike Morin.

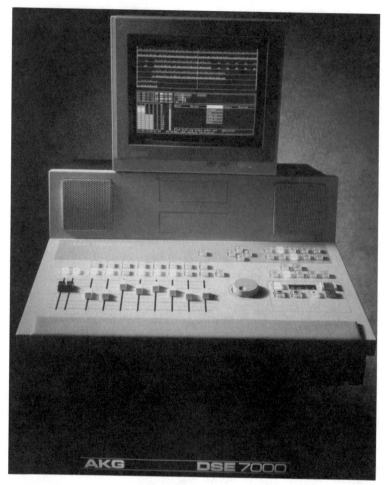

FIGURE 9.31
Digital audio
workstations mark
a new era in radio
production mixing,
editing, and
storage. Courtesy
AKG.

The duties of an announcer vary depending on the size or ranking of a station. In the small station, announcers generally fill news and/or production shifts as well. For example, a midday announcer at WXXX, who is on the air from 10 a.m. until 3 p.m., may be held responsible for the 4 and 5 p.m. newscasts, plus any production that arises during that same period. Meanwhile, the larger station may require nothing more of its announcers than the taping of voiceovers. Of course, the preparation for an airshift at a major market station can be very time consuming.

An announcer must, above all else, possess the ability to effectively read copy aloud. Among other things, this involves proper enunciation and inflection, which are improved through practice. WFYR's Bill Towery contends that

the more a person reads for personal enjoyment or enrichment, the easier it is to communicate orally. "I'd advise anyone who aspires to the microphone to read, read, read. The more the better. Announcing is oral interpretation of the printed page. You must first understand what is on the page before you can communicate it aloud. Bottom line here is that if you want to become an announcer, first become a reader."

Having a naturally resonant and pleasant-sounding voice certainly is an advantage. Voice quality still is very important in radio. There is an inclination toward the voice with a deeper register. This is true for female announcers as well as male. However, most voices possess considerable range and with training, practice, and experience even a person with a high-pitched voice can develop an appealing on-air sound. Forcing the voice into a lower register to achieve a deeper sound can result in injury to the vocal chords. "Making the most of what you already have is a lot better than trying to be something you're not. Perfect yourself and be natural," advises Morin.

Relaxation is important. The voice simply is at its best when it is not strained. Moreover, announcing is enhanced by proper breathing, which is only possible when free of stress. Initially, being "on-mike" can be an intimidating experience, resulting in nervousness that can be debilitating. Here are some things that announcers do to achieve a state of relaxation:

1. Read copy aloud before going on the air. Get the feel of it. This will automatically increase confidence, thus aiding in relaxation.

2. Take several deep breaths and slowly exhale while keeping your eyes closed.

3. Sit still for a couple of moments with your arms limp at your sides. Tune out. Let the dust settle. Conjure pleasant images. Allow yourself to drift a bit, and then slowly return to the job at hand.

4. Stand and slowly move your upper torso in a circular motion for a minute or so. Flex your shoulders and arms. Stretch luxuriously.

FIGURE 9.32
Station copy order.
Courtesy WQQQ-
AM.

PRODUCTION ORDER

(Fill in all information below.)

ACCOUNT EXECUTIVE

CLIENT _MOLLY'S DREAM_
A/E _LEO BLOOM_
DATE TO BE COMPLETED _4/19_
CASSETTE YES _✓_ NO _____
OUTGOING DUBS (LIST STATION) _____
WQQQ-AM

ANNOUNCER

DATE COMPLETED _4/19_
TIME COMPLETED _10:20 AM_
COMPLETED BY _Stephen D._

TRAFFIC

CART #   AM _468_
           FM _____

PRODUCTION DIRECTOR

ASSIGNED TO _Joyce_
DATE _4/19_

- -

DUB ORDER

AM _✓_ # OF SPOTS _12_ LENGTH _60_ START _4/20_ END _4/30_
FM _____ # OF SPOTS _____ LENGTH _____ START _____ END _____
CUTS _2_
TAGS: LIVE _✓_ PRE-RECORDED _____
TAG COPY _MOLLY'S DREAM IS OPEN
NIGHTLY UNTIL MIDNIGHT. REMEMBER,
JUST SAY 'YES' TO MOLLY._

SPECIAL INSTRUCTIONS/SPOT CHANGES _____
SOFT, SENSUOUS DELIVERY.
_____
_____

(RETURN THIS FORM TO PRODUCTION DIRECTOR)

5. When seated, check your posture. Do not slump over as you announce. A curved diaphragm impedes breathing. Sit erect, but not stiffly.

6. Hum a few bars of your favorite song. The vibration helps relax the throat muscles and vocal chords.

7. Give yourself ample time to settle in before going on. Dashing into the studio at the last second will jar your focus and shake your composure.

In most situations, an accent—regional or otherwise—is a handicap and should be eliminated. Most radio announcers in the South do not have a drawl, and the majority of announcers in Boston put the "r" in the word *car*. A noticeable or pronounced accent will almost always put the candidate for an announcer's job out of the running. Accents are not easy to eliminate, but with practice they can be overcome.

FIGURE 9.33
Production tape
storage area ("tape
morgue") contains
commercials aired
over the past year,
or in some cases
several years.
Courtesy
WHJY/WHJJ.

THE SOUND LIBRARY

Music is used to enhance an advertiser's message—to make it more
appealing, more listenable. The music

FIGURE 9.34
Index cards contain
pertinent
information about
bed music.

:60 Uptempo, piano.
Album: "Liberace's Favorites"
Cut 3, Side 1.
Acct: Dottie's Candies
Date Aired: Mar/Apr. '91

:30 Med.Tempo, Strings.
Album: "Bert Kaempfert's Hits"
Cut 4, Side 2.
Acct: Joe Paul's Paints
Date Aired: Jan.'89 - TFN.

used in a radio commercial is called a
bed simply because it backs the voice.
It is the platform on which the voice is
set. A station may bed thousands of
commercials over the course of a year.
Music is an integral component of the
production mixdown.

Bed music is derived from a couple of sources. Demonstration CDs
(demos) sent by recording companies to
radio stations are a primary source,
since few actually make it onto playlists
and into on-air rotations. These CDs are
particularly useful because the music is
unfamiliar to the listening audience.
Known tunes generally are avoided in
the mixdown of spots because they tend
to distract the listener from the copy.
However, there are times when familiar
tunes are used to back spots. Nightclubs
often request that popular music be
used in their commercials to convey a
certain mood and ambiance.

Movie soundtrack CDs are another
good place to find beds because they
often contain a variety of music, ranging
from the bizarre to the conventional.
They also are an excellent source for special audio effects, which can be used to
great advantage in the right commercial.

On-air CDs are screened for potential
production use as well. While several
cuts on an album may be placed in on-air
rotation and thereby eliminated for use in
the mixdown of commercials, there will
be cuts not programmed and therefore
available for production purposes.

Syndicated or canned bed music
libraries are available at a price and
are widely used at larger stations.
Broadcasting Yearbook contains a complete listing of production companies
offering bed music libraries. The
majority of stations choose to lift beds
from in-house CDs and LPs.

Music used for production purposes is
catalogued so that it may be located and
reused. A system once widely employed
used index cards, which could be stored
for easy access in a container or on a
rotating drum. At a growing number of
stations, computers are used to store
production library information. Using
the manual system, when an account is
assigned a certain bed, a card is made
and all pertinent data are included on it.

Should a card exist for a bed not in current use and the bed be appropriate for a new account, then either a fresh card will be prepared or the new information penciled onto the existing card.

No production studio is complete without a sound effects library. Unlike bed music, sound effects are seldom derived from in-house anymore. Sound effect libraries can be purchased for as little as one hundred dollars or can cost thousands. The quality and selection of effects vary accordingly. Specially tailored audio effects also can run into the thousands but can add a unique touch to a station's sound.

CHAPTER HIGHLIGHTS

1. The first radio commercial aired on WEAF New York in 1922.

2. Early commercials were live readings: no music, sound effects, or singing.

3. Dialogue spots, using drama and comedy to sell the product, became prominent in the 1930s. Elaborate sound effects, actors, and orchestras were employed.

4. With the introduction of magnetic recording tape and 33 LPs in the 1950s, live commercial announcements were replaced by prerecorded messages.

5. The copy, delivery, and mixdown of commercials must be adapted to match the station's format to avoid audience tune-out.

6. The production director (production manager, production chief) records voice-tracks, mixes commercials and PSAs, maintains the bed music and special effects libraries, mixes promotional material and special programs, and performs basic editing chores.

7. At smaller stations the production responsibilities are assigned part-time to on-air personnel or the program director.

8. The production director, who usually answers to the program director, also works closely with the copywriter and the traffic manager.

9. For ease of movement and accessibility, both on-air and production studio equipment are arranged in a U shape.

10. The audio console (board) is the central piece of equipment. It consists of inputs, which permit audio energy to enter the console; outputs through which audio energy is fed to other locations; VU meters, which measure the level of sound; pots (faders), which control the quantity (gain) of sound; monitor gains, which control in-studio volume; and master gains, which control general output levels.

11. When operating the console in "cue mode," the operator can listen to various audio sources without channeling them through an output.

12. Reel-to-reel tape machines are useful for recording at a variety of speeds and are necessary for editing.

13. Cartridge tape machines (cart machines) are used for recording and playback. They employ carts (plastic containers with continuous loops of magnetic tape), which are more compact and convenient than reels of tape.

14. Mini-disc machines are gradually replacing the standard analog cart deck. They let producers digitally mix and archive up to 74 minutes of audio on 2.5-inch reusable discs.

15. Cassette tape machines, although not as integral to the mixdown process,

PRODUCTION LIBRARIES

TM Century has the MOST Production Music Libraries

Trendsetter II — disc production library
Slam Dunk
Generation Three
LAZER PRODUCTION LIBRARY
Laser Lightning
Digital Director
MegaHot Country
MegaMusic

- 30 – 60 SECOND TRACKS
- PRODUCTION ELEMENTS FOR EACH LIBRARY
- MUSIC FOR ID's, CONTESTS, PROMOS, AND COMMERCIALS
- CATALOGS AND "PRODUCTION LIBRARY MANAGER" COMPUTER SOFTWARE PROVIDED
- THE LATEST MUSICAL STYLES: ADULT CONTEMPORARY ROCK/CHR, COUNTRY, SPECIALTY

FIGURE 9.35 Elaborately produced sound libraries can be a valuable tool for a radio station. Courtesy TM Communications.

MUSIC LIBRARIES

The Sound Is Simply Amazing!

● **MASTERED ENTIRELY IN THE DIGITAL DOMAIN**
   Technically superior — masters don't pass back and forth between digital and analog converters or consoles.

● **CUSTOM PRODUCED FOR RADIO BROADCAST**
   GoldDisc® is a special radio CD which gives you tight starts, consistent levels, automatic trip cues and the cleanest, most sparkling sound ever!

● **RESEARCHED TITLES — OVER 11,000!**
   To compile the best library with great depth, all songs have been extensively researched for their inclusion in the GoldDisc® Library. TM Century monitors the top rated stations, as well as hundreds of reporting stations.

● **EVERY FORMAT ON CD**
   Complete libraries are available or a variety of combination options.

ADULT CONTEMPORARY
LITE ADULT CONTEMPORARY
CLASSIC ROCK/AOR
CHR
CLASSIC HITS
COUNTRY
TM COUNTRY (Modern)
URBAN
TM MIX (Hot AC)

● **FREE CD EXCHANGE GUARANTEE**
   All damaged discs are replaced FREE during the warranty period.

● **FORMAT PROTECTION**
   If a format change becomes necessary, return the old library and we'll give you an exchange, less the pro-rated use percentage.

Whether you are changing format or upgrading your airsound, the fast, easy and superior sound of GoldDisc® is the answer to all of your music programming needs.

FIGURE 9.35
continued

are easy to handle and use convenient tape cassettes. They are often used for airchecks and actualities.

16. Audio tapes (magnetic tapes) come in a variety of thicknesses and widths. Acetate and polyester backings provide greater durability.

17. Because oxide particles from the magnetic tape, dust, and dirt accumulate on the heads of all types of tape machines, the heads should be cleaned regularly with a cotton swab and liquid head cleaner.

18. Compact disk players use a laser beam to decode the disk's surface, which eliminates stylus and turntable noises, distortion, and record damage. Recordable CDs are now in use.

19. Turntables, once the staple of control rooms, are being replaced by cart machines and compact disks. Turntables allow a record to be "cued," a process that allows the record to

reach proper speed before the sound portion is engaged.

20. Digital audio tape (DAT) offers improved sound reproduction with re-recording capability. Half the size of conventional analog tape cassettes, DAT will make inroads into the radio studio environment in the 1990s.

21. Audio processors, samplers, and MIDI enhance a radio production studio's product.

22. A patch panel is a routing device, consisting of inputs and outputs, connecting the audio console with various external sources.

23. Microphones are designed with different pickup patterns to accommodate different functions: omnidirectional (all directions), bidirectional (two directions), and unidirectional (one direction).

24. Tape editing ranges from simple repairs to complicated rearrangements of sound elements. The basic process to be mastered is called *splicing*. Today the conventional razor cut approach to tape editing is losing ground to "non-destructive" digital and multitrack methods.

25. Digital audio workstations, which rely on computer technology, are currently used in a growing number of radio production studios.

26. The station copywriter, who writes the commercials, promos, and PSAs, must be familiar with the intended audience and the product being sold. The station's format and programming approach influence the style of writing. Copy should be typed in upper case, double-spaced, with one-inch margins. Sound effects are noted in parentheses, and phonetic spellings are provided for difficult words.

27. Aspiring announcers must be able to read copy aloud with proper inflection and enunciation. A naturally resonant and pleasant-sounding voice without a regional accent is an advantage.

28. Every station maintains a sound library for use in spot mixdowns. Commercially produced sound effects and bed-music collections, and unfamiliar cuts from CDs and LPs, are common source materials.

SUGGESTED FURTHER READING

Adams, Michael H. and Massey, Kimberly. *Introduction to Radio: Production and Programming. Madison,* WI.: Brown and Benchmark, 1995.

Alten, Stanley R. *Audio in Media,* 4th ed. Belmont, Calif.: Wadsworth Publishing, 1993.

Bartlett, Bruce. *Stereo Microphone Techniques.* Boston: Focal Press, 1991.

Ford, Ty. *Advanced Audio Production Techniques.* Boston: Focal Press, 1993.

Gross, Lynne, and Reese, David E. *Radio Production Worktext: Studio and Equipment.* Boston: Focal Press, 1990.

Guidelines for Radio Continuity. Washington, D.C.: NAB Publishing, 1982.

Hilliard, Robert L. *Writing For Television and Radio,* 5th ed. Belmont, Calif.: Wadsworth Publishing, 1991.

Hoffer, Jay. *Radio Production Techniques.* Blue Ridge Summit, Pa.: Tab Books, 1974.

Hyde, Stuart W. *Television and Radio Announcing,* 7th ed. Boston: Houghton Mifflin, 1995.

Keith, Michael C. *Broadcast Voice Performance.* Boston: Focal Press, 1989.

———. *Radio Production: Art and Science.* Boston: Focal Press, 1990.

McLeish, Robert. *Technique of Radio Production,* 2nd ed. Boston: Focal Press, 1988.

Mott, Robert L. *Sound Effects: Radio, TV, and Film.* Boston: Focal Press, 1990.

National Association of Broadcasters. *Guidelines for Radio Copywriting.* Washington: NAB Publications, 1993.

Nisbet, Alec. *The Technique of the Sound Studio,* 4th ed. Boston: Focal Press, 1979.

———. *The Use of Microphones,* 3rd ed. Boston: Focal Press, 1989.

O'Donnell, Lewis B., Hauseman, Carl, and Benoit, Philip. *Announcing: Broadcast Communication.* Belmont, Calif.: Wadsworth Publishing, 1987.

Modern Radio Production, 2nd ed. Belmont, Calif.: Wadsworth Publishing, 1990.

Oringel, Robert S. *Audio Control Handbook,* 6th ed. Boston: Focal Press, 1989.

Orlik, Peter B. *Broadcast Copywriting,* 4th ed. Boston: Allyn and Bacon, 1990.

Pohlmann, Ken C. *Advanced Digital Audio.* Indianapolis: SAMS, 1991.

Rumsey, Francis. *Tapeless Sound Recording.* Boston: Focal Press, 1990.

———. *Digital Audio Operation.* Boston: Focal Press, 1991.

Runstein, Robert E. *Modern Recording Techniques,* 2nd ed. Indianapolis: Howard Sams, 1986.

Watkinson, John. *Digital Audio and Compact Disc Technology,* 3rd ed. Boston: Focal Press, 1995.

10 Engineering

PIONEER ENGINEERS

Anyone who has ever spoken into a microphone or sat before a radio receiver owes an immense debt of gratitude to the many technical innovators who made it possible. Guglielmo Marconi, a diminutive Italian with enormous genius, first used electromagnetic (radio) waves to send a message. Marconi made his historical transmission, plus several others, in the last decade of the nineteenth century. Relying, at least in part, on the findings of two earlier scientists, James Clerk Maxwell and Heinrich Hertz, Marconi developed his wireless telegraph, thus revolutionizing the field of electronic communications.

Other wireless innovators made significant contributions to the refinement of Marconi's device. J. Ambrose Fleming developed the diode tube in 1904, and two years later Lee de Forest created the three-element triode tube called the Audion. Both innovations, along with many others, expanded the capability of the wireless.

On Christmas Eve of 1906, Reginald Fessenden demonstrated the transmission of voice over the wireless from his experimental station at Brant Rock, Massachusetts. Until that time, Marconi's invention had been used to send Morse code or coded messages. An earlier experiment in the transmission of voice via the electromagnetic spectrum also had been conducted. In 1892, on a small farm in Murray, Kentucky, Nathan B. Stubblefield managed to send voice across a field using the induction method of transmission, yet Fessenden's method of mounting sound impulses atop electrical oscillations and transmitting them from an antenna proved far more effective. Fessenden's wireless voice message was received hundreds of miles away.

Few pioneer broadcast technologists contributed as much as Edwin Armstrong. His development of the regenerative and superheterodyne circuits vastly improved receiver efficiency. In the 1920s Armstrong worked at developing a static-free mode of broadcasting, and in 1933 he demonstrated the results of his labor—FM. Armstrong was a man ahead of his time. It would be decades before his innovation would be fully appreciated, and he would not live to witness the tremendous strides it would take.

Had it not been for these men, and many others like them, there would be no radio medium. Today's broadcast engineers and technologists continue in the tradition of their forebears. Without their knowledge and expertise, there would be no broadcast industry because there would be no medium. Radio is first an engineer's medium. It is engineers who put the stations on the air and keep them there.

RADIO TECHNOLOGY

Radio broadcasters utilize part of the electromagnetic spectrum to transmit their signals, and they are obliged to pay spectrum fees of up to $1,200 annually (depending on their size) for this privilege. A natural resource, the electromagnetic spectrum is comprised of radio waves at the low-frequency end and cosmic rays at the high-frequency end. In the spectrum between may be found infrared rays, light rays, x-rays, and gamma rays. Broadcasters, of course, use the radio wave portion of the spectrum for their purposes.

Electromagnetic waves carry broadcast transmissions (radio frequency) from station to receiver. It is the function of the transmitter to generate and shape the radio wave to conform to the frequency the station has been assigned by the FCC. Audio current is sent by a line from the control room to the transmitter. The current then modulates the carrier wave so that it may achieve its authorized frequency. A carrier wave that is undisturbed by audio current is called an *unmodulated carrier*.

The antenna radiates the radio frequency. Receivers are designed to pick up transmissions, convert the carrier into sound waves, and distribute them to the frequency tuned. Thus, in order for a station assigned a frequency of 950 kHz (kilohertz equals 1,000 hertz [Hz]) to reach a radio tuned to that position on the dial, it must alter its carrier wave 950,000 cycles (Hz) per second. The tuner counts the incoming radio frequency.

FIGURE 10.1 Wireless transmitter in 1918. Courtesy Westinghouse Electric.

AM/FM

Several things distinguish AM from FM. To begin with, they are located at different points in the spectrum. AM stations are assigned frequencies between 535 and 1705 kHz on the Standard Broadcast band. FM stations are located between 88.1 and 107.9

FIGURE 10.2 Early radio pioneer Dr Frank Conrad. Conrad was responsible for putting station KDKA on the air. Courtesy Westinghouse Electric.

FIGURE 10.3
1940s "Air King"
table model AM/FM
receiver.

MHz (megahertz equals 1 million hertz) on the FM band.

Ten kilocycles (kc) separate frequencies in AM, whereas 200 kilocycles is the distance between FM frequencies. FM broadcasters utilize 30 kilocycles for over-the-air transmissions and are permitted to provide subcarrier transmission (SCA) to subscribers on the remaining frequency. The larger channel width provides FM listeners a better opportunity to fine-tune their favorite stations as well as to receive broadcasts in stereo. To achieve parity, AM broadcasters developed a way to transmit in stereo, and by 1990 hundreds were doing so. The fine-tuning edge still belongs to FM, since its sidebands (15 kc) are three times wider than AM's (5 kc).

FM broadcasts at a much higher frequency (millions of cycles per second) compared to AM (thousands of cycles per second). At such a high frequency, FM is immune to low-frequency emissions, which plague AM. Whereas a car motor or an electric storm generally will interfere with AM reception, FM is static free. Broadcast engineers have

FIGURE 10.4
Antennas (towers)
propagate station
signals.

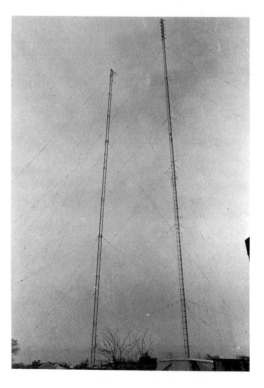

attempted to improve the quality of the AM band, but the basic nature of the lower frequency makes AM simply more prone to interference than FM. FM broadcasters see this as a key competitive advantage and refer to AM's move to stereo as "stereo with static."

Signal Propagation

The paths of AM and FM signals differ from one another. Ground waves create AM's primary service area as they travel across the earth's surface. High-power AM stations are able to reach listeners hundreds of miles away during the day. At night AM's signal is reflected by the atmosphere (ionosphere), thus creating a skywave which carries considerably further, sometimes thousands of miles. Skywaves constitute AM's secondary service area.

In contrast to AM signal radiation, FM propagates its radio waves in a direct or line-of-sight pattern. FM stations are not affected by evening changes in the atmosphere and generally do not carry as far as AM stations. A high-power FM station may reach listeners within an 80 to 100 mile radius since its signal weakens as it approaches the horizon. Since FM outlets radiate direct waves, antenna height becomes nearly as important as power. Generally speaking, the higher an FM antenna the further the signal travels.

Skywave Interference

The fact that AM station signals travel greater distances at night is a mixed blessing. Although some stations benefit from the expanded coverage area created by the skywave phenomenon, many do not. In fact, over 2,000 radio stations around the country must cease operation near sunset, while thousands more must make substantial transmission adjustments to prevent interference. For example, many stations must decrease power after sunset to ensure noninterference with others on the same frequency: WXXX-AM is 5,000

FIGURE10.5 (LEFT) 30-kw FM transmitter. Courtesy Broadcast Electronics.

FIGURE 10.6 (RIGHT) Lines leading from station transmitter to antenna base. FCC rules require fencing for safety purposes.

watts (5 kw) during the day, but at night it must drop to 1,000 watts (1 kw). Another measure designed to prevent interference requires that certain stations direct their signals away from stations on the same frequency. Directional stations require two or more antennas to shape the pattern of their radiation, whereas a nondirectional station that distributes its signal evenly in all directions only needs a single antenna. Because of its limited direct wave signal, FM is not subject to the postsunset operating constraints that affect most AM outlets.

Station Classifications

To guarantee the efficient use of the broadcast spectrum, the FCC established a classification system for both AM and FM stations. Under this system the nation's 10,000 radio outlets operate free of the debilitating interference that plagued broadcasters prior to the Radio Act of 1927.

AM Classifications

Class A: These are clear channel stations with power not exceeding 50 kw. Their frequencies are protected from interference up to 750 miles. Among the pioneer, or oldest, stations in the country are KDKA, WBZ, WSM, WJR.

Class B: Stations in this class are assigned power ranging from a minimum of 250 watts to a maximum of 50 kw. They must protect Class I outlets by altering their signals around sunset. As a Class B station, WINZ-AM in Miami is required to reduce power from 50 kw to 250 watts so as not to intrude upon other stations at 940 kHz. These stations also operate on Regional channels.

Class D: Also operating on Regional channels with less coverage area than Class A and B operations, Class Ds are assigned up to 5 kw ERP (effective radiated power) and are intended to service the city or town and adjacent areas of their license. Regionals outnumber clear channel stations and are located all across the Standard band.

Class C: Local channel Class C stations operate with the lowest power (1 kw and under), often must sign off around sunset, and generally are found at the upper end of the dial— 1,200 to 1,500 kHz. Class C stations

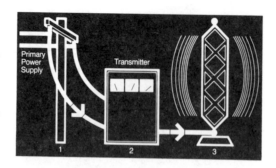

FIGURE 10.7
Stations receive
their power from
conventional utility
companies. From
*FCC Broadcast
Operator's
Handbook,* fig. 3–1.

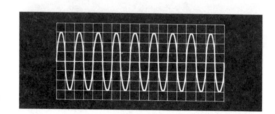

FIGURE 10.8
Unmodulated
(undisturbed)
carrier. From *FCC
Broadcast
Operator's
Handbook,* fig. 5–1.

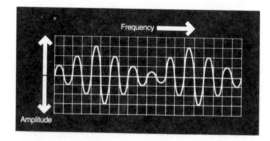

FIGURE 10.9
Amplitude
modulated (AM)
carrier. From *FCC
Broadcast
Operator's
Handbook,* fig. 5–2.

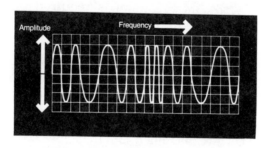

FIGURE 10.10
Frequency
modulated (FM)
carrier. From *FCC
Broadcast
Operator's
Handbook,* fig. 5–4.

FIGURE 10.11
Standard AM and
FM band. From *FCC
Broadcast
Operator's
Handbook,* figs.
4–7 and 4–8.

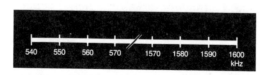

are particularly prevalent in small
and rural markets.

Check section 73.21 of the *Code of
Federal Regulations* (Part 73) for more
details on AM station classifications.

New AM band space (1,605 to 1,705
kHz) is being allocated. At present, the
FCC is encouraging existing AM license
holders to shift to the new space as a
means of reducing interference on the
clogged band.

FM Classifications

Class C: The most powerful FM outlets
with the greatest service parameters
are Class C's. Stations in this class
may be assigned a maximum ERP of
100 kw and a tower height of up to
2,000, feet. Class C radio waves
carry on the average 70 miles from
their point of transmission.

Class B: Class B stations operate with
less power—up to 50 kw—than Class
Cs and are intended to serve smaller
areas. The maximum antenna height
for stations in this class is 500 feet.
Class B signals generally do not
reach beyond 40 to 50 miles.

Class A: Class A stations are the least
powerful of commercial FM stations;
they seldom exceed 3 kw ERP (except
in select cases where a ceiling of 6 kw
is imposed) and 328 feet in antenna
height. The average service contour
for stations in this category is 10 to
20 miles.

Class D: The Class D category is set
aside for noncommercial stations
with 10 watts ERP. This type of station is most apt to be licensed to a
school or college.

In the 1980s the FCC introduced
three new classes of FM stations under
Docket 80-90 in an attempt to provide
several hundred additional frequencies,
and more sub-classes were added later.
They are as follows:

Class C1: Stations granted licenses to
operate within this classification may
be authorized to transmit up to 100
kw ERP with antennas not exceeding

FIGURE 10.12
Radio spectrum
table.

VLF (Very Low Frequency) 30 kHz and below	—Maritime use
LF (Low Frequency) 30 kHz to 300 kHz	—Aeronautical/maritime
MF (Medium Frequency) 300 kHz to 3000 kHz	—AM, amateur, distress, etc.
HF (High Frequency) 3 MHz to 30 MHz	—CB, fax, international, etc.
VHF (Very High Frequency) 30 MHz to 300 MHz	—FM, TV, satellite, etc.
UHF (Ultra High Frequency) 300 MHz to 3000 MHz	—TV, satellite, CB, DAB (proposed), etc.
SHF (Super High Frequency) 3 GHz to 30 GHz	—Satellite, radar, space, etc.
EHF (Extreme High Frequency) 30 GHz to 300 GHz	—Space, amateur, experimental, etc.

FIGURE 10.13
Coverage maps
show where a
station's signal
reaches. Courtesy
WLS.

984 feet. The maximum reach of stations in this class is about 50 miles.

Class C2: The operating parameters of stations in Class C2 are close to Class Bs. The maximum power granted Class C2 outlets is 50 kw, and antennas may not exceed 492 feet. Class C2 stations reach approximately 35 miles.

Class C3: These stations operate with shorter antennas and with power which typically exceeds 6kw ERP.

Class B1: The maximum antenna height permitted for Class B1 stations (328 feet) is identical to Class As; however, Class B1s are assigned at least 25 kw ERP. Class B1 signals carry 25 to 30 miles.

Due to the seemingly constant revisions made to FM classifications, it is suggested that the reader consult section 73.210 of the current *Code of Federal Regulations* (Part 73).

DIGITAL AUDIO BROADCASTING

Radio is undergoing a true metamorphosis as analog signal processing is being supplanted by digital processing. The reason for the transformation is simple—the demand for better and more evolved sound is at an all-time high. Stations must convert to digital, or they will not be competitive with home audio products such as CD and DAT players.

The conversion to digital broadcasting is being planned and is likely to be undertaken this decade. At the 1992 World Administrative Radio Conference (WARC), conducted by the International Telecommunications Union (ITU) in

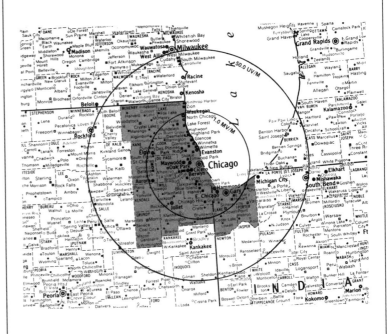

WLS FM COVERAGE

The map below indicates the area "covered" by the WLS FM signal. The inner circle represents 1 millivolt per meter. The outer circle represents the total reception range—50 microvolts per meter. Our transmitter move to the Sears Tower in June of 1983 provided a 22% increase in square mile coverage—the maximum allowable under present FCC rules.

Eleven County Metro Area (effective Fall '84) includes: Cook, IL, DuPage, IL, McHenry, IL, Kane, IL, Lake, IL, Grundy, IL, Will, IL, Kendall, IL, Lake, IN Porter, IN and Kenosha, WI (AREA SHADED ON MAP).

FIGURE 10.14 WNBC's (now WFAN) coverage maps illustrate the difference between daytime and nighttime reach. Courtesy WNBC-AM.

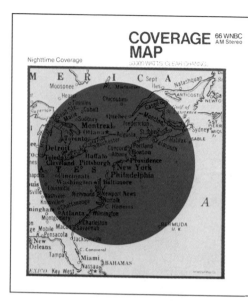
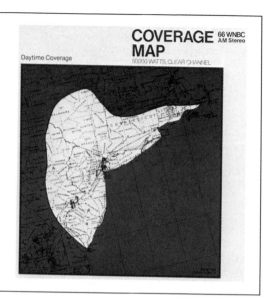

FIGURE 10.15 AM signal radiation. From *FCC Broadcast Operator's Handbook,* fig. 3–2.

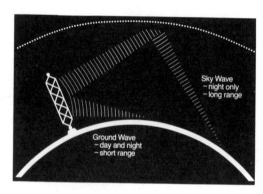

FIGURE 10.16 Nondirectional and directional antenna radiation. From *FCC Broadcast Operator's Handbook,* fig 7–2.

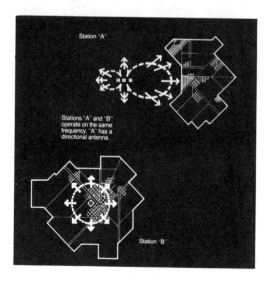

Spain, the FCC proposed use of the S-Band (2,310-2,360 MHz) for the propagation of DAB signals.

Much still remains to be resolved, especially the question of whether U.S. DAB will be a terrestrial or a satellite service and whether it will find in-band placement. The NAB supports an in-band on-channel (IBOC) terrestrial DAB system. Its philosophy is based on a desire to maintain a strong local broadcast industry, and that is likely to be the outcome.

Whereas the present system of analog broadcasting essentially replicates sound waves (this with inherent shortcomings), digital converts sound waves into a bitstream of 1s and 0s for processing into a low-band width. In digital, sound waves are assigned numeric values and become coded pulses.

Simply put, in digital, sounds are quantified. This allows for a more accurate representation of audio signals. Unlike analog, which is limited in what it can reproduce, digital provides greater frequency response and dynamic range. Thus more audio information is conveyed to the listener, who hears more. Another positive feature from the broadcast operator's perspective is the fact that digital signals do not require as much power as does analog.

Obviously when DAB is implemented, new receivers will be introduced to the consumer public. Part of their appeal, according to telecommunications professor Ernest Hakanen, is the fact that they "will allow for much more faithfulness of signal reproduction. DAB receivers will be designed to use reflected signals as alternative sources of information when the primary signal deteriorates. Using receivers that correct the fading and interference problems associated with AM and FM broadcasts, DAB signals that include specific information that can 'tell' the receiver how to compensate for information lost between transmitter and receiver can be received."

Eventually the existing system of AM and FM broadcasting will be passé. It is not likely, however, that the conversion to DAB will occur overnight. Some predict that analog broadcasting will be around for many years and that, even when digital is the preeminent broadcasting system, AM and FM stations will still be out there—that is, until the FCC no longer perceives them as providing a viable service.

SMART RECEIVERS

It is now possible to get more than just audio from a radio receiver. In fact, consumers can format scan without actually having to listen to stations. These so-called smart receivers may also feature emergency alerting capabilities, traffic announcements, advertisements and promos, and other informational services via a built-in LCD display panel. This is made possible by a technology known as RDS or RBDS (radio broadcast data system). The system, which originated in Europe as RDS, employs special signaling codes generated by stations.

Some programmers oppose the idea because they feel that it is difficult to categorize a format given the existing options, especially with a limited number of letters. The thought of a quasi-teletext component to radio inspires mixed emotions in many broadcasters. Will people be watching radio, and what exactly will that mean? In the main,

however, RBDS is perceived as an important value-added feature for radio.

Although it is difficult at this time to predict the impact and role of these innovations, it is certain that eventually receivers will do more than simply tune frequencies.

One other plus offered by smart-receiver technology is that it will allow car radios to retune automatically a different station offering the same format when a vehicle leaves the coverage area of the first station.

Several hundred stations are already providing "sight radio" to users. However, the RBDS receiver market is still in its infancy. According to media experts, it will grow quickly.

BECOMING AN ENGINEER

Most station managers or chief engineers look for experience when hiring technical people. Formal training such as college ranks high but not as high as actual hands-on technical experience. "A good electronics background is preferred, of course. This doesn't necessarily mean ten years of experience or an advanced degree in electronic engineering, but rather a person with a solid foundation in the fundamentals of radio electronics, perhaps derived from an interest in amateur radio, computers, or another hobby of a technical nature. This is a good starting point. Actually, it has been my experience that people with this kind of a background are more attuned to the nature of this business. You don't need a person with a physics degree from MIT, but what you do want is someone with a natural inclination for the technical

FIGURE 10.17
In digital processing of sound, an analog waveform is quantified, that is, given a numeric/binary value.

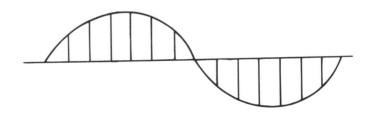

FIGURE 10.18 Features of a smart receiver turn radio medium into a visual data system. Courtesy RDS.

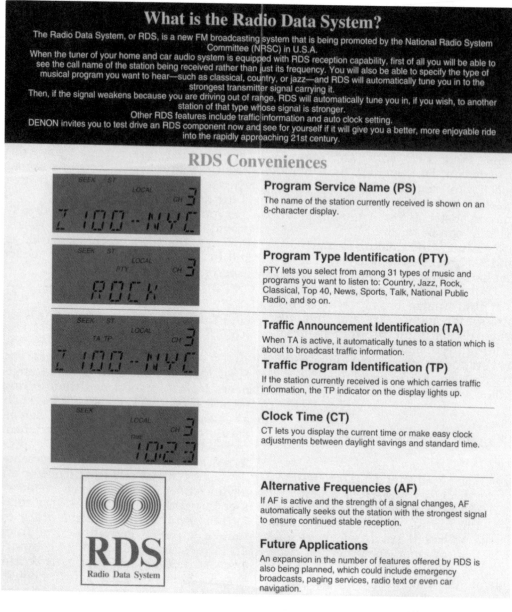

What is the Radio Data System?

The Radio Data System, or RDS, is a new FM broadcasting system that is being promoted by the National Radio System Committee (NRSC) in U.S.A.

When the tuner of your home and car audio system is equipped with RDS reception capability, first of all you will be able to see the call name of the station being received rather than just its frequency. You will also be able to specify the type of musical program you want to hear—such as classical, country, or jazz—and RDS will automatically tune you in to the strongest transmitter signal carrying it.

Then, if the signal weakens because you are driving out of range, RDS will automatically tune you in, if you wish, to another station of that type whose signal is stronger.

Other RDS features include traffic information and auto clock setting.

DENON invites you to test drive an RDS component now and see for yourself if it will give you a better, more enjoyable ride into the rapidly approaching 21st century.

RDS Conveniences

Program Service Name (PS)
The name of the station currently received is shown on an 8-character display.

Program Type Identification (PTY)
PTY lets you select from among 31 types of music and programs you want to listen to: Country, Jazz, Rock, Classical, Top 40, News, Sports, Talk, National Public Radio, and so on.

Traffic Announcement Identification (TA)
When TA is active, it automatically tunes to a station which is about to broadcast traffic information.

Traffic Program Identification (TP)
If the station currently received is one which carries traffic information, the TP indicator on the display lights up.

Clock Time (CT)
CT lets you display the current time or make easy clock adjustments between daylight savings and standard time.

Alternative Frequencies (AF)
If AF is active and the strength of a signal changes, AF automatically seeks out the station with the strongest signal to ensure continued stable reception.

Future Applications
An expansion in the number of features offered by RDS is also being planned, which could include emergency broadcasts, paging services, radio text or even car navigation.

side. Ideally speaking, you want to hire a person with a tech history as well as some formal in-class training," contends Kevin McNamara, director of engineering, Beasley Broadcasting Group.

Jim Puriez, chief engineer of WARA-AM, Attleboro, Massachusetts, concurs. "A formal education in electronics is good, but not essential. In this business if you have the desire and natural interest, you can learn from the inside out. You don't find that many broadcast engineers with actual electronics degrees. Of course, most have taken

basic electronics courses. The majority are long on experience and have acquired their skills on the job. While a college degree is a nice credential, I think most managers hire tech people on the basis of experience more than anything else."

Station engineer Sid Schweiger also cites experience as the key criterion for gaining a broadcast engineer's position. "When I'm in the market for a tech person, I'll check smaller market stations for someone interested in making the move to a larger station. This way,

I've got someone with experience right from the start. The little station is a good place for the newcomer to gain experience."

Formal training in electronics is offered by numerous schools and colleges. The number shrinks somewhat when it comes to those institutions actually providing curricula in broadcast engineering. However, a number of technical schools do offer basic electronics courses applicable to broadcast operations.

Before August 1981 the FCC required that broadcast engineers hold a First Class Radiotelephone license. In order to receive the license, applicants were expected to pass an examination. An understanding of basic broadcast electronics and a knowledge of the FCC rules and regulations pertaining to station technical operations were necessary to pass the lengthy examination. Today a station's chief engineer (also called *chief operator*) need only possess a Restricted Operator Permit. Those who held First Class licenses prior to their elimination now receive either a Restricted Operator Permit or a General Radiotelephone license at renewal time.

It is left to the discretion of the individual radio station to establish criteria regarding engineer credentials. Many do require a General Radiotelephone license or certification from associations, such as the Society of Broadcast Engineers (SBE) or the National Association of Radio and Telecommunications Engineers (NARTE), as a preliminary means of establishing a prospective engineer's qualifications. Appendices at the end of this chapter contain reproductions of SBE and NARTE membership application forms.

Communication skills rank highest on the list of personal qualities for sta-

tion engineers, according to McNamara. "The old stereotype of the station 'techhead' in white socks, chinos, and shirtpocket pen holder weighed down by its inky contents is losing its validity. Today, more than ever, I think, the radio engineer must be able to communicate with members of the staff from the manager to the deejay. Good interpersonal skills are necessary. Things have become very sophisticated, and engineers play an integral role in the operation of a facility, perhaps more now than in the past. The field of broadcast engineering has become more competitive, too, with the elimination of many operating requirements."

Due to a number of regulation changes in the 1980s, most notably the elimination of upper-grade license requirements, the prospective engineer now comes under even closer scrutiny by station management. The day when a "1st phone" was enough to get an engineering job is gone. There is no direct "ticket" in anymore. As in most

FIGURE 10.19
Station engineer at the work bench. Courtesy WMJX-FM.

FIGURE 10.20
A station engineer must be knowledgeable about the sophisticated state-of-the-art audio processing equipment (such as the limiter and processor shown here) used by many stations, especially in metro markets where great sound gives a station an important competitive edge. Courtesy CRL Audio and Broadcast Electronics.

FIGURE 10.21
Solid-state circuitry
has become state-
of-the-art in
broadcasting.
Courtesy IGM.

other areas of radio, skill, experience, and training are what open the doors the widest.

THE ENGINEER'S DUTIES

The FCC requires that all stations designate someone as chief operator. This individual is responsible for a station's technical operations. Equipment repairs and adjustments, as well as weekly inspections and calibrations of the station transmitter, remote control equipment, and monitoring and metering systems, fall within the chief's area of responsibility.

Depending on the makeup and size of a station, either a full-time or part-time engineer will be contracted. Many small outlets find they can get by with a weekly visit by a qualified engineer who also is available should a technical problem arise. Larger stations with more studios and operating equipment often employ an engineer on a full-time basis. It is a question of economics. The small station can less afford a day-to-day engineer, whereas the larger station usually finds that it can ill-afford to do without one.

Beasley Broadcast Group's McNamara considers protecting the station's license

his number one priority. "A station is only as good as its license to operate. If it loses it, the show is over. No other area of a station is under such scrutiny by the FCC as is the technical. The dereg movement in recent years has affected programming much more than engineering. My job is to first keep the station honest, that is, in compliance with the commission's rules. This means, keep the station operating within the assigned operating parameters, i.e., power, antenna phase, modulation, and so on, and to take corrective action if needed."

Steve Church, chief engineer of WFBQ/WNDE, Indianapolis, says that maintenance and equipment repairs consume a large portion of an engineer's time. "General repairs keep you busy. One moment you may be adjusting a pot on a studio console and the next replacing a part on some remote equipment. A broadcast facility is an amalgam of equipment that requires care and attention. Problems must be detected early or they can snowball. The proper installation of new equipment eliminates the chance of certain problems later on. The station's chief must be adept at a whole lot."

Other duties of the chief engineer include training techs, monitoring radiation levels, planning maintenance schedules, and handling a budget. Many stations hire outside engineering firms to conduct performance proofs, but it is ultimately the responsibility of the chief operator to ensure that the outlet meets its technical performance level. Proofs ascertain whether a station's audio equipment performance measurements fall within the prescribed parameters. A station's frequency response, harmonic distortion, FM noise level, AM noise level, stereo separation, crosstalk, and subcarrier suppression are gauged. If found adequate, the proof is passed. If not, the chief sees to it that necessary adjustments are made. Although the FCC no longer requires that Proof of Performance checks be undertaken, many stations continue to observe the practice as a fail-safe measure.

The duties of a station engineer are wide ranging and demanding. It is a position that requires a thorough grasp of electronics relative to the broadcast environment, knowledge of FCC rules and regulations pertaining to station technical operations, and, especially in the case of the chief engineer, the ability to manage finances and people.

STATION LOG

In 1983 the FCC dispensed with its requirement that radio stations keep maintenance and operating logs. In their place the commission created a new and considerably modified document called the Station Log, which stations must maintain. The new log requires that information pertaining to tower light malfunctions, Emergency Alert System (EAS) tests, and AM directional antenna systems be entered. Station Logs are kept on file for a period of two years.

Despite the fact that the FCC has eliminated the more involved logging procedures, some stations continue to employ the old system. "I like the accountability that Maintenance and Operating logs provide. We still use them here, and they are inspected daily. Despite the elimination of certain requirements, namely the tech logs, a station is still required to meet the operating stipulations of their license. Actually, enforcement action has been on the rise at the FCC, perhaps in reaction to the deregs. The commission is really interested in station technical operations. Keeping daily logs ensures compliance," says McNamara.

§ 73.1820 Station log.

(a) Entries must be made in the station log either manually by a properly licensed operator in actual charge of the transmitting apparatus, or by automatic devices meeting the requirements of paragraph (b) of this section. Indications of operating parameters that are required to be logged must be logged prior to any adjustment of the equipment. Where adjustments are made to restore parameters to their proper operating values, the corrected indications must be logged and accompanied, if

FIGURE 10.22
The luxury of solid-state electronics—pull out the bad, put in the good. Courtesy IGM.

any parameter deviation was beyond a prescribed tolerance, by a notation describing the nature of the corrective action. Indications of all parameters whose values are affected by the modulation of the carrier must be read without modulation. The actual time of observation must be included in each log entry. The following information must be entered:

(1) All stations: (i) Entries required by § 17.49 of this chapter concerning any observed or otherwise known extinguishment or improper functioning of a tower light:

(A) The nature of such extinguishment or improper functioning.

(B) The date and time the extinguishment or improper operation was observed or otherwise noted.

(C) The date, time and nature of adjustments, repairs or replacements made.

(ii) Any entries not specifically required in this section, but required by the instrument of authorization or elsewhere in this part.

FIGURE 10.23
Maintaining prescribed technical parameters is one of many engineering responsibilities.

(iii) An entry of each test of the Emergency Broadcast System procedures pursuant to the requirement of Subpart G of this part and the appropriate EBS checklist. All stations may keep EBS test data in a special EBS log which shall be maintained at any convenient location; however, such log should be considered a part of the station log.

(2) Directional AM stations without an FCC-approved antenna sampling system (See § 73.68): (i) | An entry at the beginning of operations in each mode of operation, and thereafter at intervals not exceeding 3 hours, of the following (actual readings observed prior to making any adjustments to the equipment and an indication of any corrections to restore parameters to normal operating values):

(A) Common point current.

(B) When the operating power is determined by the indirect method, the efficiency factor F and either the product of the final amplifier input voltage and current or the calculated antenna input power. See § 73.51(e).

(C) Antenna monitor phase or phase deviation indications.

(D) Antenna monitor sample currents, current ratios, or ratio deviation indications.

(ii) Entries required by § 73.61 performed in accordance with the schedule specified therein.

(iii) Entries of the results of calibration of automatic logging devices (see paragraph (b) of this section), extension meters (see § 73.1550) or indicating instruments (see § 73.67) whenever performed.

(b) Automatic devices accurately calibrated and with appropriate time, date and circuit functions may be utilized to record entries in the station log provided:

(1) The recording devices do not affect the operation of circuits or accuracy of indicating instruments of the equipment being recorded;

(2) The recording devices have an accuracy equivalent to the accuracy of the indicating instruments;

(3) The calibration is checked against the original indicators as often as necessary to ensure recording accuracy;

(4) Provision is made to actuate automatically an aural alarm circuit located near the operator on duty if any of the automatic log readings are not within the tolerances or other requirements specified in the rules or station license;

(5) The alarm circuit operates continuously or the devices which record each parameter in sequence must read each parameter at least once during each 30 minute period;

(6) The automatic logging equipment is located at the remote control point if the transmitter is remotely controlled, or at the transmitter location if the transmitter is manually controlled;

(7) The automatic logging equipment is located in the near vicinity of the operator on duty and is inspected periodically during the broadcast day. In the event of failure or malfunctioning of the automatic equipment, the employee responsible for the log shall make the required entries in the log manually at that time.

(8) The indicating equipment conforms to the requirements of § 73.1215 (Indicating instru-

ments—specifications) except that the scales need not exceed 2 inches in length. Arbitrary scales may not be used.

(c) In preparing the station log, original data may be recorded in rough form and later transcribed into the log. [43 FR 45854, Oct. 4, 1978, as amended at 44 FR 58735, Oct. 11, 1979; 47 FR 24580, June 7, 1982; 48 FR 38481, Aug. 24, 1983; 48 FR 44806, Sept. 30, 1983; 49 FR 33603, Aug. 23, 1984]

§ 73.1835 Special technical records.

The FCC may require a broadcast station licensee to keep operating and maintenance records as necessary to resolve conditions of actual or potential interference, rule violations, or deficient technical operation. [48 FR 38482, Aug. 24, 1983]

§ 73.1840 Retention of logs.

(a) Any log required to be kept by station licensees shall be retained by them for a period of 2 years. However, logs involving communications incident to a disaster or which include communications incident to or involved in an investigation by the FCC and about which the licensee has been notified, shall be retained by the licensee until specifically authorized in writing by the FCC to destroy them. Logs incident to or involved in any claim or complaint of which the licensee has notice shall be retained by the licensee until such claim or complaint has been fully satisfied or until the same has been barred by statute limiting the time for filing of suits upon such claims.

(b) Logs may be retained on microfilm, microfiche or other data-storage systems subject to the following conditions:

(1) Suitable viewing—reading devices shall be available to permit FCC inspection of logs pursuant to § 73.1226, availability to FCC of station logs and records.

(2) Reproduction of logs, stored on data-storage systems, to full-size copies, is required of licensees if requested by the FCC or the public as authorized by FCC rules. Such reproductions must be completed within 2 full work days of the time of the request.

(3) Corrections to logs shall be made:

(i) Prior to converting to a data-storage system pursuant to the requirements of § 73.1800 (c) and (d), (§ 73.1800, General requirements relating to logs).

(ii) After converting to a data-storage system, by separately making such corrections and then associating with the related data-stored logs. Such corrections shall contain sufficient information to allow those reviewing the logs to identify where corrections have been made, and when and by whom the corrections were made.

(4) Copies of any log required to be filed with any application; or placed in the station's local public inspection file as part of an application; or filed with reports to the FCC must be reproduced in fullsize form when complying with these requirements. [45 FR 41151, June 18, 1980, as amended at 46 FR 13907, Feb. 24, 1981; 46 FR 18557, Mar. 25, 1981; 49 FR 33663, Aug. 24, 1984]

WFBQ-FM
operating
log

DAY _____

DATE _____

Time	Plate Volts	Plate Amps	Power Out

1. All times are EST.
2. Unless otherwise noted, all operations are continued from previous
   day.
3. WFBQ leases its subcarrier to
   Comcast, Inc. for the operation of
   a background music service. Unless
   otherwise noted, subcarrier program
   material is background music provided by the Muzak Corp., subcarrier
   is on continuously when modulation
   present, and, with the exception of
   short breaks between music segments
   modulation is continuous.
4. Responsibility for tower maintenance
   and light checks delegated to WRTV
5. Po = Ep x Ip x Eff
   Eff= 74.12% as ind. in mfrs instr.

Comments:

POWER TOLERANCE: Low Limit = 18,000 w / High Limit = 21,000 w

EBS TEST SENT

_____ _____
time   initial

EBS TEST RECEIVED

VIA _____

_____ _____
time   initial

EBS RECEIVER CHECK

_____ _____
time   initial

Operator on	Time	Operator off	Time

Ch.Op.Review:

_____

Date _____

FIGURE 10.24
Operating log
currently in use at
WFBQ-FM. Courtesy
WFBQ-FM.

Allen Myers

The FCC and Radio

In evaluating the service to the radio medium provided by the Federal Communications Commission, it is necessary to understand that the Commission was created by Congress in the Communications Act of 1934 and that the agency, therefore, carries out the wishes of that body. If the Commission wants to implement regulations for which it lacks the statutory authority, it must first obtain the approval of Congress.

The Commission's service to the radio medium is two-fold. First, it sets the technical standards under which the medium operates. Second, it ensures an adequate and equitable distribution of radio services throughout the United States. The Commission was established by Congress with these specific objectives in mind. Prior to the existence of the Commission and its predecessor, The Federal Radio Commission, radio broadcasting in this country was in a state of chaos. There was no spectrum planning. Operators put the stations on the air wherever they wanted. If a new station caused interference to a station already on the air, the operator of the older station often would just increase power—sort of an electronic shouting match. There were also no standards for radio receivers to prevent them from causing radio frequency interference. So when Congress created the FCC, it charged it with making the radio medium "serve the public interest need and necessity."

The Commission's principal missions are accomplished with this objective in mind. Some of the agency's rules set minimum and maximum power requirements for radio stations; others set interference and distance standards —all with the objective of making sure that when a listener tunes to a radio station, he or she will be able to hear it clearly. The Commission's role in setting technical standards also extends to the equipment used in the transmission and reception of radio signals. Manufacturers of transmitters and receivers are required to receive FCC "type acceptance" approval before putting their products on the market. This insures that a clean signal is transmitted and received and that the equipment does not cause interference to other stations or electronic services. Other Commission rules deal with spectrum planning and are intended to ensure an equitable distribution of radio stations throughout the country so that as many communities as possible will have a local radio station and possibly access to several different stations to provide a multitude of voices.

The Commission recognizes the different types of services that radio stations provide to listeners. To this end, it will often establish rules to foster the growth of a group of stations providing a unique service. For example, in 1945 the Commission reserved the first 20 channels in the FM band (88.1 to 91.9 MHz) for radio stations licensed to nonprofit, educational institutions and organizations to be operated as noncommercial, educational radio stations. The Commission then established a set of rules for this type of station, including both technical standards and spectrum planning.

Finally, in looking at the Commission's service to the radio medium, one must realize that the Communications Act is a living document. It has been amended many times to allow for new technologies in the radio medium, and the Commission has implemented regulations to carry out these changes. There is no doubt that the Communications Act will continue to be amended to take into consideration future changes in radio service and the Commission will proscribe regulations implementing these changes.

The views expressed by the author are not necessarily those of the Federal Communications Commission.

FIGURE 10.25

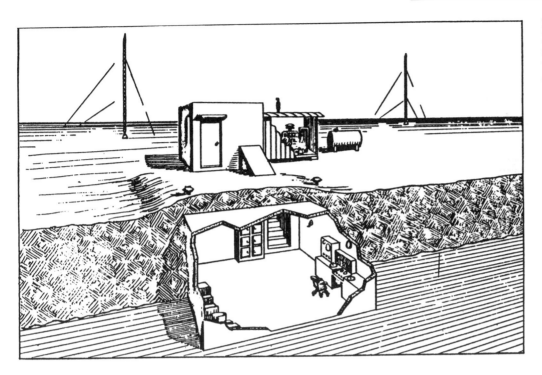

FIGURE 10.26
FMEA rendering of
a protected
underground
shelter for an EBS
station.

THE EMERGENCY ALERT SYSTEM (FORMERLY THE EMERGENCY BROADCAST SYSTEM)

In 1994, the FCC established the Emergency Alert System (EAS), which replaced the old Emergency Broadcast System (EBS). The latter came into existence following World War II as the nation and the world entered the nuclear age. The system was designed to provide the president and heads of state and local government with a way to communicate with the public in the event of a major emergency.

In the 1990s, EBS was viewed as outmoded due to the revolution in technology, and it was significantly revamped. EAS is intended to upgrade the effectiveness of broadcast warnings by employing digital equipment and sophisticated automation. Its speed and timeliness are greatly enhanced under the new protocol. Stations are expected to have the new EAS system fully installed by mid-1997. At present, stations take the following steps should the president and/or heads of state and

local government agencies deem it necessary to alert the public of a potential or imminent disaster. (Check the new rules pertaining to EAS in the *Code of Federal Regulations* (Part 73). At the time of this revision, the 1995 CFR containing the updates on the EAS procedure were unavailable. Therefore, the language which follows remains essentially unaltered:

1. Receive Emergency Action Notification (EAN) via AP/UPI teletype, network feed, or EBS (EAS) monitor receiver. Continue to monitor for further instructions.
2. Refer to EBS (EAS) Checklist. Each station has this folder on hand and must post it so that it is readily accessible to on-duty operators. The folder contains procedural information pertaining to a station's participation in the Emergency Action Notification System.
3. Authenticate EAN. This applies to AP/UPI subscribers and network affiliates only.
4. Discontinue normal programming and broadcast the first short announcement given in the Checklist.
5. Transmit attention signal.

Stations not designated to remain in operation in the event of an EAN then remove their carriers from the air after advising listeners where to tune for further information. Those participating stations continue to broadcast information as it is received from the nation's base of operations. Every radio station is required to install and operate an EBS (EAS) monitor. Failure to do so can result in a substantial penalty imposed by the FCC.

Stations are also required to test the Emergency Broadcasting System (Emergency Alert System) by airing both an announcement and an attention signal. Here is the text (changes may have been implemented as this edition publishes) that must be broadcast once a week on a rotating basis between 8:30 a.m. and local sunset:

EBS [EAS] TEST MESSAGE TEXT

THIS IS A TEST. (YOUR CALL LETTERS OR "THIS STATION") IS CO NDUCTING A TEST OF THE EMERGENCY BROADCAST SYSTEM [EMERGENCY ALERT SYSTEM]. THIS IS ONLY A TEST. (THEN BROADCAST THE "ATTENTION SIGNAL" GENERATED BY YOUR EBS EQUIPMENT FOR 8–25 SECONDS.) THIS IS A TEST OF THE EMERGENCY BROADCAST SYSTEM [EMERGENCY ALERT SYSTEM]. THE BROADCASTERS OF YOUR AREA IN VOLUNTARY COOPERATION WITH FEDERAL, STATE, AND LOCAL AUTHORITIES HAVE DEVELOPED THIS SYSTEM TO KEEP YOU INFORMED IN THE EVENT OF AN EMERGENCY. IF THIS HAD BEEN AN ACTUAL EMERGENCY, THE ATTENTION SIGNAL YOU JUST HEARD WOULD HAVE BEEN FOLLOWED BY OFFICIAL INFORMATION, NEWS, OR INSTRUCTIONS. (YOUR CALL LETTERS OR "THIS STATION") SERVES THE (OPERATIONAL AREA NAME) AREA. THIS CONCLUDES THIS TEST OF THE EMERGENCY BROADCAST SYSTEM [EMERGENCY ALERT SYSTEM].

EBS (EAS) tests are documented in the Station Log when they are broadcast. The entry must include the time and the date of the test.

The Federal Emergency Management Agency (FEMA) makes funds available to stations designated to remain on the air during an authentic emergency through the Broadcast Station Protection Plan. Under this provision the government provides financial assistance to EBS stations for the purpose of constructing and equipping a shelter designed to operate for at least fourteen days under emergency conditions.

In the 1990s, the FCC began an inquiry into whether the system needed updating or replacement. Critics of EBS claimed that the system had become obsolete.

In late 1992, proposed EBS revisions included the following:

Replacement of the existing emergency alerting system
Updating of EBS equipment
Cable media involvement in emergency alerting
Self-testing of the system
Mandated equipment standards
Rules to prohibit false and deceptive use of the system
Revised EBS test script

The new EAS embraces many of the above, as well as additional innovations and procedures.

AUTOMATION

The FCC's decision in the mid-1960s requiring that AM/FM operations in markets with populations of more than 100,000 originate separate programming 50 percent of the time provided significant impetus to radio automation. Before then *combo* stations, as they were called, simulcast their AM programming on FM primarily as a way of curtailing expenses. FM was still the poor second cousin of AM. (In the late 1980s the FCC dropped most of its simulcast requirements. Since then many stations have resorted to simulcasting as a means of dealing with the realities of fierce competition and a declining AM market.)

Responding to the rule changes, many stations resorted to automation systems as a way to keep expenses down. Interestingly enough, however, automation for programming, with its emphasis on music and deemphasis on chatter, actually helped FM secure a larger following, resulting in increased revenue and stature.

Today, over a quarter of all commercial stations are automated. Some are fully automated, for while others rely on automation for part of their broadcast day. Automation is far more prevalent on FM, but in the late 1970s and 1980s many AM outlets were employing automation systems to present Nostalgia and Easy Listening programming. The advent of AM stereo also has generated more use of automation on the Standard Broadcast band.

Although a substantial initial investment usually is necessary, the basic purpose of automation is to save a station money, and this it does by cutting staffing costs. Automation may also reduce the number of personnel problems. However, despite early predictions that automation eventually would replace the bulk of the radio work force, very few jobs have actually been lost. In fact, new positions have been created.

Automated stations employ operators as well as announcers and production people. The extent to which a station uses automation often bears directly on staffing needs. Obviously, a fully automated station will employ fewer programming people than a partially automated outlet.

A simple automation system consists of a small memory bank that directs the sequence of program elements, whereas a more complex system may be run by a computer that also produces logs, music sheets, invoices, affidavits, and so on. In either case, an automation system primarily consists of reel-to-reel tape decks, with 10- and/or 12-inch reel capacity for longer play, and cartridge units, known as carousels or *stack racks*. Many stations have incorporated CDs, as well as DAT decks, into their automation chains.

Programming elements are aired when a trip mechanism in a tape machine is activated by a cue tone. All

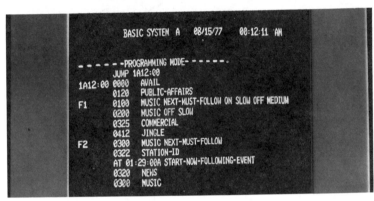

reels and carted elements, such as commercials, promos, PSAs, newscasts, features, and the weather must be impressed with a cue tone in order that the programming chain be maintained. A tape source without a cue tone will not signal the next program item in the sequence. To prevent extended periods of dead air, automation systems are equipped with silence sensors which, after a period of two to 40 seconds, depending on how they are set, automatically trigger the next element in the sequence.

Automation systems often are housed in a separate room that may be adjacent to the on-air studio. Remote start and stop switches located in the control studio permit the operator to go live whenever necessary. The engineer is responsible for the repair and maintenance of automation equipment, and at some stations he or she actually operates the system.

Most station automation systems are driven by computers and many are linked to satellite program services

FIGURE 10.27 Computers were used by automated stations three decades ago as well as today. Many automated systems are run by computers. The operator simply types in the programming routine for the system to follow. Courtesy IGM.

FIGURE 10.28
(LEFT)
Satellite networks
may interface with
station automation
systems. Courtesy
SMN.

FIGURE 10.29
(RIGHT)
Automation
carousel unit
containing carted
music. Courtesy
IGM.

(networks and suppliers) in this age of the high-tech, "turn-key" operation.

The day of in-house loading of format elements is on the wane. Satellite syndicators using computers control local station ingredients (news, weather, promos, spots) remotely from the uplinks.

POSTING LICENSES AND PERMITS

The FCC requires that a station's license and the permits of its operators be posted. What follows are the rules pertaining to this requirement as outlined in Subpart H, Section 73.1230 of the FCC's regulations.

§ 73.1230 Posting of station and operator licenses.

(a) The station license and any other instrument of station authorization shall be posted in a conspicuous place and in such a manner that all terms are visible at the place the licensee considers to be the principal control point of the transmitter. At all other control or ATS monitoring and alarm points a photocopy of the station license and other authorizations shall be posted.

(b) The operator license of each station operator employed full-time or part-time or via contract, shall be permanently posted and shall remain posted so long as the operator is employed by the licensee. Operators employed at two or more stations, which are not co-located, shall post their operator license or permit at one of the stations, and a photocopy of the license or permit at each other station. The operator license shall be posted where the operator is on duty, either:

(1) At the transmitter; or

(2) At the extension meter location; or

(3) At the remote control point, if the station is operated by remote control; or

(4) At the monitoring and alarm point, if the station is using an automatic transmission system.

(c) Posting of the operator licenses and the station license and any other instruments of authorization shall be done by affixing the licenses to the wall at the posting location, or by enclosing them in a binder or folder which is retained at the posting location so that the documents will be readily available and easily accessible. [43 FR 45847, Oct. 4, 1978, as amended at 49 FR 29069, July 18, 1984]

CHAPTER HIGHLIGHTS

1. Guglielmo Marconi first used electromagnetic (radio) waves to send a message at the end of the nineteenth century. Marconi used earlier findings by James C. Maxwell and Heinrich Hertz.

2. J. Ambrose Fleming developed the diode tube (1904), and Reginald Fessenden transmitted voice over the wireless (1906).

3. Edwin Armstrong developed the regenerative and superheterodyne circuits, and first demonstrated the static-free FM broadcast signal (1933).

4. Broadcast transmissions are carried on electromagnetic waves. The transmitter creates and shapes the wave to correspond to the "frequency" assigned by the FCC.

5. Receivers pick up the transmissions, converting the incoming radio frequency (RF) into sound waves.

6. AM stations are assigned frequencies between 535 and 1,705 kHz, with ten kilocycle separations between frequencies. AM is disrupted by low-frequency emissions, can be blocked by irregular topography, and can travel hundreds (along surface-level ground waves) or thousands (along nighttime skywaves) of miles.

7. Because AM station signals travel greater distances at night, in order to avoid "skywave" interference, over 2,000 stations around the country must cease operation near sunset. Thousands more must make substantial nighttime transmission adjustments (decrease power), and others (directional stations) must use two or more antennas to shape the pattern of their radiation.

8. FM stations are assigned frequencies between 88.1 and 107.9 MHz, with 200-kilocycle separations between frequencies. FM is static free, with direct waves (line-of-sight) carrying up to 80–100 miles.

9. To guarantee efficient use of the broadcast spectrum and to minimize station-to-station interferences, the FCC established four classifications for AM stations and seven classifications for FM. Lower-classification stations are obligated to avoid interference with higher-classification stations. Recent FCC actions have created more sub-classifications.

10. Analog is being replaced by digital audio (DAB) because the latter provides superior frequency response and greater dynamic range. New spectrum space may be allocated to accommodate the digital service.

11. A station's chief engineer (chief operator) needs experience with basic broadcast electronics, as well as a knowledge of the FCC regulations affecting the station's technical operation. The chief must repair and adjust equipment, as well as perform weekly inspections and calibrations. Other duties may include installing new equipment, training techs, planning maintenance schedules, and handling a budget.

12. A Proof of Performance involves checking the station's frequency response, harmonic distortion, FM noise level, AM noise level, stereo separation, crosstalk, and subcarrier suppression.

13. Although the FCC dispensed with the maintenance and operating log requirements (1983), a Station Log must be maintained. The log lists information about tower light malfunctions, Emergency Broadcast System tests, and AM directional antenna systems.

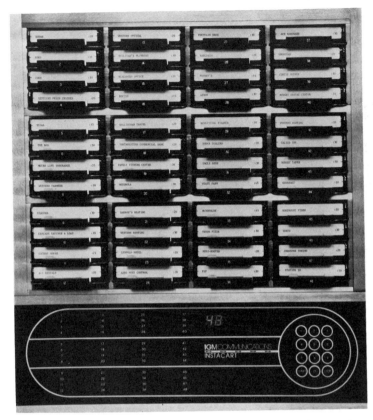

FIGURE 10.30 Automation stack cart system containing carted commercials. Courtesy IGM.

FIGURE 10.31
PC-based program controller running an automation system.
Courtesy IGM Communications.

14. The Emergency Broadcast System (today revamped as the Emergency Alert System), implemented after World War II, provides the government with a means of communicating with the public in an emergency. Stations must follow rigid instructions both during periodic tests of the system and during an actual emergency.

15. Over one-quarter of today's commercial stations are fully or partially automated. More prevalent in FM stations, automation reduces staffing costs but requires a significant equipment investment. Automated programming elements are aired when a trip mechanism is activated by a cue tone, which is impressed on all program material. Either an operator or a computer can maintain the programming chain. At many stations, satellite programming services use computers (at both uplink and downlink sites) to control station automation systems.

16. The FCC requires that a station's license and the permits of its operators be accessible in the station area.

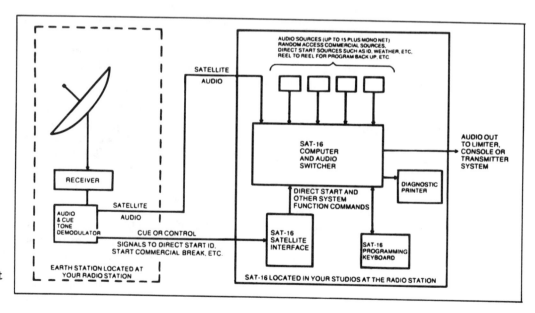

FIGURE 10.32
Satellite-linked automation systems are available in certain formats. Diagram shows how the system works.
Courtesy Broadcast Electronics.

FIGURE 10.33
Appendix 10A: SBE
Application.

An invitation to join SBE

Who are we?

The Society of Broadcast Engineers, formed in 1963 as the Institute of Broadcast Engineers, is a non-profit organization serving the interests of broadcast engineers. We are the only society devoted to all levels of broadcast engineering.

Our membership, which is international in scope, is made up of studio and transmitter operators and technicians, supervisors, announcer-technicians, chief engineers of large and small stations and of commercial and educational stations, engineering vice presidents, consultants, field service and sales engineers, broadcast engineers from recording studios, schools, CCTV and CATV systems, production houses, advertising agencies, corporations, audio-visual departments, and all other facilities that utilize broadcast engineers.

What can we do for you?

Help you keep pace with our rapidly changing industry through educational seminars, and a look at new technology through industry tours and exhibits at monthly chapter meetings, regional conventions, and our national meeting held in conjunction with the National Association of Broadcasters (NAB).

Give you national representation. To serve as a voice for you in the industry; a liaison for you with governmental agencies as well as other industry groups.

To provide a forum for the exchange of ideas and sharing of information with other broadcast engineers and industry people.

To promote the profession of broadcast engineering.

To establish standards of professional education and training for broadcast engineering, and to recognize achievement of these standards.

In addition to the intangible benefits of membership in the SBE, the tangible benefits of an insurance program, com-

munications through **The SBE Signal**, certification and re-certification opportunities, and a readily available network of specialized professionals.

All this adds up to an increase in your worth as a broadcast engineer to your employer.

Where does your money go?

A small office staff handling membership, certification, and the day-to-day business of the Society. Many duties of the SBE are handled by officers and board members who volunteer their time with no remuneration.

A library of videotape training material for loan from the national headquarters.

The production of our bi-monthly newsletter, **The SBE Signal**.

Allows SBE representation—through a professionally designed informational booth at state and regional meetings as well as NAB and NRBA.

A portion of your annual dues returns to subsidize the local chapter.

Supplements expenses for special events such as NAB, Chapter Chairman and Certification Chairman Meetings, and invitational opportunities to represent the SBE.

How to be one of us

Membership categories include Student, Associate Member, Member, Senior Member, Honorary Member, and Fellow.

Qualification for Member grade requires that the individual be actively engaged in broadcast engineering or have an academic degree in electrical engineering or its equivalent in scientific or professional experience in broadcast engineering or a closely related field or art.

The cost of membership is $20 annually for member and associate member grades, and $10 for student memberships.

What do we want from you?

First, we'd like to have your name on the SBE roster.

Strength in numbers gives us additional clout. And we want and need your participation and input at the regional and national levels as well as at the local level.

Certification Program

The program issued its first certificates on January 1, 1977, and now conducts tests at various times and places for those people, either members or non-members, who wish to have a certificate attesting to their competence as broadcast engineers. The certificates are issued for two different levels of achievement in either radio or TV and are valid for five years from date of issue. Recertification may be accomplished by earning professional credits for activities which maintain competence in the state-of-the-art or by re-examination.

Emphasis in the tests is on practical working knowledge rather than general theory. The tests are as valid for people in related industries as they are for broadcasters.

An entry-level certificate was added to the certification program in January 1982 to attract new technical talent to the broadcast industry and provide incentive for them to grow with technology.

The certification program is conducted by the SBE to benefit everyone in the industry. The program recognizes professional competence as judged by one's peers, and encourages participation in seminars, conventions, and meetings to help keep abreast of the constantly changing technology in broadcast engineering.

If you ever wanted to meet the people who design the equipment you use,
- to talk with your fellow engineers and technicians,
- to tour the many facilities that employ engineers and technicians,
- to keep abreast of the state-of-the-art equipment,
- to upgrade your skills for certification,
- here is the opportunity to become a member of the most prestigious society in its field.

SOCIETY OF BROADCAST ENGINEERS, INC.
P.O. Box 50844, Indianapolis, Indiana 46250

Group Insurance Program

When you join the SBE, you have the opportunity to participate in the Group Insurance Program for SBE members and their dependents, which offers a wide range of coverage to suit your individual needs. The low rates are made possible through the economics of group administration and by the fact that SBE does not profit from the insurance program. Please note that requests for coverage under some of the plans are subject to insurance company approval.

- Term Life Insurance Plan offers options of up to $195,000 for eligible members, with lesser amounts for dependents.

- High-Limit Accident Insurance provides protection wherever you go, 24 hours a day, and eliminates the need for special accident insurance every time you travel.

- Disability Income Plan protects your income by providing monthly benefit payments when you are unable to work due to a disabling illness or accident.

- Excess Major Medical Plan supplements your regular hospital/medical coverage in the event of a catastrophic illness or accident, paying up to $1,000,000 after you satisfy your deductible.

- In-Hospital Plan pays up to $100 per day for every day you spend in the hospital—up to 365 days—directly to you, to spend as you wish.

- Major Medical Expense Insurance is designed for members who have little or no basic medical coverage.

For further information concerning membership, certification, application, regional meetings and conventions, contact the Society of Broadcast Engineers, Inc., P.O. Box 50844, Indianapolis, IN 46250, (317) 842-0836.

MEMBERSHIP APPLICATION

SOCIETY OF BROADCAST ENGINEERS
P.O. Box 50844 • Indianapolis, Indiana 46250 • 317/842-0836

(Please type or print)

Application For:
- New Member $20.00
- Associate Member $20.00
- Student Member $10.00
  - Change in Grade:
  - To Member
  - Sr. Member

Reinstate:
- Former Member #_____

Name: _____

Full home Address: (don't abbreviate)
_____ ____ Receive SBE Mail here?
_____
_____ Home Phone ( )_____

Full Company Name and Address: ____ or here?
_____
_____
_____ Business Phone ( )_____

If accepted, please consider me a member of _____ Chapter

SBE Certification # _____ (If applicable)   Date of Birth_____

Current Job Title: _____   Date Employed:_____

Type of Facility: _____

Description of Duties: _____

Total years of responsible   Field of   Radio _____
Engineering Experience:_____   Activity: Television _____
                                            Other _____

PROFESSIONAL LICENSES OR CERTIFICATES
_____
_____

Additional Information Requested on Reverse Side

ADMISSIONS COMMITTEE ACTION

Date:_____

Action deferred for more information _____

Admissions Committee Chairman's
Signature:_____

Approved for Grade_____
Candidate Notified_____
Entered in Records_____

EXPERIENCE RECORD

List in chronological order, beginning with the most recent, all formal experience in Broadcast Engineering or related employment. Indicate field or fields of specialization under 'Position.' Please do not limit yourself to the four spaces below. ATTACH A BRIEF DESCRIPTION OF DUTIES.

From Mo. Yr.	To Mo. Yr.	Company Name and Location	Position or Title	Type of Facility

EDUCATION

College, University, or Technical Institute	From Mo. Yr.	To Mo. Yr.	Credits or Yrs. Compl.	Course or Major	Degree

List Short Courses, Seminars Related to Broadcast-Communications Technology

SPECIAL ACHIEVEMENTS
List awards, patents, books, articles, etc.

REFERENCES
List two references - familiar with your work

Name	Company	Address	Phone

Have you ever been convicted of a violation of the Communications Act of 1934, as amended. Yes ☐ No ☐. If so, describe in full.
(Use additional space if necessary) _____
_____ I have enclosed the required application fee.

Signed _____ Date_____ 19__
I agree to abide by the By-Laws of the Society if admitted.

FIGURE 10.33
continued

FIGURE 10.34
Appendix 10B:
NARTE.

APPLICATION FOR CERTIFICATION AND ENDORSEMENT
THE NATIONAL ASSOCIATION
OF
RADIO & TELECOMMUNICATIONS ENGINEERS, INC.

P.O. BOX 15029 • SALEM, OR 97309
503-581-3336

☐ CHECK HERE IF YOU ARE A NARTE MEMBER. NUMBER _____
☐ CHECK HERE IF YOU WANT A NARTE MEMBERSHIP APPLICATION.

Date of Application _____

FOR CERTIFICATION COMMITTEE ONLY
Date Recieved _____ 19____
Date Reviewed _____ 19____
Date Returned _____ 19____
Date Granted _____ 19____
Date Denied _____ 19____
Certificate No. Assigned _____

1. I, the undersigned, hereby apply for NARTE Certification and Endorsement under the **Grandfather** provisions outlined in the NARTE Administrative Rules Handbook on Certification and Endorsement, and submit the following information to substantiate my experience in the relevant disciplines. *My check for $_____ is enclosed.*

2. FULL NAME _____
   LAST   FIRST   MIDDLE OR INIT   (AS TO BE SHOWN ON CERTIFICATE)

3. DATE OF BIRTH _____ PLACE OF BIRTH _____
   CITY   STATE/PROVINCE   COUNTRY

   I AM A CITIZEN OF _____
   COUNTRY

   PREFERRED _____
   MAILING
   ADDRESS   P O BOX OR STREET

   CITY   STATE   ZIP

   TELEPHONE _____

   CHECK IF THIS IS
   ☐ HOME OR
   ☐ BUSINESS

4. Existing professional licensing (include FCC)

ISSUING AGENCY	FIRST DATE OF ISSUE	LIC. OR CERT. NO.	TYPE	CHECK IF CURRENT OR LAPSED

ATTACH COPY OF PRESENT LICENSES, IF ANY.

5. Education

A GIVE THE HIGHEST ELEMENTARY OR HIGH SCHOOL GRADE COMPLETED _____		B IF YOU COMPLETED HIGH SCHOOL. GIVE DATE _____				
C NAME AND LOCATION OF COLLEGE OR UNIVERSITY		DATES ATTENDED		YEARS COMPLETED		CREDIT HOURS	DEGREES RECEIVED
		FROM	TO	DAY	NIGHT	SEMESTER	

D CHIEF GRADUATE COLLEGE SUBJECTS	CREDIT HOURS SEMESTER	E CHIEF UNDERGRADUATE COLLEGE SUBJECTS	CREDIT HOURS SEMESTER

USE ADDITIONAL SHEETS IF NECESSARY

FORM 002 6-84

APPENDIX 10C

FEDERAL COMMUNICATIONS COMMISSION

FACT SHEET

Hints on Filing Comments with the FCC

The FCC is interested in any experiences, knowledge, or insights that outside parties may have to shed light on issues and questions raised in the rule-making process. The public and industry have the opportunity to comment upon Petitions for Rule Makings, NOI's, NPRM's, Further NPRM's, Reports and Orders, and others' comments on the aforementioned documents. It is a common misconception that one must be a lawyer to be able to file comments with the FCC, but all that is necessary is an interest in an issue and the ability to read and follow directions.

Prior to drafting comments it is crucial to read and understand fully the item you wish to comment on. Usually, the NPRM, NOI, or other item will specify and invite comment upon the issue(s) that the Commission is interested in studying further. Examination of the issue(s) and relevant documents is the most important part of the comment process. Comments may take any form, but below are some hints to assist you in writing them.

Format: There is no required format for informal comments, although if you plan to file formally, it is required that they be typed, double-spaced, and on 8.5˝ x 11˝ paper. Additional requirements for formal filings are set forth in Sections 1.49 and 1.419 of the FCC Rules. The Docket Number or Rule Making Number of the item at hand should be included on your comments, and can be found on the front page of the Commission document or public notice. You should also include your name and complete mailing address.

Content: Your comments should state who you are and what your specific interest is. (You do not need to represent yourself in an official capacity. You may, for example, express your opinion as a concerned consumer, concerned parent, etc., and sign your name.) State your position and the facts directly, as thoroughly but as briefly as possible. Explain your position as it relates to your experience and be explicit. Make clear if the details of a proposed rule or only one of several provisions of the rule are objectionable. If the rule would be acceptable with certain safeguards, explain them and why they are necessary.

Support: Statements of agreement or dissent in comments should be supported to the best extent possible by factual (studies, statistics, etc.), logical, and/or legal information. Support should illustrate why your position is in the public interest. The more support made, the more persuasive the comments will be.

Length: Comments may be any length, although it is preferred that they be succinct and direct. If formal comments are longer than ten pages, it is required that they include a summary sheet.

Time frame: Your comments should be submitted well within the time frame designated on the original document or public notice. It is almost always included on the first page of an NPRM or NOI. However, if the deadline has passed, you can still submit your views informally in a permissible ex parte presentation.

Filing: Send your written comments to Secretary, Federal Communications Commission, 1919 M Street, N.W., Washington, D.C. 20554. If you wish your comments to be received as an informal filing, submit the original and one copy. If you want your comments to be received as a formal filing, you should submit an original and four copies. For more specific filing information, please refer to the FCC Public Notice "Guidelines for Uniform Filings" available from the same address.

Reply comments: As the name implies, reply comments are used to respond to comments filed by other parties. You may file reply comments even if you did not submit comments initially. When drafting reply comment, use the same guidelines expressed above regarding content and be careful not to raise additional or irrelevant issues.

Tracking your comments: After you have properly filed your comments with the FCC, they will be part of the official Commission record. To track the progress of proceedings in which you have filed comments, you may check the Daily Digest or Federal Register for releases and notices. The Daily Digest can be obtained from the Office of Public Affairs, 1919 M Street, N.W., Washington, D.C. 20554 or from a daily recorded listing of texts and releases at (202) 418-2222.

For further information: For further information, you may directly contact the Secretary's office at the FCC, (202) 418-0300. Explicit information about filings in rule making proceedings can be found in Sections 1.49 and 1.419 of the FCC Rules. Copies of any FCC documents can be obtained through the FCC's duplicating contractor, ITS, 1919 M Street, N.W., Room 246, Washington, D.C. 20554, (202) 857-3800, or from one of the private distributors of FCC releases. A list of distributors is available from the Public Service Division, 1919 M Street, N.W., Room 254, Washington, D.C. 20554, (202) 418-0190.

SUGGESTED FURTHER READING

Abel, John D., and Ducey, Richard V. *Gazing into the Crystal Ball: A Radio Station Manager's Technological Guide to the Future.* Washington, D.C.: NAB, 1987.

Antebi, Elizabeth. *The Electronic Epoch.* New York: Van Nostrand Reinhold, 1982.

Butler, Andy. *Practical Tips for Choosing and Using Consulting and Contract Engineers.* Washington: NAB Publications, 1994.

Cheney, Margaret. *Tesla: Man out of Time.* Englewood Cliffs, N.J.: Prentice-Hall, 1983.

Considine, Douglas M., ed. *Van Nostrand's Scientific Encyclopedia.* New York: Van Nostrand Reinhold, 1983.

Davidson, Frank P. *Macro: A Clear Vision of How Science and Technology Will Shape Our Future.* New York: William Morrow, 1983.

Ebersole, Samuel. *Broadcast Technology Worktext.* Boston: Focal Press, 1992.

Finnegan, Patrick S. *Broadcast Engineering and Maintenance Handbook.* Blue Ridge Summit, Pa.: Tab, 1976.

Grant, August E. *Communication Technology Update.* Boston: Focal Press, 1995.

Mirabito, Michael, and Morgenstern, Barbara. *The New Communication Technologies*, 2nd ed. Boston: Focal Press, 1994.

National Association of Broadcasters. *Understanding DAB*, 2nd ed. Washington: NAB Publications, 1994.

Noll, Edward M. *Broadcast Radio and Television Handbook*, 6th ed. Indianapolis: Howard Sams, 1983.

Reitz, John R. *Foundations of Electromagnetic Theory.* Reading, Mass.: Addison-Wesley, 1960.

Roberts, Robert S. *Dictionary of Audio, Radio, and Video.* Boston: Butterworths, 1981.

Starr, William. *Electrical Wiring and Design: A Practical Approach.* New York: John Wiley and Sons, 1983.

Watkinson, John. *The Art of Digital Audio.* Boston: Focal Press, 1992.

11 Consultants and Syndicators

RADIO AID

Two things directly contributed to the rise of radio consultants: more stations—from 2,000 in the 1950s to 12,000 in the 1990s—and more formats—from a half dozen to several dozen during the same period. Broadcast consultants have been around almost from the start, but it was not until the medium set a new course following the advent of television that the field grew to real prominence. By the 1960s consultants were directing the programming efforts of hundreds of stations. In the 1970s over a third of the nation's stations enlisted the services of consultants. Today, radio consultants work with an even greater number of stations, and the growing level of competition in the radio marketplace should increase station involvement with consultants into the 1990s and beyond. In fact, some industry experts predict that the day will come when the overwhelming majority of stations, regardless of size, will solicit the aid of an outside consultant or consultancy firm in some way, shape, or form. Whether or not this comes to pass, consultants do play an important role in the shaping and management of the medium today.

Observes prominent radio consultant George Burns, "The principle role of radio consultants has evolved considerably since Mike Joseph started the whole thing in the 1950s. We began by being very specifically task oriented. A consultant was assumed to have greater expertise at the job than anyone that the station could afford full-time. Currently consultants serve primarily

as outside (and, it is hoped, impartial) monitors of station progress. The job is to assure management that everything possible is being done to maximize the station's potential. If something is not functioning properly or needs to be changed, consultants are expected to give voice to these concerns. Over the years, the job has become infinitely more complex. Musical and nonmusical aspects of programming have spread widely apart. Research has become a separate discipline. And lately, the marketing side of radio has achieved a 'life of its own.' Different consultants approach each station's progress from varying points of view. Specialization was inevitable."

Stations use consultants for various reasons, says Fred Jacobs, president of Jacob's Media: "Stations realize that they need an experienced, objective ear to make intelligent evaluations. Consultants are also exposed to ideas and innovations from around the country that they can bring to their client stations. As radio has become more competitive, stations understand that their need for up-to-date information about current trends in programming and marketing has increased."

Dave Scott, president of Century 21 Programming, Dallas, Texas, adds that a lack of research expertise on the local station level prompts many stations to use consultants. "We're well into the information age, the age of highly sophisticated research techniques and computerized data. It takes a lot of resources to assess a market and prescribe a course of action. Most stations do not have the wherewithal. At Century 21 Programming, each of our consultants goes through more ratings surveys and research data than most station owners, managers, or program directors will in a lifetime. The way the marketplace is today, using a consultant generally is a wise move. Radio stations that attempt to find their niche by trial and error make costly mistakes. A veteran consultant can accelerate a station's move on the road to success."

Boston-based radio consultant Donna Halper agrees with Scott and adds, "Consultants give their client stations an objective viewpoint and another experienced person's input. Consultants are support people, resource people, who bring to a situation a broader vision rather than the purely local perspective. Consultants, and not just out-of-work PDs who call themselves consultants but in reality aren't, have a lot of research, information, and expertise they can make available to a client with an ailing station."

Mikel Hunter of Mikel Hunter Broadcast Services, Las Vegas, says consultants help stations develop a distinctiveness that they need in order to succeed. "Unfortunately most station PDs are bandwagon riders. Many watch what other stations do around the country and clone them in their markets. Sometimes this works. Often it doesn't. It likely was a consultant who helped design the programming of that successful station being copied, and the consultant did so based on what was germane to that particular market, not one a thousand miles away. Therein lies the problem. Simply because a station in Denver is doing great book by programming a certain way does not guarantee that a station in Maryland can duplicate that success. A good consultant brings originality and creativity to each new situation, in addition to the knowledge and experience he possesses. The follow-the-leader method so

FIGURE 11.1
There are an abundance of radio consulting services for stations to choose from. Courtesy Shane Media and Lund Consultants.

FIGURE 11.1
continued

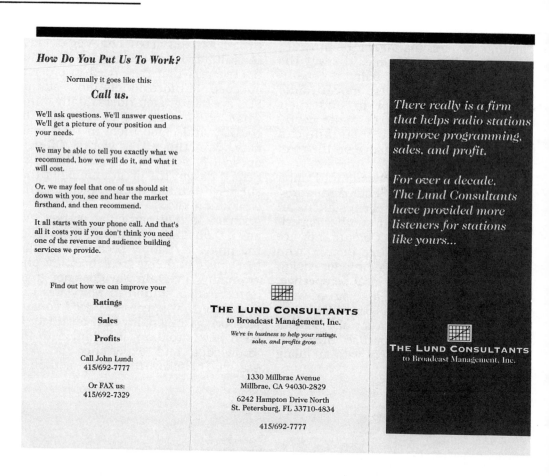

prevalent among programmers actually creates a lot of the problems that consultants are called on to remedy."

There are over a hundred broadcast consultants listed in the various media directories around the country. More than half of this number specialize in radio. Generally, consultancy companies range in size from 20 to 50 employees, but may be composed of as few as two or three. Many successful program directors also provide consultancy to stations in other markets in addition to their regular programming duties. A growing number of station rep companies provide their client stations consultancy services for an additional fee.

CONSULTANT SERVICES

Stations hire program consultants to improve or strengthen their standings in the ratings surveys. An outside consul-

tant may share general program decisions with the station's PD or may be endowed with full control over all decisions affecting the station's sound, contends Donna Halper. "I have as little or as much involvement as the client desires. Depending on the case, I can hire and train staff (or fire staff), design or fine-tune a format, or simply motivate and direct deejays, which is actually anything but simple. Whatever a station wants, as a professional consultant I can provide. Usually, I make recommendations and then the owner or GM decides whether or not I will carry them out. At some of my stations, I've functioned as the acting PD, for all intents and purposes. At other client stations, I've been sort of the unofficial mother figure, providing support, encouragement, and sometimes a much-needed kick in the behind."

Among other services, Fred Jacobs says his company offers "In-market visits for monitoring and strategizing; ongoing monitors of client competition from airchecks or station 'listen lines'; critiques of on-air talent, assistance/design of music scheduling and selection; computer programs that assist with promo scheduling, data-base marketing, and morning show preparation; design of off-air advertising and coordination with production; and design/implementation of market research for programming, image, and music."

Most consultant firms are equipped to provide either comprehensive or limited support to stations. "In some cases, consultants offer a packaged 'system for success' in the same way a McDonald's hamburger franchise delivers a 'system for success' to an investor. The consultant gets control. In other instances, consultants deliver objective advice or research input to a station more on a one-to-one basis. This parallels the role of most accountants or attorneys in that the decisions are still made by the station management, not the consultant," notes Century 21's Dave Scott.

In the mid-1990s, niche consultants have come into vogue. For example, a consulting service called Air Support focuses on improving the ratings of station morning shows by working on "talent development, preparation, creativity, and performance," reports *Radio Ink*.

Program consultants diagnose the problems that impair a station's growth and then prescribe a plan of action designed to remedy the ills. For example, WXXX, located in a 20-station market, is one of three that programs current hits, yet it lags behind both of its competitors in the ratings. A consultant is hired to assess the situation and suggest a solution. The consultant's preliminary report cites several weaknesses in WXXX's overall programming. The consultant's critique submitted to the station's general manager may be written like this:

Dear GM:

Following a month-long analysis of WXXX's on-air product, here are some initial impressions. A more extensive report on each of the areas cited herein will follow our scheduled conference next week.

K-ONE
AM 1450

KOZZ 105.7
CLASSIC ROCK & ROLL

The #1 Combo

August 13, 1991

Burkhart, Douglas and Associates
6500 River Chase Circle, East
Atlanta, Georgia 30328

I wanted to take a few minutes to thank your staff for the hours you've put in to help make KOZZ a winner!

Greg Gillespie gives us a great deal of his time, makes himself available and always gets back to me even when he's on the road. Greg works from the time he gets up till the time he turns in when he's visiting KOZZ -- it's nice to know he's never too busy to answer our questions or offer input. Greg played a very significant part in our success. We came in #1 Adults 25-54 in both the Birch and the other survey that we don't subscribe to. When you consider that we did it with $9,000.00 and 10 billboards it's down right astounding!

Looking forward to bigger and better numbers in the Fall and continuing our relationship for years to come.

Yours truly,

Harry Reynolds
Program Director

HR/ca

P.O. Box 9870 • 2900 Sutro Street • Reno, Nevada 89507 • FAX 323-1450 • Telephone (702) 329-9261

FIGURE 11.2 Consultants make a dramatic difference at many stations. Courtesy Burkhart/Douglas & Associates.

---

1. Personnel: Morning man Jay Allen lacks the energy and appeal necessary to attract and sustain an audience in this daypart. While Allen possesses a smoothness and warmth that would work well in other time slots, namely midday or evenings, he does not have the "wake-up and roll" sound, nor the type of humor listeners have come to expect at this time of day. The other "hot hit" stations in the market offer bright and lively morning teams. Allen does not stand up against the competition. His contrasting style is ill-suited for AM drive, whereas midday man Mike Curtis would be more at home during this period. His upbeat, witty, and casual style when teamed with newspeople Chuck Tuttle and Mark Fournier would strengthen the morning slot.

Tracy Jessick and Michelle Jones perform well in their respective time periods. Overnight man Johnny

BURKHART/DOUGLAS & ASSOCIATES: "DON'T ROCK WITHOUT US"

6500 River Chase Circle, East, Atlanta, GA 30328 (404) 955-1550/FAX: (404) 955-6220

For 17 years, Burkhart/Douglas & Associates has been an integral force in the development of AOR radio. Through consistent execution of unique marketing and promotional concepts, researched music and features, and talent development, B/D & A clients have attained proven results. And the Spring '88 Arbitron shows Burkhart/ Douglas & Associates clients performing better than ever!

Here are the facts:

❑ Six stations scored their best 12+ numbers in their history. These include WHCN/Hartford, WRXK/Ft. Myers, KFMF/Chico, WBLM/Portland, KEZO/ Omaha, and WYMG/Springfield, which has the distinction of being the nation's highest rated AOR with a 23.3 share and a cume equal to 1/3 of the metro population!

❑ Eight stations scored number one 12+, including WYMG, KFMF, WBLM, KEZO, WPYX/Albany, WFBQ/Indianapolis, WRDU/Raleigh-Durham, and KOZZ/Reno. WKLS/ Atlanta and WIMZ/Knoxville were both a strong number two 12+.

❑ In the all important target demographics, the success of B/D & A clients shows even more strength. Fifty-three percent of B/D & A clients are number one 18-34 Adults, including KKDJ/Fresno, WPYX/Albany, WKLS/ Atlanta, WZZO/Allentown, WRQK/ Canton, KFMF/Chico, WRXK/Ft. Myers, WHCN/Hartford, KLOL/ Houston, WIMZ/Knoxville, KEZO/ Omaha, WBLM/Portland, WRDU/ Raleigh-Durham, WFBQ/Indianapolis, KYYS/Kansas City, and KOZZ/Reno.

❑ Thirty percent of B/D & A's clients are number one 25-54 Adults, including KKDJ/Fresno, WPYX/Albany, KFMF/Chico, WRXK/Ft. Myers, WIMZ/Knoxville, KEZO/Omaha, WBLM/ Portland, WRDU/Raleigh-Durham, WFBQ/Indianapolis, and KOZZ/ Reno. WKLS/Atlanta, KYYS/ Kansas City and WZZO/Allentown came in number two 25-54 Adults.

❑ Overall, the B/D & A scorecard shows 67% of our AOR clients are up, 30% down, and 3% stayed the same.

And now with fragmentation a reality, Burkhart/Douglas & Associates is proud to offer three rock based formats that address the fragmentation issue! The formats available are:

SOLID ROCK

Solid Rock is B/D & A's mainstream AOR approach. Solid Rock is a proven winner, as evidenced by the aforementioned statistics. To insure the continued growth and success of the Solid Rock approach, B/D & A recently applied the latest available research and trend data to redefine the mainstream approach for the first time in a decade. Now featuring even better demographic/psychographic controls, Solid Rock will help you maximize your audience potentials throughout the various dayparts.

MAC

MAC, or Male AC, is B/D & A's newly designed and researched Adult Rock format. MAC was developed to meet the needs of 25-49 Adults who find typical AOR too hard, new or weird, and AC too soft and boring. MAC is also the perfect evolution for Classic Rock stations who find their approach becoming a stagnated mood service. While MAC is library-intensive, it also utilizes a customized proportion of in-sync, current music to insure freshness and forward momentum.

Male CHR

Male CHR is B/D & A's newly developed and researched approach designed to reach the growing number of disenchanted 16-25 year olds. These young adults are finding AOR too old and CHR too urban or teeny-bop. Male CHR mixes the best cutting edge music with mainstream and hard rock hits. The new to old percentages are very similar to CHR — 75 to 80 percent current versus 15 to 20 percent library. The presentation features high-profile personalities and quick-paced, forward moving, state of the art production elements.

As we enter a new decade, Burkhart/Douglas & Associates has made the commitment to the future of AOR radio. Whether you're an established AOR looking for increased ratings and revenue or you're considering a format change to AOR, contact B/D & A first.

Burkhart/Douglas & Associates: Don't rock without us!

FIGURE 11.3
One of the nation's top consultancy firms getting the word out. Courtesy Burkhart/Douglas & Associates.

Christensen is very adequate. Potential as midday man should Curtis be moved into morning slot.

Weekend personnel uneven. Better balance needed. Carol Mirando, 2 to 7 p.m. Sunday, is the strongest of the part-timers. Serious pacing problems with Larry Coty in 7 to midnite slot on Saturday. Can't read copy.

2. Music: Rotation problems in all dayparts. Playlist narrowing and updating necessary. Better definition needed. As stands, station verges on Adult Con-temporary at certain times of the day, especially during a.m. drive. On Monday the 14th, during evening daypart, station abandoned currents and assumed Oldies sound. More stability and consistency within format essential. Computerized music scheduling possible solution. Separate report to follow.

3. News Programming: General revamping necessary. Too heavy an emphasis during both drive periods. Cut back by 20 to 30 percent in these two dayparts. Fifteen-minute "Noon News" needs to be eliminated. Tune-out factor in targeted demos. Same holds true for half-hour, 5 to 5:30 p.m., "News Roundup." Hourly five-minute casts reduced to minute headlines after 7 p.m. Both content and style of newscasts presently inappropriate for demos sought. Air presentations need adjusting to better, more compatibly suit format. Tuttle and Fournier of morning show are strong, whereas p.m. drive news would benefit from a comparable team. Ovitt, Hart, and Lexis do not complement each other. Van Sanders is effective in evening slot. More sounders and actualities in hourly newscasts. Greater "local" slant needed, especially on sports events.

4. General Programming: Too much clutter! A log-jam in drive dayparts. Spots clustered four deep in spot sets, sometimes at quarter hour. So much for "maintenance." Rescheduling needed for flow purposes as well. "Consumer Call" at 8 a.m., noon, and 5 p.m. not suitable for demos. "Band News" good, but too long. One-minute capsule versions scheduled through day would be more effective. Friday evening "Oldies Party" too geriatric, breaks format objective. Sends target demos off to competition by appealing to older listeners with songs dating back to 1960s. Public Affairs programs scheduled between 9 a.m. and noon on Sundays delivers teens to competition that airs music during same time period. Jingles and promos dated. Smacks of decade ago. New package would add contemporary luster needed to sell format to target demo.

5. Promotions: "Bermuda Triangle" contest aimed at older demos. Contest prizes geared for 25- to 39-year-old listener. Ages station. Concert tie-in good. Album "giveaway" could be embellished with other prizes. Too thin as is. Response would indicate lack of motivation. True also of "Cash Call." Larger sums need to be awarded. Curtis's "Rock Trivia" on target. Hits demos on the money. Expand into other dayparts. Bumper stickers and "X-100" calendar do not project appropriate image. New billboards and bus-

boards also need adjusting. Paper ads focus on weak logo. Waiting to view TV promo. Competition promos are very weak. A good "X-100" TV promo would create advantage in this area. Opportunity.

6. Technical: Signal strong. Reaches areas that competition does not. Significant null in Centerville area. Competition's signals unaffected. Occasional disparity in levels. Spots sometimes very hot. Promos and PSAs, especially UNICEF and American Cancer Society, slightly muddy. In general, fidelity acceptable on music. Extraneous noise, possibly caused by scratches or dirt, on some power rotation cuts. Stereo separation good. Recommend compressor and new limiter. Further plant evaluation in progress. A more detailed report to follow.

Following an extensive assessment of a station's programming, a consultant may suggest a major change. "After an in-depth evaluation and analysis, we may conclude that a station is improperly positioned in its particular market and recommend a format switch. Sometimes station management disagrees. Changing formats can be pretty traumatic, so there often is resistance to the idea. A critique more often recommends that adjustments be made in an existing format than a change over to a different one. There are times when a consultant is simply called upon to assist in the hiring of a new jock or newsperson. Major surgery is not always necessary or desired," says Donna Halper.

Today the majority of stations in major and medium markets switching formats do so with the aid of a consultant. According to the NAB, 3 to 5 percent of the nation's stations change formats each year. Consultant fees range from five hundred to over twelve hundred dollars a day, depending on the complexity of the services rendered and the size of the station.

CONSULTANT QUALIFICATIONS

Many consultants begin as broadcasters. Some successfully programmed stations before embarking on their own or joining consultancy firms. According to Rick Sklar, deceased president of Sklar Communications, consultants who have a background in the medium have a considerable edge over those who do not. "The best way to fully understand and appreciate radio is to work in it. As you might imagine, radio experience is very helpful in this business." Jacobs agrees with Sklar. "Ideally a consultant should have a successful

FIGURE 11.4
Consultant's response to commonly asked questions. Courtesy Donna Halper and Associates.

Donna Halper & Associates
Radio Programming Consultants
304 Newbury St., #506
Boston, MA 02115
(617) 786-0666

QUESTIONS I AM OFTEN ASKED ABOUT HIRING A CONSULTANT

1. What kind of station would hire a consultant?

All kinds! From major market #1 stations that want to stay that way to new stations that need help choosing a format or hiring staff.

2. Aren't most consultants just out-of-work Program Directors?

Not today. Competition is too intense. Most of us who have stayed in the consulting field have years of experience in one thing: CONSULTING.

3. Should I hire a 'big name' consultant?

Since the majority of consultants today are experienced, you should choose one based on what his/her areas of expertise are. Interview a few consultants and you will see that each has some specialty-- whether it's a certain format (some consultants prefer to do only one format) or a certain market size. Choosing the right consultant for your station is an important decision, and you shouldn't do it on name alone.

4. What can a consultant offer my station that my own people can't provide?

First, consultants aren't there to replace your people, nor do they want them to look bad. While staff changes may result from the recommendations of a consultant, our first purpose is to offer you an UNBIASED, outside overview of how your station sounds, both its strengths and its weaknesses. We work WITH your people, providing research, guidance, training, market studies, etc. Often, because we are not caught up in the day-to-day circumstances, we can offer a fresh, objective point of view.

5. What are the benefits of HALPER & ASSOCIATES?

I'm glad you asked. We've been in business since 1980. (Before that, Donna Halper spent 13 years in major markets as an announcer, Music Director, PD, news reporter, and writer/producer of special programming.) Our specialties include critiques/positioning studies, staff training and motivation, and talent development. We work in markets of all sizes, but we are best-known for our ability to turn around failing small and medium market stations. We also do motivational seminars, and are expert at handling morale problems. Unlike some consultants who only do one format, Halper & Associates can show success stories in AC, Gold, CHR, Urban, Classic Rock, Full-Service/M-O-R, and Country. SINCE 1980, OVER 90% OF OUR CLIENTS HAVE SHOWN RATINGS GROWTH. And, our critiques and market studies have been used by some of the biggest and best companies. Also, Halper and Associates has experience with Canadian radio, and we have consulted in Puerto Rico.

6. Can you promise results for every client?

No consultant wins 'em all, although we'd like to. But, our slogan has always been "NO PROMISES...JUST RESULTS." We are proud of our many satisfied clients and our renewal rate is quite high. Many of our clients say they would never use another consultant. So, when it's time to think about a consultant, choose DONNA HALPER & ASSOCIATES. We can get results for you. To find out more, call us at 617-786-0666. We'll give you the attention you might not get from the "big names," affordable rates, and, most important, you can count on us to make a positive impact on your station and its staff! DONNA HALPER GETS RESULTS!!!

background in programming, with expertise in a number of areas, including research, sales, marketing, and promotion. The key word is *success*—a solid track record in a number of different market situations is invaluable. Consultants also need to have strong communication and tracking skills to best work with a variety of clients in markets around the country." Not all consultants have extensive backgrounds in the medium. Most do possess a thorough knowledge of how radio operates on all its different levels, from having worked closely with stations and having acquired formal training in colleges offering research methodology,

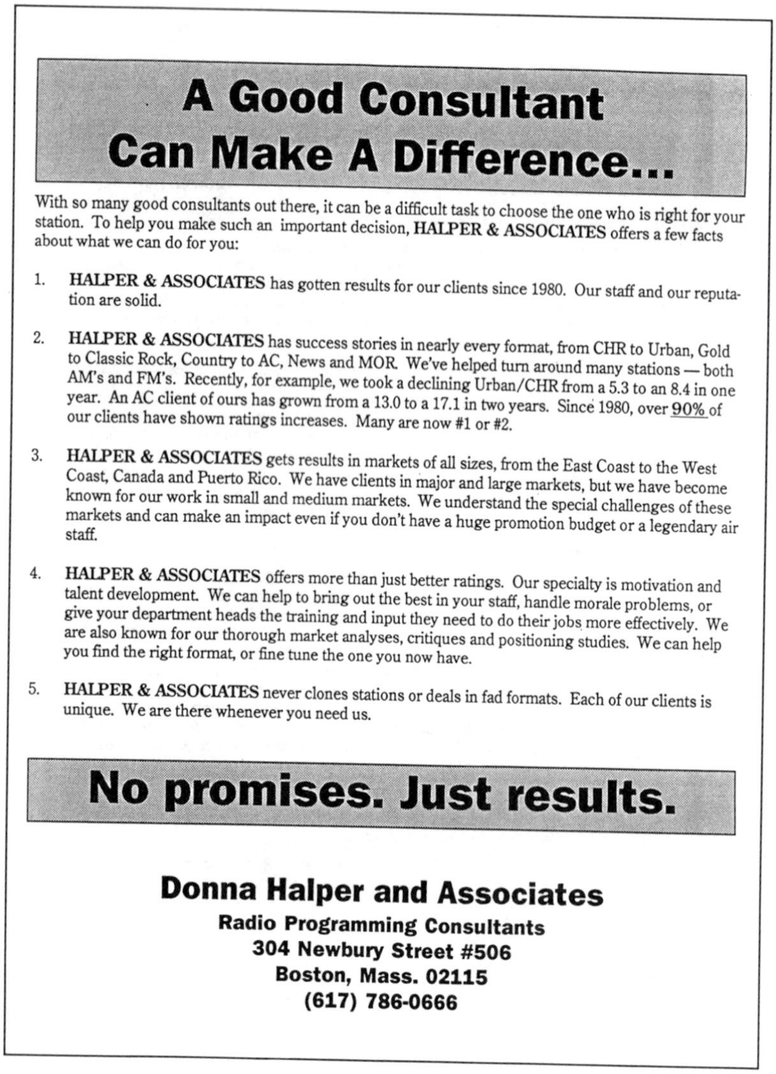

FIGURE 11.5 Consultant promotional piece. Courtesy Donna Halper and Associates.

audience measurement, and broadcast management courses. "A solid education is particularly important for those planning to become broadcast consultants. It is a very complex and demanding field today, and it is becoming more so with each passing day. My advice is to load up. Get the training and experience up front. It is very competitive out there. You make your own opportunities in this profession," says Century 21's Dave Scott.

Both Halper and Scott rate people skills and objectivity highly. "Consulting requires an ability to deal with people. Decisions sometimes result in drastic personnel changes, for example, changing formats. A consultant must be adept at diplomacy but must act with conviction when the diagnosis has been made. Major surgery invariably is traumatic, but the idea is to make the patient, the station, healthy again. You can't let your own personal biases or tastes get in the way of what will work in a given market," observes Halper. Dave Scott shares Halper's sentiments. "A consultant, like a doctor, must be compassionate and at the same time maintain his objectivity. It is our intention and goal as a program consultant service to make our client stations thrive. As consultants, we're successful because we do what we have to do. It's not a question of being mercenary. It's a question of doing what you have to do to make a station prosper and realize its potential."

Consultant company executives also consider wit, patience, curiosity, sincerity, eagerness, competitiveness, and drive, not necessarily in that order, among the other virtues that the aspiring consultant should possess.

CONSULTANTS: PROS AND CONS

There are as many opponents of program consultants within the radio industry as there are advocates. Broadcasters who do not use consultants argue that local flavor is lost when an outsider comes into a market to direct a station's programming. Donna Halper contends that this may be true to some degree but believes that most professional consultants are sensitive to a station's local identity. "Some consultants do clone their stations. Others of us do not. In fact, I'd say most do not. For those of us who recognize local differences, there need not be any loss whatsoever as a consequence of consultant-recommended changes. But the hits are pretty much the hits, and good radio is something that Tulsa deserves as well as Rochester. So I do try to localize my music research and acquire a good feel for the market I'm working in. But as far as basic rules of good radio are concerned, those don't vary much no matter what the market is. It's important for a station to reflect the market it serves, and I support my clients in that. Because I work out of Boston doesn't mean that my AOR client in Duluth should sound like a Boston album rocker. It should sound like a solid AOR station that could be respected in any city but fits the needs of Duluth."

Consultant Dwight Douglas says that localization is essential for any radio station and that consultants are amply aware of this fact. "It is an industry axiom that a station must be a part of its environment. An excellent station will be uniquely local in relating to its audience. That tends to take the form of news, weather, sports, public service, general information, and jock talk. A good consultant will free a station from music worries and allow it to concentrate on developing local identity. We work hard at customizing formats to suit the geodemographics or lifestyles of the audiences of our client stations."

A station has an obligation to retain its sense of locality regardless of what a consultant may suggest, contends Mikel Hunter. "No station should simply turn itself over body and soul to a consultant. Local flavor does not have to be sacrificed if a station has a strong PD and a general manager who doesn't insist that the PD merely follow the consultant's suggestions. A station should not let itself become a local franchise. Consultants are a valuable resource, but both the station and the consultant must pool their wisdom to make the plan work."

Jacobs strikes a similar note of caution regarding the importance of local connection. "With a consultant, a station can conceivably lose some of its localness if there isn't adequate effort to give it a hometown flavor. But the loss of local presence is far more likely with satellite-delivered formats. Consultants need to work closely with station management (and vice versa) to find local ties and signposts, because listeners care most about what's happening in their town. It's always important to understand that there are regional differences in taste, personalities, and music. Many high-powered on-air personalities would be hard-pressed to duplicate their success in another market."

The cost factor is another reason why some stations do not use consultants. "Consultants can be expensive, although most consultants scale their fees to suit the occasion, that is, the size of the market. A few hundred dollars a day can be exorbitant for many smaller stations. But the cost of the research, analysis, and strategy usually is worth the money. I believe that a station, in most cases, gets everything it pays for when it uses a consultant. It's worth investing a few thousand to make back a million," contends Dwight Douglas.

Dave Scott believes that certain stations can become too dependent on consultants. "A consultant is there to provide support and direction when needed. If a station is infirm, it needs attention, perhaps extensive care. However, when a station regains its health, an annual or semi-annual checkup is usually sufficient. A checkup generally can prevent problems from recurring."

Mikel Hunter agrees with Scott, adding, "A radio doctor needs the cooperation of his client. On the other hand, a station must insist that a consultant do more than diagnose or critique. Positive input, that is, a remedying prescription, is what a consultant should provide. Conversely, a station should be willing to use the aid that the consultant provides."

```
Pro's and con's of using a consultant:

   +'s                              -'s

* Objective, experienced view    * Overreliance on consultant
* Exposure to new ideas            and not enough local input
* Ongoing evaluation of the      * Program director gets too
  station                          much advice from too many
* Input about stations from        sources
  around the country
* National research and information
* Experience/assistance in a wide
  range of areas including music,
  promotion, marketing, talent
  management, etc.
```

FIGURE 11.6
Courtesy Jacobs Media.

Statistically, those stations that use programming consultants more often than not experience improved ratings. In case after case consultants have taken their client stations from bottom to top in many of the country's largest markets. Of course, not all succeed quite so dramatically. However, a move from eleventh place to sixth in a metro market is considered a noteworthy achievement and has a very invigorating effect on station revenue. "The vast majority of consultants benefit their clients by increasing their position in the book. This means better profits," notes Halper, who has improved the ratings of 90 percent of her client stations.

About the future of radio consultancy, George Burns says, "I see the role of consultants undergoing considerable change in the next few years. The rules of ownership and the very principles under which our industry is organized are altering radically. Consultants will probably take even more of an advisory role and have less involvement in the day to day operations of a station. The new and larger broadcasting companies, in all size markets, will keep expertise 'in-house' and rely less on outside input in these areas. I see consultants operating at 'higher levels' in the future. They will be working on organization, continuing education, motivation, compensation, human resources and other 'top management' concerns. Consultants, I believe, will become more policy oriented and less concerned with ground-level activities."

PROGRAM SUPPLIERS

The widespread use of automation equipment commencing in the 1960s sparked significant growth in the field of programming syndication. Initially the installation of automation systems motivated station management to seek out syndicator services. Today the highly successful and sophisticated program formats offered by myriad syndicators often inspires stations to invest in automation equipment. It is estimated that nearly half of the country's radio outlets purchase syndicated programming, which may consist of as little as a series of one- or two-minute features or as much as a 24-hour, year-round station format.

Program specialist Dick Ellis cites economics as the primary reason why stations resort to syndicators. "When I programmed for Peters Productions they supplied high-quality programming and engineering at a relatively low cost. For instance, for a few hundred dollars a month a small market operator gets a successful program director, a highly skilled mastering engineer, all the music he'll ever need (no service problems with record companies) recorded on the highest quality tapes available. It takes a programmer eight hours to program one 24-hour cut reel. It takes a mastering engineer eight hours to remove all the pops and clicks found on even brand-new records, plus place the automation tones. All of this frees the local operator to concentrate his efforts on promotion and, of course, sales."

William Stockman of Schulke Radio Productions (SRP was purchased by Bonneville Broadcasting System in the mid-1980s) says that stations are attracted to syndicators because of the highly professional, major market sound they are able to provide. "By using SRP's unique programming service, a smaller station with limited resources can sound as polished and sophisticated as any metro station."

Both economics and service motivate radio stations to contract syndicators, contends Satellite Music Network programmer Lee Abrams. "Stations are attracted to our affordable, high-quality programming. Its just that simple. We provide an excellent product within a cost-effective context. Our expertise in delivering niche concepts is very appealing to radio operators."

The late and great Rick Sklar observed, "In today's cost-conscious economic climate, more and more radio station operators are turning to suppliers of 24-hour formats for their programming. Whether delivered via satellite, conventional tape, CD or DAT, these increasingly sophisticated products are not only penetrating new markets but larger markets as well, where until now, traditional thinking has held that locally originated programming was the only way to go."

The demand for syndicator product has paralleled the increase in the number of radio outlets since the 1960s. Current projections indicate that this trend should continue.

SYNDICATOR SERVICES

The major program syndicators usually market several distinctive, fully packaged radio formats. "Peters Productions makes available a complete format service with each of their format blends. They're not merely a music service. Their programming goal is the emotional gratification of the type of person attracted to a particular format," says Dick Ellis, whose former company offers a dozen different formats, including Beautiful Music, Easy Listening, Standard Country, Modern Country, Adult Contemporary, Standard MOR, Super Hits, Easy Contemporary, and a country and contemporary hybrid called *Natural Sound.*

Century 21 Programming also is a leader in format diversity, explains Dave Scott. "Our inventory includes everything from the most contemporary super hits sound to several Christian formats. We even offer a full-time Jazz format. We have programming to fit any need in any market."

Drake-Chenault Enterprises, Satellite Music Network, and TM Century also are among the most successful of those syndicators marketing several program formats. Some syndicators prefer to specialize in one or two programming areas. For example, Bonneville Broadcasting and Churchill Productions primarily specialize in the adult Easy Listening format.

Each format is fully tested before it is marketed, explains Stockman. "At Schulke our strategy has been to reorient the music from essentially a producer-oriented to a consumer-oriented product. Music is tested on a cut-by-cut basis in several markets coast-to-coast. Using patented and proven methodology, music is carefully added or selectively deleted. By determining what songs the listeners like to hear and which songs they dislike, SRP has assembled a totally researched library that has been on the air via our subscriber stations since March 1983. Every song played on our stations has been rated by the listeners as a 'winner,' and all the 'stiffs' that have a high dislike factor have been eliminated altogether."

Customized sound hours are designed for each format to ensure consistency and compatibility on the local station level. "An exact clock is tailored for our client station after our market study. The format we provide will perfectly match the station in tempo, style, music mix, announcing, promos, news, weather, and commercial load," says Century 21's Dave Scott.

Audience and market research and analysis are conducted by syndicators before implementing a particular for-

FIGURE 11.7
The president of Century 21 Programming David Scott (standing) oversees the packaging of syndicated material. Courtesy Century 21.

FIGURE 11.8
Century 21's
catalogue of
syndicated formats.
Courtesy Century
21.

Eighteen Successful Sounds: One's Right For You

The Z Format.

Since 1973, Century 21's contemporary hit Z Format has delivered the best track record in the business!

The Hot Z Format.

Today's hottest hits give you the most popular music, so your station will be the most popular in your market!

Album Oriented Z.

Century 21's album rock format is available either unannounced or uniquely custom-voiced.

The A-C Format.

Adult contemporary music goes hand-in-hand with a big, responsive audience.

Good Ol' Rock & Roll.

Here are the top hit oldies of the 50's, 60's & 70's that still sound good today.

You can choose from our *three* different formats: modern, traditional, or pop/cross-over country.

The C-C Format.

This Country-Crossover Format blends country with adult contemporary music.

The E-Z Format.

Century 21's E-Z Format moves "middle-of-the-road" music into *your* lane.

MORE BEAUTIFUL.

MORe Beautiful blends middle-of-the-road vocals with the finest instrumentals.

Simply Beautiful.

Beautiful music is the favorite format of broadcasters seeking trouble-free, stable operations.

MEMORY MUSIC.

Here's the ideal mixture of M-O-R and nostalgia. Music vintages are tailored to your station.

SACRED SOUNDS.

Century 21 delivers four Christian formats: beautiful, adult contemporary, traditional & country gospel.

The Jazz-Z Format.

As either a full- or part-time format, Century 21's Jazz-Z sound gives your station unique popularity.

mat. "Our clients receive comprehensive consulting services from our seasoned staff. We begin with a detailed study of our client station's market. We probe demographics, psychographics, and population growth trends of a station's available audience. We analyze a client's competition quantitatively through available ratings and qualitatively from airchecks. Then the programming Century 21 provides is professionally positioned to maximize our client's sales, ratings, and profits. All of our programming is solidly backed by systematic studies of the listening tastes of each format's target audience. Our research includes call-out and focus group studies, in-depth market analysis, attitudinal audience feedback, psychographic patterns and tests, and several in-house computers with ratings data on-line," says Dave Scott.

Format programming packages include hundreds of hours of music, as well as breaks, promos, and IDs, by seasoned metro market announcers. Customized identity elements, such as jingles and other special formatic features (taped time checks), are made available by the majority of syndicators. "We try to cover all bases to ensure the success of our clients. We back each of our formats a dozen different ways. For example, image builders in the form of promotions, contests, and graphics also are an element of our programming service at Radio Arts," says programmer Larry Vanderveen.

To stay in step with the ever-changing marketplace, syndicators routinely update the programming they provide their subscribers. "When you want people to listen to a station a lot, you've got to keep them interested in it. To do so you have to air a sound that's always fresh and current. Tape updates from Century 21 are plentiful. We give stations the most extensive initial collection of music tapes available. Then we follow them up with hundreds more throughout the year. For instance, our CHR, AOR, and Country subscribers receive over 100 updates annually. All categories have frequent updates, so our client's sound stays fresh and vital," says Dave Scott, who adds that

the lines of communication are kept open between the client and syndicator long after the agreement has been signed. "Since the success of our clients is very important to us, continuing consultation and assistance via a toll-free hotline is always available 24 hours a day. Automation-experienced broadcasters are in our production studios around the clock, and consultants can be reached at work or home any time. Help is as close as the phone."

Syndicators assist stations during the installation and implementation stage of a format and provide training for operators and other station personnel. Comprehensive operations manuals are left with subscribers as a source of further assistance.

Fees for syndicator services vary according to a sliding scale that is based on station revenues for a specific market. "Our fee structure at Bonneville ranges from something under $1,000 per month to over $10,000. It

FIGURE 11.9 Sample customized sound hour. Courtesy Century 21.

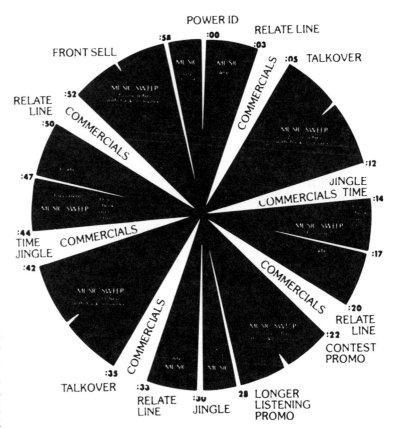

very much depends on the size of the market in which a station is located," explains Marlin Taylor, Bonneville Broadcasting's president. The range for syndicator formats falls somewhere between several hundred and several thousand dollars per month, depending on the nature of the service and market size.

Leasing agreements generally stipulate a minimum two-year term and assure the subscriber that the syndicator will not lease a similar format to another station in the same market. Should a station choose not to renew its agreement with the syndicator, all material must be returned unless otherwise stipulated.

The majority of format syndicators also market production libraries, jingles, and special features for general market consumption.

FIGURE 11.10 Satellite program services. Courtesy WFMT.

HARDWARE REQUIREMENTS

Syndicated programming typically is designed for automation systems. "All Century 21 programming airs smoothly with a minimum of equipment. Of course, sophisticated systems can be fully utilized by our formats, but you can run our material with the most basic automation setup," notes Dave Scott, whose company, as well as most other format syndicators, require that stations possess the following automation equipment:

Four reel-to-reel tape decks capable of holding ten-and-a-half-inch NAB reels. Some formats will work with only two or three reel-to-reel decks (or CD or DAT players when the station has upgraded to this new technology).

End-of-music sensors also are recommended to alert operators of necessary tape changes.

Sufficient multiple cart playback systems for spot load, news, PSAs, weather, promos, IDs, and so forth.

Two single-cartridge playback units or use of extra trays in spot players.

One time-announce unit with two size C cartridge players and controller.

Scott also notes that his company now provides clients with the newest in sound reproduction technology. "Century 21 offers 160 hours of full-form programming on CD and DAT. Right now we have hundreds of clients on CD product as compared to 350 on reel-to-reel. We also provide subscribers with high-quality DAT players when they choose to go in that direction."

Technical and engineering assistance is made available to subscribers. "We provide periodic on-site station technical reviews together with ongoing technical consultation, continuing new technology development, and new equipment evaluation," explains Bonneville's Marlin Taylor.

The use of satellites by syndicators has grown enormously since 1980. A recent NAB survey concluded that over three-quarters of the nation's stations receive some form of satellite programming. The majority of stations with satellite dishes use them to draw network feeds. However, the percentage of stations receiving product from syndicators and other programming services has more than doubled over the past decade, and the use of station hardware (other than computers) for syndicator programming is winding down. It is more cost-effective and efficient to catch the digital satellite signals than it is to handle actual product on the local station level. In fact, the majority of program syndicators have ceased to mail material to stations, opting to beam it to them instead.

SYNDICATOR FIDELITY

Syndicators are very particular about sound quality and make every effort to ensure that their programming meets or exceeds fidelity standards. "Our company uses the finest quality recording studio equipment. Actually, it's far superior to any broadcast-grade gear. Therefore, it is quite important that subscribers have adequate hardware, too. We utilize a number of highly regarded audio experts to make our sound and our client's the very best possible. In fact, we use special audiophile 'super disks' master tapes from record companies, noise reduction, click editing, and precise level control or slight equalization, if needed," says Century 21's Scott.

Periodic airchecks of subscriber stations are analyzed from a technical perspective to detect any deficiencies in sound quality.

CHAPTER HIGHLIGHTS

1. The significant increase in stations and formats created a market for consultants.

2. Consultants provide various services, including market research, programming and format design, hiring and training of staff, staff motivation, advertising and public relations campaigns, news and public affairs

FIGURE 11.10
continued

restructuring, and technical evaluation (periodic airchecks of sound quality).

3. Aspiring consultants should acquire background experience in the medium, solid educational preparation, and strong interpersonal skills.

4. Station executives opposed to using consultants fear losing the station's local flavor, becoming a clone of other stations, and the substantial cost.

5. Statistically, stations using programming consultants more often than not experience improved ratings.

6. Increased use of programming syndication is related to the increased use of automation. Nearly half of the nation's stations purchase some form of syndicated programming.

7. Syndicated programs are generally cost-effective, of high quality, and reliable, thus allowing smaller stations to achieve a metro station sound.

8. Program syndicators provide a vari-

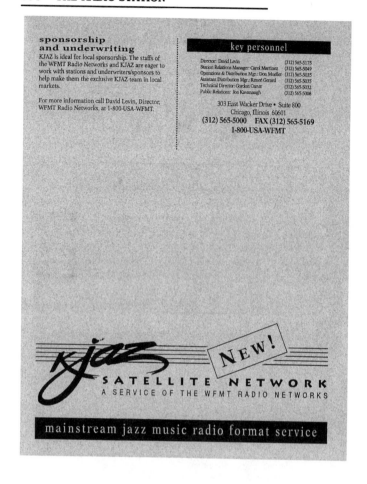

FIGURE 11.10
continued

ety of test-marketed, packaged radio formats—from Country to Top 40 to Religious. Packages may include music, breaks, promos, customized IDs, and even promotions. Package updates are frequent.

9. Most nonsatellite syndicated programs/formats require the use of reel-to-reel tape decks, end-of-music sensors, single-cartridge playback units, and a time-announce unit.

10. The number of syndicators using satellites to deliver programming is at an all-time high. Many only deliver programming via satellite.

APPENDIX: STATION CRITIQUE

To: GM/B- Radio
Fr: Donna L. Halper, Halper & Associates
Re: Critique of tapes of B-

Thanks for sending along the latest batch of tapes for me to critique. I do hear some improvements since I last visited the station. On the other hand, I am still hearing some areas that we need to work on. Most of what I noticed is problems with formatics, although a few little things stood out. In no particular order:

FIGURE 11.11
SMN offers an explanation of their system which requires little equipment on the client-subscriber level. This is one of the many cost advantages of satellite programming.
Courtesy SMN.

1. Bet is back to being too close to the mike, causing her to pop her *P*s again. The good news is that on this tape, her voice is now VERY mid-range—and not high-end or "cutesy," and she sounds more natural. The bad news is that she seems determined to use verbal cliches (however, she is not the only one with this habit). For example, when she reads the liner about "playing the music that made FM great," she repeatedly says "and here's another great example" when she introduces the next song. She is also trying too hard to make simple format elements sound enthusiastic: the school lunch menu basically just needs to be read, rather than embellished upon for two minutes. I know she is trying, but it just sounds artificial to get *that* excited about school lunches. The entire staff seems stuck on the phrases *keep it locked* and *music from . . .* ("that was music from the Beatles; now here's music from Steely Dan"). In real life, do we really talk that way, or do we talk about great SONGS, or use phrases like "a classic from" . . . We need to VARY what we say, or else we sound like robots. Thus, if I just said "now, here's a classic from Bob Dylan," I don't want to use that phrase for every front-sell. Ditto for "keep it locked"—it's a rather over-used AOR phrase as it is, but boy, do our jocks say it a lot . . . can't we find some other ways to invite people to listen?

2. Virtually everyone seems to have acute liner-itis. It's a disease where you want to read a lot of liners all at once. I heard "the all-new B-, the station that plays the classic hits and the music that made FM great!!!" ONE LINER PER BREAK IS JUST FINE, THANKS. If we just used "the all new B-" or "your station for Classic Hits," we don't need to add in two more liners. A good front-sell might be as simple as "on the home of classic hits, B-, here's Cat Stevens." Or "playing the music that made FM great, we're B-, with Fleetwood Mac." Simple is better, in other words.

3. How much weather do we need in mid-days? While in morning drive, the announcer should give time, temp, and say good morning to the audience each break; that isn't necessary the rest of the day. Unless there are major storms coming, I wouldn't have so many weather forecasts during the workday—folks already got there, the kids are in school, etc. But we should still be friendly: I seldom heard Dave say his name (I assume it was Dave?), and I like a liner that thanks people for listening or invites them to tell a friend about the All New B-. . . .

4. While I agree that call letters are crucial, I heard them used way too much at some times: it is not conversational to say "on the all new B-, here's Bob Seger on the all-new B-." That just sounds repetitious. My training has been to use call letters going into a song, and use them when coming out of a long music set ("That was Traffic on the All-New B-, and we also heard Joan Baez and Dan Fogelberg.") But to use them two or three times within the same front-sell strikes me as too much of a good thing. Also, I'm hearing B- a lot more than I'm hearing Classic Hits. And a final grammatical note—you CAN'T say "the classic hits of all time"—you can play the greatest hits, but *classic hits* is a format description and a

FIGURE 11.12
Many syndicators use satellites to feed programming to client stations. This greatly simplifies the distribution process. Handling of tapes and mailing is eliminated. Satellite syndication also keeps station equipment costs down. Courtesy IDB.

positioning statement. I'd suggest "playing ALL the classic hits" or "playing nothing but classic hits" or "playing the classic hits of the past and the classic hits of today"—you get the idea.

5. More cliches—why is Wednesday called *hump day*? Again, do we really talk that way? And in several of the forecasts I heard various people do, I heard about *scattered rain*—I knew that showers could be scattered, but rain? I remind everyone to keep being CONVERSATIONAL—how would you talk to a FRIEND? Would you really say "it's 52 minutes past the hour of 6 o'clock"? I also like consistency—some jocks, as mentioned earlier, said their name often, some said it repeatedly, some seldom said it at all. Some used one liner over and over, some used a variety. These elements must be formatted in so that all the liners are rotated evenly. Also, I'm not sure if it was the skimmer, but I heard some problems with levels—when a song ended, the jock's voice sounded much louder or much softer than the song at certain times. A request—could you tape an hour of each jock and NOT scope it down—in other words, just put a tape in a boom-box and let it run, so that I can hear an uninterrupted hour, commercials and all? I'd like to hear what the audience heard.

More critique later—I'll go over the music in more detail in my next memo. What I heard sounded basically hit-oriented, which is good, but let's stress CLASSIC HITS too, because we are getting to a point where the variety of the music is right on target! I'll be in touch.

P.S.—I meant to mention that the PSA or Street-Sheet outro may be too long—although it may also be the jock ad-libbing. The ending of a PSA should be brief and to the point. What I heard was over a MINUTE of numbers and advice—"if your church, social club, or nonprofit organization has a message you want to be publicized, just send the who-what-when-where-how-and-why, etc."—boy, that's wordy!!! It's better to be simple—without tons of addresses and phone numbers. You are wiser to advise those with something to send to call B- for our fax number, rather than taking up so much time telling them they can phone it or fax it or mail it, then giving the address

FIGURE 11.13
Satellite Music
Network program
clock. Courtesy
SMN.

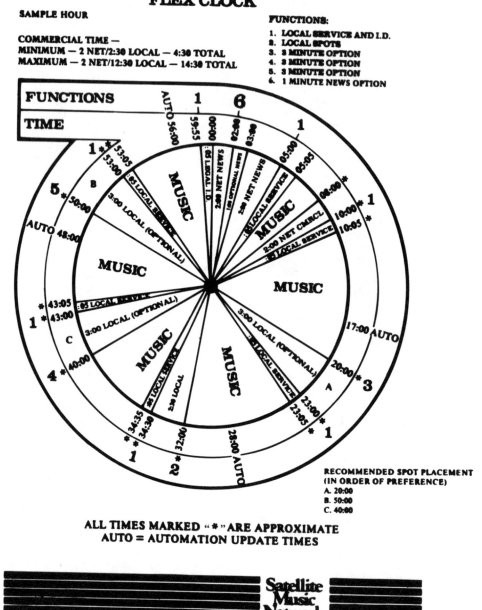

FIGURE 11.14
Many stations have come to rely on program suppliers. Courtesy Shane Media and TM Century.

on top of everything. This slows the station down too much. . . It also goes without saying that the way something is sent to us on a press release may not sound good read verbatim on the air. What I heard B- reading sounded as if we had just put the press release right into the studio with no rewrite. Local news and local PSAs should be rewritten for clarity, and it helps that they be conversational. (I would also like to hear our local news, by the way, plus how Chris sounds.) Perhaps the announcers were nervous because they were doing an aircheck, but they do need to become accustomed to taping themselves regularly so that they can eliminate the verbal crutches they use and make their show sound smoother.

SUGGESTED FURTHER READING

Broadcasting Yearbook. Washington, D.C.: Broadcast Publishing, 1935 to date, annually.

Fornatale, Peter, and Mills, Joshua. *Radio in the Television Age.* Woodstock, N.Y.: Overlook Press, 1980.

Hall, Claude, and Hall, Barbara. *This Business of Radio Programming.* New York: Billboard Publishing, 1977.

Howard, Herbert H., and Kievman, Michael S. *Radio and Television Programming.* Columbus, Ohio: Grid Publishing, 1983.

Inglis, Andrew F. *Satellite Technology.* Boston: Focal Press, 1991.

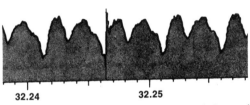

Waveform of commercial CD with clicks and tape hiss.

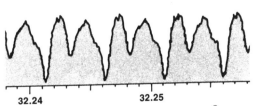

Same song as above on GoldDisc[3] after **NoNOISE** computer eliminates click and digitally removes hiss without changing music.

FIGURE 11.15
Syndicators are especially sensitive to quality control. Here a commercial CD is enhanced by a syndicator's computer prior to shipment to a client. Courtesy Century 21.

Keith, Michael C. *Radio Programming: Consultancy and Formatics*. Boston: Focal Press, 1987.

Mirabito, Michael M., and Morgenstern, Barbara L. *The New Communication Technologies*, 2nd ed. Boston: Focal Press, 1994.

The Radio Programs Sourcebook, 2nd ed. Syosset, N.Y.: Broadcast Information Bureau, 1983.

Series, Serials, and Packages. Syosset, N.Y.: Broadcast Information Bureau, annually.

Vane, Edwin T., and Gross, Lynne S. *Programming for TV, Radio, and Cable*. Boston: Focal Press, 1994.

Wasserman, Paul. *Consultants and Consulting Organization Directory*, 3rd ed. Detroit: Gale Research, 1976.

Glossary

ABC American Broadcasting Company; network.

Account executive Station or agency salesperson.

Actives Listeners who call radio stations to make requests and comments or in response to contests and promotions.

Actuality Actual recording of news event or person(s) involved.

ADI Area of Dominant Influence; Arbitron measurement area.

Adjacencies Commercials strategically placed next to a feature.

Ad lib Improvisation. Unrehearsed and spontaneous comments.

Affidavit Statement attesting to the airing of a spot schedule.

AFTRA American Federation of Television and Radio Artists; union comprised of broadcast performers: announcers, deejays, newscasters.

Aircheck Tape of live broadcast.

AM Amplitude Modulation; method of signal transmission using Standard Broadcast band with frequencies between 535 and 1,605 (1,705) kHz.

AMAX Enhanced AM receiver developed by the NAB.

Announcement Commercial (spot) or public service message of varying length.

AOR Album-Oriented Rock radio format.

AP Associated Press; wire and audio news service.

Arbitron Audience measurement service employing a seven-day diary to determine the number of listeners tuned to area stations.

ASCAP American Society of Composers, Authors, and Publishers; music licensing service.

Audio Sound; modulation.

Audition tape Telescoped recording showcasing talents of air person.

Automation Equipment system designed to play prepackaged pro-gramming.

Average quarter-hour (AQH) persons See the research glossaries in Chapter 6.

AWRT American Women in Radio and Television.

Back announce Recap of preceding music selections.

Barter Exchange of airtime for programming or goods.

BEA Broadcast Education Association.

Bed Music behind voice in commercial.

Blasting Excessive volume resulting in distortion.

Blend Merging of complementary sound elements.

Book Term used to describe rating survey document; "Bible."

BM Beautiful Music radio format.

BMI Broadcast Music Incorporated; music licensing service.

BPME Broadcast Promotion and Management Executives.

Bridge Sound used between program elements.

BTA Best Time Available, also Run Of Schedule (ROS); commercials logged at available times.

Bulk eraser Tool for removing magnetic impressions from recording tape.

Call letters Assigned station identification beginning with "W" east of the Mississippi and "K" west.

Capstan Shaft in recorder that drives tape.

Cart Plastic cartridge containing a continuous loop of recording tape.

CFR *Code of Federal Regulations.*

Chain broadcasting Forerunner of network broadcasting.

CHR Contemporary Hit Radio format.

Clock Wheel indicating sequence or order of programming ingredients aired during one hour.

Cluster   See *Spot set.*

Cold   Background fade on last line of copy.

Combo   Announcer with engineering duties; AM/FM operation.

Commercial   Paid advertising announcement; spot.

Compact disk (CD)   Digital recording using laser beam to decode surface.

Console   Audio mixer consisting of inputs, outputs, toggles, meters, and pots; board.

Consultant   Station advisor or counselor; "radio doctor."

Control room   Center of broadcast operations from which programming originates; air studio.

Cool out   Gradual fade of bed music at conclusion of spot.

Co-op   Arrangement between retailer and manufacturer for the purpose of sharing radio advertising expenses.

Copy   Advertising message; continuity; commercial script.

Cost Per Point (CPP)   See the research glossaries in Chapter 6.

Cost Per Thousand (CPM)   See the research glossaries in Chapter 6.

CPB   Corporation for Public Broadcasting.

CRMC   Certified Radio Marketing Consultant.

Crossfade   Fade out of one element while simultaneously introducing another.

Cue   Signal for the start of action; prepare element for airing.

Cue burn   Distortion at the beginning of a record cut resulting from heavy cueing.

Cume   See the research glossaries in Chapter 6.

DAB   Digital Audio Broadcasting.

DARS   Digital Audio Radio Service

DAT   Digital audio tape.

Dayparts   Periods or segments of broadcast day: e.g., 6 to 10 a.m., 10 a.m. to 3 p.m., 3 to 7 p.m.

Daytimer   AM station required to leave the air at or near sunset.

Dead air   Silence where sound usually should be; absence of programming.

Deejay   Host of radio music program; announcer; "disk jockey."

Demagnetize   See *Erase.*

Demographics   Audience statistical data pertaining to age, sex, race, income, and so forth.

Direct Broadcast Satellite (DBS)   Powerful communications satellites that beam programming to receiving dishes at earth stations.

Directional   Station transmitting signal in a pre-ordained pattern so as to protect other stations on similar frequency.

DMX   Digital music satellite service.

Donut spot   A commercial in which copy is inserted between segments of music.

Double billing   Illegal station billing practice in which client is charged twice.

Drivetime   Radio's primetime: 6–10 a.m. and 3–7 p.m.

Dub   Copy of recording; duplicate (dupe).

EBS   Emergency Broadcast System.

Edit   To alter composition of recorded material; splice.

ENG   Electronic news gathering.

Erase   Wipe clean magnetic impressions; degause, bulk, deflux, demagnetize.

ERP   Effective radiated power; tape head configuration: erase, record, playback.

ET   Electrical transcription.

Ethnic   Programming for minority group audiences.

Fact sheet   List of pertinent information on a sponsor.

Fade   To slowly lower or raise volume level.

FCC   Federal Communications Commission; government regulatory body with authority over radio operations.

Fidelity   Trueness of sound dissemination or reproduction.

Fixed position   Spot routinely logged at a specified time.

Flight   Advertising air schedule.

FM   Frequency Modulation; method of signal transmission using 88 to 108 MHz band.

FMX   System used to improve FM reception.

Format   Type of programming a station offers; arrangement of material, formula.

Frequency   Number of cycles-per-second of a sine wave.

Fulltrack   Recording utilizing entire width of tape.

Gain   Volume; amplification.

Generation   Dub; dupe.

Grease pencil   Soft-tip marker used to inscribe recording tape for editing purposes.

Grid   Rate card structure based on supply and demand.

Gross Rating Points (GRP)   See the research glossaries in Chapter 6.

Ground wave   AM signal traveling the earth's surface; primary signal.

Headphones   Speakers mounted on ears; headsets, cans.

Hertz (Hz)   Cycles per second; unit of electromagnetic frequency.

HLT   Highly Leveraged Transaction.

Hot   Overmodulated.

Hot clock   Wheel indicating when particular music selections are to be aired.

Hype   Exaggerated presentation; high intensity, punched.

IBEW   International Brotherhood of Electrical Workers; union.

ID   Station identification required by law to be broadcast as close to the top of the hour as possible; station break.

Input   Terminal receiving incoming current.

Institutional   Message promoting general image.

IPS   Inches per second; tape speed: 1, 3, 15, 30 IPS.

ITU   International Telecommunications Union; world broadcasting regulatory agency.

Jack   Plug for patching sound sources; patch-cord, socket, input.

Jingle   Musical commercial or promo; signature, logo.

Jock   See *Deejay*.

KDKA   Radio station to first offer regularly scheduled broadcasts (1920).

Kilohertz (kHz)   One thousand cycles per second; AM frequency measurement, kilocycles.

Leader tape   Plastic, metallic, or paper tape used in conjunction with magnetic tape for marking and spacing purposes.

Level   Amount of volume units; audio measurement.

Licensee   Individual or company holding license issued by the FCC for broadcast purposes.

Line   Connection used for transmission of audio; phone-line.

Line-of-sight   Path of FM signal; FM propagation.

Liner cards   Written on-air promos used to ensure adherence to station image; prepared ad-libs.

Live copy   Material read over air; not prerecorded.

Live tag   Postscript to taped message.

LMA   Local Marketing (or Management) Agreement.

Local channels   Class IV AM stations found at high end of band: 1,200 to 1,600 kHz.

Make-good   Replacement spot for one missed.

Market   Area served by a broadcast facility; ADI.

Master   Original recording.

Master control   See *Control room*.

MBS   Mutual Broadcasting System; radio network.

Megahertz (MHz)   Million cycles per second; FM frequency measurement, megacycles.

Mini-disc Machines   Digital cart decks employing floppy disc technology for audio reproduction and archiving.

Mixdown   Integration of sound elements to create desired effect; production.

Monitor   Studio speaker; aircheck.

Mono Single or fulltrack sound; monaural, monophonic.

MOR Middle-Of-the-Road radio format.

Morning Drive Radio's primetime daypart: 6:00 to 10:00 a.m.

MSA Metro Survey Area; geographic area in radio survey.

Multitracking Recording sound-on-sound; overdubbing, stacking tracks.

Music sweep Several selections played back-to-back without interruption; music segue.

NAB National Association of Broadcasters.

NAEB National Association of Educational Broadcasters.

Narrowcasting Directed programming; targetting specific audience demographic.

NBC National Broadcasting Company; network.

Network Broadcast combine providing programming to affiliates: NBC, CBS, ABC, MBS.

Network feed Programs sent via telephone lines or satellites to affiliate stations.

News block Extended news broadcast.

NPR National Public Radio.

NRSC National Radio Systems Committee.

O and Os Network or group owned and operated stations.

OES Optimum Effective Scheduling.

Off-mike Speech outside normal range of microphone.

Out-cue Last words in a line of carted copy.

Output Transmission of audio or power from one location to another; transfer terminal.

Overdubbing See *Multitracking*.

Overmodulate Exceed standard or prescribed audio levels; pinning VU needle.

Packaged Canned programming; syndicated, prerecorded, taped.

Passives Listeners who do not call stations in response to contests or promotions or to make requests or comments.

Patch Circuit connector; cord, cable.

Patch panel Jack board for connecting audio sources: remotes, studios, equipment; patch bay.

PBS Public Broadcasting System.

Pinch roller Rubber wheel that presses recording tape against capstan.

Playback Reproduction of recorded sound.

Playlist Roster of music for airing.

Plug Promo; connector.

Popping Break-up of audio due to gusting or blowing into mike; blasting.

Positioner Brief statement used on-air to define a station's position in a market.

Pot Potentiometer; volume control knob, gain control, fader, attentuator, rheostat.

PSA Public Service Announcement; noncommercial message.

Psychographics Research term dealing with listener personality, such as attitude, behavior, values, opinions, and beliefs.

Production See *Mixdown*.

Punch Emphasis; stress.

Quadraphonic Four speaker/channel sound reproduction.

RAB Radio Advertising Bureau.

Rack Prepare or set up for play or record: "rack it up"; equipment container.

RADAR Nationwide measurement service by Statistical Research, Inc.

Rate card Statement of advertising fees and terms.

Rating Estimated audience tuned to a station; size of listenership, ranking.

RBDS Technology that enables AM and FM stations to send data to "smart" receivers, allowing them to perform several automatic functions.

RCA Radio Corporation of America; NBC parent company.

Recut Retake; rerecord, remix.

Reel-to-reel Recording machine with feed and take-up reels.

Remote Broadcast originating away from station control room.

Reverb Echo; redundancy of sound.

Rewind Speeded return of recording tape from takeup reel.

Ride gain Monitor level; watch VU needle.

Rip 'n' read Airing copy unaltered from wire.

rpm Revolutions per minute: 33⅓, 45, and 78 rpm.

RTNDA Radio and Television News Directors Association.

Run-of-station (ROS) See *BTA*.

Satellite Orbiting device for relaying audio from one earth station to another; DBS, Comsat, Satcom.

SBE Society of Broadcast Engineers.

SCA Subsidiary Communication Authority; subcarrier FM.

Secondary service area AM skywave listening area.

Segue Uninterrupted flow of recorded material; continuous.

SESAC Society of European Stage Authors and Composers; music licensing service.

SFX Abbreviation for sound effect.

Share Percentage of station's listenership compared to competition; piece of audience pie.

Signal Sound transmission; RF.

Signature Theme; logo, jingle, ID.

Simulcast Simultaneous broadcast over two or more frequencies.

Spec tape Specially tailored commercial used as a sales tool to help sell an account.

Splice To join ends of recording tape with adhesive; edit.

Splicing bar Grooved platform for cutting and joining recording tape; edit bar.

Sponsor Advertiser; client, account, underwriter.

Spots Commercials; paid announcements.

Spot set Group or cluster of announcements; stop set.

Station Broadcast facility given specific frequency by FCC.

Station identification See *ID*.

Station log Document containing specific operating information as outlined in Section 73.1820 of the FCC Rules and Regulations.

Station rep Company acting in behalf of local stations to national agencies.

Stereo Multichannel sound; two program channels.

Stinger Music or sound effect finale preceded by last line of copy; button, punctuation.

Straight copy Announcement employing unaffected, nongimmicky approach; institutional.

Stringer Field or on-scene reporter.

Subliminal Advertising or programming not consciously perceived; below normal range of awareness, background.

Sweep link Transitional jingle between sound elements.

Syndicator Producer of purchasable program material.

Tag See *Live tag*.

Talent Radio performer; announcer, deejay, newscaster.

Talk Conversation and interview radio format.

TAP Total Audience Plan; spot package divided between specific dayparts: AAA, AA, A.

Tape speed Movement measured in inches per second: 3, 7, 15 IPS.

Telescoping Compressing of sound to fit a desired length; technique used in audition tapes and concert promos, editing.

TFN Till Further Notice; without specific kill date.

Trade-out Exchange of station airtime for goods or services.

Traffic Station department responsible for scheduling sponsor announcements.

Transmit To broadcast; propagate signal; air.

TSA Total Survey Area; geographic area in radio survey.

Underwriter See *Sponsor*.

Unidirectional mike Microphone designed to pick up sound in one direction; cardioid, studio mike.

UPI   United Press International; wire and audio news service.

VOA   Voice of America.

Voiceover   Talk over sound.

Voice-track   Recording of announcer message for use in mixdown.

Volume   Quantity of sound; audio level.

Volume control   See *Pot*.

VU   Meter Gauge measuring units of sound.

WARC   World Administrative Radio Conference; international meeting charged with assigning spectrum space.

Wheel   See *Clock*.

Windscreen   Microphone filter used to prevent popping and distortion.

Wireless telegraphy   Early radio used to transmit Morse code.

Wow   Distortion of sound created by inappropriate speed; miscue.

Afterword: Marking the Occasion of the Birth of a Radio Station

by Norman Corwin

THE TIME AT THE TONE

The time at the tone . . . . the weather
will be . . . . and now for a word . . . .

The time at the tone is early autumn
    in the ebb of a millenium
the numbers to the next one
    programmed to fall from the
    calendar like the leaves of a
    thousand oaks
And here we speak unwalled on a
    campus
And you listen at home or school or at
    work, or in a car, or out walking
    with a walkman
And between and over and around us,
    oceans of air flowing in currents,
    bearing vapors and scents and
    flotsam of pollens and dust . . . .
And by a miracle now ordinary, as
    common and set in its ways as a
    hinged door or an old slipper
We demolish space on the instant; and
    the noise we make and the music
    we make and the sense we make
    when we make it, as flashy in the
    run as light

For this is Radio.

We are coded and listed among the
    K's: KLCU.
Born this morning, but already
    tracking on a timetable set to run
    down decades

Transacting speech and sound and
    music to be heard everywhere in
    the spread of its shadow,
    obtainable at the push of a
    button, the twist of a wrist

The weather will be . . .

The weather will be what we make it.
Not the kind cooked by the sun or
    chilled by the poles,
Not hemispheric highs and lows, jet
    streams, isobars
Drizzles and blizzards, impetuous
    lightnings and the extreme
    rudeness of hurricanes
But climates—climates of peace, of
    war, of fear, winters of discontent,

315

springtimes of fresh beginnings
and awakened hope, winds of
commerce, trade winds becalmed
or brisk,
Flurries of public opinion, squalls of
faction—
But overarching all the weathers, best
of all, devoutly to be wished,
sublime, empyrean,
A climate of freedom that knows no
seasons.

And now for a word . . .

And now for a word, a sprinkle of
words among millions to be cast
abroad for as long as the tower
stands:
Good morning and good evening and
good night: and between them,
news from all the spots and
flexions of the only planet that
carries passengers, news foreign
and domestic, local and express,
good, bad, mixed
Announcements, bulletins, drama,
comedies, psalms and
syncopation.
Words, yea, words. As John wrote the
moment his quill touched down,
In the Beginning was the Word
And in the middle will be the word,
whenever that middle comes, and
words will serve to the end of time
And the greatest of those words is and
always will be Peace—
Call it pax, paix, pax, friedlich, salaam,
shalom—whatever the tongue
Whatever:
A peace on earth that will pass all
misunderstanding.

*The time at the tone will be 4:22 on
Thursday, October 20, 1994.*

—Norman Corwin

Transmitted on the date given, as the
first broadcast over KCLU,
California Lutheran University's
FM station in Thousand Oaks,
and rebroadcast as its sign-on
every morning since then.

Index